Medical Monopoly

synthesis

A series in the history of chemistry, broadly construed,
edited by Angela N. H. Creager, Ann Johnson, John E. Lesch,
Lawrence M. Principe, Alan Rocke, E. C. Spary, and Audra J. Wolfe,
in partnership with the Chemical Heritage Foundation

Medical Monopoly

*Intellectual Property Rights
and the Origins of the Modern
Pharmaceutical Industry*

JOSEPH M. GABRIEL

The University of Chicago Press   Chicago and London

The University of Chicago Press, Chicago 60637
The University of Chicago Press, Ltd., London
© 2014 by The University of Chicago
Published 2014
Paperback edition 2020
Printed in the United States of America

29 28 27 26 25 24 23 22 21 20   1 2 3 4 5

ISBN-13: 978-0-226-10818-6 (cloth)
ISBN-13: 978-0-226-71022-8 (paper)
ISBN-13: 978-0-226-10821-6 (e-book)
DOI: https://doi.org/10.7208/chicago/9780226108216.001.0001

Library of Congress Cataloging-in-Publication Data

Gabriel, Joseph M. (Joseph Michael)
    Medical monopoly: intellectual property rights and the origins of the modern pharmaceutical
industry / Joseph M. Gabriel.
        pages cm. — (Synthesis)
    Includes bibliographical references and index.
    ISBN 978-0-226-10818-6 (cloth: alk. paper) — ISBN 978-0-226-10821-6 (e-book)
    1. Pharmaceutical policy—United States—History. 2. Drug development—United States—History.
    3. Pharmaceutical industry—Law and legislation—United States. 4. Drugs—Patents—United
    States. 5. Intellectual property—United States. I. Title.
    RA401.A3G33 2014
    338.4'76151—dc23
                                                                                        2014002923

♾ This paper meets the requirements of ANSI/NISO Z39.48-1992 (Permanence of Paper).

For Jacob

Contents

A Note about Terms

In this book I have generally italicized scientific names to distinguish them from commercial names. In some cases, however, there is some ambiguity because the two categories did not always operate in the same way that they do today. I have also capitalized commercial names but left nonproprietary names lowercase. There is ambiguity in some of these cases, since the boundary between commercial and nonproprietary names was both fluid and frequently contested. In both cases I hope the context makes the situation clear, but in some cases my choices in these matters are necessarily arbitrary.

There are also a number of terms that are easy to misunderstand, and while I have tried to explain them in the body of the text, it is worth calling attention to them here to avoid confusion. Before 1890, the term "officinal" was widely used to refer to products that were included in the *Pharmacopeia of the United States*. (After 1890 the term was replaced with "official.") "Patent medicines" refers to a class of pharmaceutical products that were typically made with secret ingredients but were in fact rarely patented. In the first chapter, the term "botanic" is used to refer to an alternative medical system developed by Samuel Thomson and the broader health movement that grew out of his work, not to the use of plants in medicine more generally. "Botanical," on the other hand, is used both as a noun to refer to raw plants that were sold on the drug market and as an adjective in discussion of plants and the science of botany.

Finally, the term "ethical" when applied to manufacturers is used in a nominal sense to refer to a segment of the pharmaceutical industry that consciously conformed to the ethical norms of the orthodox medical community. I do not claim that "ethical" manufacturers hewed to these norms in all cases. More important, I do not use the term in a normative sense. In other words, I am not arguing that these firms or their scientific and business practices were more ethical than their competitors in the so-called patent medicine industry. In most cases I have avoided using the modifier "so-called" or scare quotes around the term. Thus, when I describe a manufacturer as belonging to the ethical wing of the industry, readers should be aware that I am not making an assertion about its ethical or moral standing.

Introduction

When we open a bottle of medicine and consume its contents, we connect ourselves to forces far beyond us. Pharmaceuticals are the product of highly complex scientific and technological networks, the organization of expert knowledge and practice, and the efforts of countless men and women who go about their daily work researching, manufacturing, and distributing these powerful substances. They are also, of course, the vehicle for generating immense profits for those who make and sell them. As such, pharmaceuticals can be understood as congealed moments of massively complex systems dedicated to both the promotion of health and the accumulation of economic value. When we take our pills we hope—sometimes desperately—that they will do their job and help us. Whether or not they do, simply because we have purchased them, they have already fulfilled their other function.

At first glance, it seems obvious that drug companies should patent their products so that they can recoup their considerable investment in the scientific process and earn a profit. While critics sometimes decry the length of time patents can be extended, the relationship between patents and prices, and other aspects of how patenting affects the commercial development of new drugs, the simple idea that innovation and profit should be linked together is a fundamental assumption in the way we think about the role of the pharmaceutical industry in contemporary society. Patents are a key mechanism through which the investment of resources in the scientific process is translated into economic value, and few critics would suggest that patent rights have no role to play in commercial drug development. Despite sometimes contentious debates over the proper extent of these rights, their impact on the increasingly complex landscape of scientific innovation and other issues, few doubt their fundamental place in our drug development system. Even those who criticize pharmaceutical patenting in strong terms rarely suggest that it should be abolished altogether.

The same can be said about trademarks and, more broadly, advertising and other promotional efforts that depend on forging a distinctive brand identity for new drugs. The pharmaceutical industry spends an immense amount of money advertising its products, and critics frequently decry promotional campaigns targeted at the public, the methods used to market drugs to physicians, the sometimes fuzzy boundary between scientific research and commercial promotion, and the simple fact that the industry spends such a large amount of money selling its goods. Yet despite these concerns, few question the basic right of manufacturers to assign trademarked names to their products. Assigning trademarked names to new drugs and working to develop the reputation of products by promoting these names seem to be natural and reasonable parts of commercial drug development. Few critics would suggest that product branding should be eliminated.

Yet none of this is obvious from a historical perspective. Before the Civil War, patenting pharmaceuticals was considered deeply unethical by many physicians, pharmacists, and drug manufacturers. Indeed, reputable companies refrained from patenting their products, and those few manufacturers that did were denounced by the medical community as quacks. As is well known among historians of medicine, the early drug industry was bifurcated into two main sections, with the so-called patent medicine industry manufacturing goods with secret ingredients and advertising its products in the popular press. The so-called ethical segment of the industry, on the other hand, manufactured well-known goods, refrained from keeping ingredients secret, and marketed only to physicians. Neither segment of the industry patented goods before the Civil War to any meaningful extent, but those who did were loudly denounced as violating the spirit and ethics of scientific medicine. Yet in the decades following the Civil War, drug manufacturers began to patent a few of their products, and by the outbreak of World War I the "ethical" segment of the industry had cautiously embraced patenting as a part of its overall strategy of corporate growth. Medical patenting was no longer considered a form of quackery. It had become a legitimate part of scientific drug development.

The fact that physicians no longer consider medical patenting to be, by definition, both unethical and unscientific points to a central argument of this book. Historians of medicine have in recent years suggested that the complex changes that characterized American medicine between the collapse of Reconstruction and World War II should be understood as part of the broader story of the rise of corporate capitalism.[1] As Martin Sklar reminds us, the corporate reconstruction of American capitalism was something that industrialists, jurists, and other actors *did*—it was not just something that happened *to* them—and as such it should not be understood as some sort of external force or supposedly "objective" economic or organizational process to which people conformed.[2] One implication of this is that the values and beliefs of the men and women who built the new economy need to be taken seriously in our descriptions of the past. I thus argue that the emergence of the modern pharmaceutical industry was both a cause and a function of a profound transformation in the ethical sensibilities of physicians and other actors toward medical patenting. Over the course of the late nineteenth and early twentieth centuries, what had once been the mark of unethical quackery was reinterpreted as an ethically legitimate component of scientific drug development. This transformation in values was an essential component of the corporate reconstruction of the American pharmaceutical industry and its subsequent growth in the decades following World War I.

Physicians' attitudes toward trademarks followed a different, though related, trajectory. Today, when a drug is approved for commercial sale by the Food and Drug Administration (FDA), it is given an official nonproprietary, or generic, name by the United States Adopted Name Council, a nonprofit organization sponsored by the American Medical Association, the United States Pharmacopeia Convention, and the American Pharmacists Association. A new drug is also given a chemical name, which is usually long and complex, and the drug's manufacturer coins a brand name for the new product that the company trademarks. Once patent protection on a drug expires, other manufacturers can then (with approval from the FDA) begin to market it under either the generic name or different brand names. This system works because we assume that all pharmaceuticals that share the same generic name are instances of the same drug—in other words, we assume that as long as they share the same generic name, the pills sold under different trade names are made from the same chemical substance and will have the same effects on us when we take them. We assume that these pills are equivalent to one another and thus interchangeable—or at least we assume that they are similar enough to one another to be interchangeable.[3]

Yet before the Civil War, the relationship between names and things operated in a much different way, and there was no real concept of a *generic* name at all. Medicinal substances were typically referred to by common names, such as "opium" and "rhubarb," and drug companies did not assign additional brand names to their own particular versions of these goods. At the same time, trademark law was of little use to manufacturers who did assign distinctive names to their products—such as "Dr. Robertson's Infallible Worm Destroying Lozenges"—because they were assumed to indicate the origins of the good in question and therefore could be freely adopted by any manufacturer as long as the competitor *also* indicated the origin of his or her product. As a result, common names and commercial names were not juxtaposed in the same way that they are today. Indeed, there was no clear distinction between nonproprietary and proprietary drug names at all.

Following the Civil War, however, trademarks became increasingly important to manufacturers, and courts began to recognize that they had value in and of themselves. As a result, trademarks began to acquire the power to monopolize the sale of goods, to influence medical practice, and to otherwise shape the therapeutic market. Both physicians and pharmacists grew alarmed about their growing use, and by the end of the nineteenth century trademarks had emerged as a significant area of concern among therapeutic reformers. The result was the emergence of a new way of understanding the relationship between names and things based on the idea that equivalent products should share a common name that could not be monopolized by any one manufacturer. The *generic* thus emerged as a sort of parallel form of property to both trademarked names and patented medicines, a type of social property belonging to all that resisted commodification and thereby came to occupy a central place in debates about monopoly rights in the pharmaceutical industry. Perhaps ironically, the emergence of the generic also worked to legitimize privately held trademarked names.

These complex processes were also deeply intertwined with the history of therapeutic reform. This study builds on the work of Harry Marks and others who have traced the efforts of twentieth-century reformers to rationalize medicine by linking the methods of scientific research to the practice of clinical care. According to Marks, what united these therapeutic reformers was "the shared belief that better knowledge about the effects and uses of drugs will lead directly to better therapeutic practice."[4] Nineteenth-century reformers shared this belief as well, and in this volume I trace the effort to improve the practice of medicine by investigating drugs to the early decades of that century. I also argue that in the decades

following the Civil War therapeutic reformers embraced both laboratory and clinical science as a way to improve medical care and that this
emergent therapeutic framework was deeply intertwined with the possibilities of commercially developing new remedies in ways that would
not offend the ethics of conservative physicians. Efforts to develop new
remedies along ethical lines were, in turn, part of a broader framework
that assumed that a rationally operating therapeutic market would lead
to legitimate profits for those who rightfully deserved them, that state
governments (and then the federal government) had an important role
to play in promoting this market, and that impositions on the therapeutic market—whether unethical manufacturers, quack doctors, irrational
forms of behavior, accidents, or other dangers—should be suppressed.
I thus use the phrase "therapeutic reform" more broadly than Marks to
refer not only to the physicians and scientists who sought to link the
study of drugs through laboratory and clinical science to the practice of
medicine but also to physicians who sought to suppress what they saw
as quackery, to pharmacists who sought to improve the practice of selling drugs, to government officials who sought to enforce laws regulating
the market, and to pharmaceutical manufacturers who sought to simultaneously improve medicine and earn a profit. From my perspective, all
these actors—and many more—were engaged in the process of therapeutic reform because they each sought to improve medicine by promoting
what they saw as scientific standards governing the manufacture, distribution, and use of pharmaceuticals. Shifting attitudes toward patents and
trademarks were a tremendously important part of this process.

As a I hope is clear from this brief overview, this project draws on a
variety of scholarly fields to tell a story about intellectual property, therapeutic reform, medical ethics, and the nineteenth-century origins of the
American pharmaceutical industry. It should be of interest to scholars
of intellectual property, to historians and others who study the pharmaceutical industry, and to researchers interested in the history of science,
medicine, and technology more broadly. I certainly hope that this book
contributes to the dynamic conversations currently under way about the
scope, meaning, and ethics of intellectual property. I also hope that it will
be of interest to researchers who study the pharmaceutical industry. Over
the past decade historians and other scholars have increasingly turned
their attention to the industry, examining the relationship between the
introduction of new pharmaceuticals and shifting definitions of disease,
public policy related to the industry and its products, and other important topics.[5] Finally, I hope that this project will be of interest to clinicians, bioethicists, and members of the public concerned about the power

of the industry to shape both biomedical science and medical practice toward its own ends. A growing number of ethicists, academic physicians, and other critics have argued that as a result of pharmaceutical marketing practices the boundary between commerce and science has become dangerously blurred.[6] From this perspective, industry support for biomedical research may be important, but the tendency of the profit motive to corrupt both science and medical practice needs to be guarded against and the many problems that currently plague the system reformed. As I hope to show, these concerns are not new. They date back to the very origins of the modern pharmaceutical industry, to the moment when the pursuit of profit and the advancement of medical science were first linked to one another. Perhaps ironically, at that moment preserving the boundary between the two also became a matter of immense importance.

Medical Science and Property Rights in the Early Republic

In the summer of 1807 a young medical student named Caspar Eddy visited the country residence of the noted physician Samuel L. Mitchill on Long Island. Mitchill had been trained at the University of Edinburgh and was one of the leading physicians of his day; at the time he met Eddy, he was both the editor of the nation's first medical journal, the *Medical Repository*, and a United States senator. Eddy was on a "botanical excursion" as part of his medical studies. Mitchill encouraged him in his endeavors by asking him to compile a list of the plants growing in the neighborhood of his home. Eddy happily complied, walking about the land and writing down the names of the numerous plants he saw, many of which had medicinal properties. His long inventory was then published by Mitchill and appeared alongside articles on topics such convulsions, temporary insanity, smallpox vaccination, and the preparation of mineral waters, which, thanks to the "beautiful applications of chemistry," could now be prepared by anyone with the requisite skills and proper ingredients. "The laboratory thus becomes a manufactory of these rare and far-sought liquids," notes the essay, "and the weak, the sick and the poor, who cannot conveniently travel to the fountain head, may, nevertheless, regale themselves with exact imitations of those health-inspiring draughts."[1]

I begin with this anecdote because it illustrates both the diversity and the quickly changing nature of elite medical practice in the early nineteenth century. Trained either abroad or at one of the small handful of American medical schools, men such as Samuel Mitchill and Casper Eddy operated in a tradition of medical care that stretched back to the ancient Greeks. As Eddy's inventory indicates, the use of healing plants was central to this tradition, which, for lack of a better term, I shall refer to as "orthodox" medicine. The fact that Mitchill deemed Eddy's list worthy of publication indicates that he considered it a useful contribution to the expanding body of scientific knowledge about plants that underlay orthodox medical practice in the young country. At the same time, other articles that he published point to the growing power of scientific and technological innovation to transform medical practice away from its traditional reliance on plants. Chemists had known how to manufacture diethyl ether since at least 1540, for example, but by the turn of the nineteenth century, advances in chemistry had made the production of ether commercially feasible on a large scale. Purification became easier as well, and as a result physicians began to investigate the drug's therapeutic properties and incorporate it into medical practice as a treatment for asthma, "windy disorders of the stomach," and other problems.[2]

Of course, in addition to elite physicians such as Mitchill, there were also physicians of a more common—and affordable—stock, having trained by apprenticing with a master or simply by studying on their own. There were also a wide variety of other healers in the early republic, including herb and root doctors, midwives, and itinerant peddlers who traveled through towns selling medicines that were often made with mysterious ingredients. Equally important, medical knowledge infused people's understanding of the world in which they lived, and as they went about the business of living, they did what they could to address the pains and ailments of daily life. Friends, family members, and even the occasional stranger dispensed advice about what to take and how to prepare it; newspapers, pamphlets, and almanacs published recipes for those who could read them; and for people who had the money, there were herbals and other guidebooks for sale, many of which were imported from England. Such knowledge grew out of complex and overlapping traditions. The extent to which Indian knowledge influenced the healing practices of whites is open to debate, for example, but it is clear that there was at least some transmission of medical knowledge between the two. Slaves and the small communities of free blacks had their traditions as well, traditions that combined with Indian and white practices in fruitful ways and were undoubtedly complex and internally variegated in their own right.[3] Medical care in the early republic thus encompassed a wide variety of practices, overlapping sets of knowledge, and therapeutic innovation. How could it have been otherwise?

The diversity of healing practices in the early republic fueled the growth of a vibrant market in therapeutic goods. Most of the goods available for purchase were grown or gathered locally, but many also came from different regions of the country or even more distant realms. Medicinal herbs, roots, and flowers were grown in family gardens or gathered in the wild, but they were also commodities that were bought and sold in local markets that were themselves part of extended networks of trade. Cinchona bark, for example, played an important role in the practice of healing in early America, as did cinnamon, camphor, nutmeg, Indian and Russian rhubarb, and other plants that came from foreign lands.[4] Manufactured products were also important, although less so than raw botanicals. These included tinctures, elixirs, and other goods made from plants; animal and mineral products; and chemical preparations such as sulfuric ether and mercury. Manufactured products also included preassembled remedies made from what were frequently secret ingredients. These were essentially ready-made formulas of various medicinal plants compounded together and sold as pills or powdered mixtures. During the colonial period these types of products had been imported from England and were referred to as "patent medicines" because they were supposedly protected by patents granted by the king of England.[5] Following the revolution patent medicines began to be produced domestically in large numbers, retaining the familiar designation despite the fact that they were not, in general, actually patented.

Elite members of the orthodox medical community such as Mitchill and Eddy responded to the diverse and changing therapeutic market with significant concern. In the early decades of the nineteenth century, orthodox physicians began to consolidate their authority and establish the formal structures needed to fully distinguish themselves from other healers and practices. They did so against the backdrop of a changing market in medicinal substances, competing claims of expertise made by members of other groups, and their own long-standing traditions and self-understanding. Orthodox physicians thus constituted themselves into a distinct class not only through the establishment of laws and institutional structures but also through the formalization of their own ethical sensibilities and the marginalization of other forms of healing. The self-identity of orthodox physicians was based on a profound belief in their own supposed benevolence and dedication to the advancement of scientific knowledge. It was also fundamentally based on the rejection of what they considered quackery, in all its numerous forms. A critique of patenting, and monopoly more broadly, was an important part of this process.

Patent Rights and the Early Drug Industry

In 1796 a physician named Samuel Lee Jr., of Windham, Connecticut, obtained a patent for a formula for "bilious pills." He soon began selling his pills up and down the Eastern Seaboard. He was quite successful, and within a year "Dr. Lee's Windham Bilious Pills" could be found in stores from Vermont to South Carolina. In 1798, however, a druggist named Samuel H. P. Lee from New London, Connecticut, began marketing his own preparation, which he called "Lee's New-London Bilious Pills." The two Lees soon became involved in a bitter dispute. The first Lee, of Windham, argued that his competitor's products were a "wicked" effort to defraud him and to "impose on the public" by selling a counterfeit good.[6] The second Lee responded by claiming that his pills were made from different ingredients, and in 1799 he secured his own patent for them. In the same year, he earned a medical degree and began advertising his products under the name "Doctor Lee," no doubt adding to the confusion. Samuel Lee of Windham renewed his patent in 1810, shortly before his death, and the New London Lee obtained a second patent for his medicine in 1814. Other competitors entered the "bilious pill" market around the same time, several of whom patented their own preparations.[7] The brisk trade in these pills was one part of the growing and sometimes contentious markets of the young country.

Before the ratification of the Constitution in 1789, patent laws had varied from colony to colony, and in many areas there had been no patent law at all.[8] The deliberations of the framers of the Constitution about protecting the interests of inventors are beyond the scope of this book, but article 1, section 8, gave Congress the authority to "Promote the Progress of Science and useful Arts, by securing for limited Times to Authors and Inventors the exclusive Right to their respective Writings and Discoveries." Congress passed the first federal patent law in 1790. It established an examination system for applications, and if the item in question was found "sufficiently useful and important," the inventor was granted the exclusive right to the invention for a period of fourteen years. In exchange, the inventor was required to describe the invention in detail and to make this information available to the public by submitting it to the Patent Office. The goal was both to encourage invention by rewarding private effort and to make knowledge about new inventions available to all by balancing the competing interests of the inventor and the public in a way that would maximize the good to all concerned. However, the initial patent law was quickly attacked for making patents too difficult to obtain, and in 1793 a revised law was enacted that abolished the examination system and allowed registration of patents without significant review. The law was followed by a number of minor revisions over the next several years, but the basic framework stayed in place until the next major revision in 1836.[9]

By all accounts the inhabitants of the young republic were an inventive lot. In 1811, for example, William Thornton, the first superintendent of the Patent Office, noted with astonishment the inventiveness of his countrymen, claiming that "no nation on earth surpasses them in genius. Even the unlettered inhabitants of the forest have perfected inventions that would have done honor to Archimedes."[10] Yet inhabitants of the young country also had complex attitudes toward patenting. On the one hand, they inherited a deep skepticism of patents based on the English tradition. The British patent system was highly arcane and difficult to navigate, courts seemed unfair and arbitrary in their judgments, and expensive fees meant that patenting tended to be a privilege of the elite.[11] As a result, many Americans inherited a tendency to view patents through the lens of class privilege, and for many critics they seemed to border on an oppressive form of monopoly. Antimonopoly sentiments were deeply rooted in American public life and animated much of the political discourse in the early republic. Patents were one early target of this antimonopoly critique, denounced by those hostile to class privilege as a form of "knavery."[12] At the same time, however, in the American political context patents were also frequently thought of as a form of property to which inventors had rights of ownership rather than as a privilege granted by an elite authority. Inventors and their lawyers also argued that patenting played an important role in promoting national economic development. These complicated attitudes translated into a significant amount of debate about the role of patents in society and whether they promoted or hindered the developing fortunes of the young country. Still, by the 1830s patent law had become a vibrant part of a dynamic legal system that served the needs of an expanding economy. Inhabitants of the new republic thus secured an impressive number of patents: between 1790 and 1836 almost ten thousand patents were issued for a wide variety of inventions.[13] Along with this burst of inventive activity came complex attitudes toward patenting that blended dreams of economic progress and personal gain with fears of class hierarchy and the concentration of economic and political power.

The development of drug manufacturing was one part of this economic and social dynamism. Following the revolution, American patriotism and the growth of domestic markets encouraged the development of a variety of medicinal products, including preassembled remedies and tinctures, elixirs, and other derivative products made from raw botanicals (commonly referred to as the "essence" of the plant in question). Easily assembled from readily available ingredients, these types of products made a sensible business venture, and by the second decade of the nineteenth century the market in manufactured remedies had become strongly competitive. To take just one example, in 1820 one apothecary in New York advertised a variety of manufactured products, including Doctor Church's Cough Drops, Hooper's Female Pills, and Scotch Ointment, "a safe, pleasant, effectual cure for the itch, without mercury." The apothecary also sold Restorative Balsam (for nervous disorders), Anodyne Essence (for headaches), Essences of Peppermint, Spruce, and Mustard (for rheumatism and other ailments), Worm Lozenges (for worms), Lee's Windham Bilious Pills, and several products that had long been sold in the colonies, such as Bateman's Drops (made from alcohol, opium, camphor, and anise oil) and Duffy's Elixir (made from senna leaves, jalap root, coriander seeds, and alcohol).[14]

Chemical products were also an important part of the early therapeutic market. During the eighteenth century chemical preparations such as sulfuric ether and quicksilver (now known as mercury) were generally imported from England, but in the early decades of the nineteenth century a small but growing number of domestic manufacturers also began to produce these types of goods. The alkaloids are a particularly important example. Morphine was first isolated in 1804 by the German pharmacist Friedrich Sertürner and was being imported and sold by pharmacists in the United States as early as 1826.[15] A long series of other alkaloids, or substances that were thought at the time to be alkaloids, were soon isolated by European scientists, including strychnine (1818), piperine (1819), and quinine (1820). By the late 1820s American manufacturers had begun to produce these and other chemical goods for the domestic market in relatively substantial amounts. In 1822, for example, Rosengarten & Sons was established in Philadelphia. One of the first firms to produce quinine sulfate in the United States, within fifteen years the company was manufacturing an impressive line of chemical preparations, including morphine salts, piperine, strychnine, bismuth, and silver salts. Other manufacturers produced tartaric acid, muriate of ammonia, corrosive sublimate, and numerous other chemical products intended to heal.[16]

Yet despite the early interest in patenting so-called bilious pills, only a small number of patents were taken out for medicinal products in the first decades of the nineteenth century. In December of 1836 a fire destroyed the records of the Patent Office, but according to two major sources that list patents issued before the fire, only about 110 patents were granted between 1793 and 1836 for medicinal goods. There is some ambiguity in the sources about the exact number of patents that were issued, but whatever the actual number it is clear that very few patents were in fact granted for so-called patent medicines and other medicinal products.[17] As far as I have been able to determine, this was not due to an unwillingness on the part of the Patent Office; since there was no examination system, patents were generally granted in a routine manner with minimal review. Simple formulas for preassembled remedies had long been patentable in England, and there was no particular reason for them to be turned down. Indeed, formulas seemed akin to the description of a new machine in that they were instructions for how to make something useful; as such they fell under both the traditional idea of manufacture and the statutory language of "composition of matter" in the 1793 law.[18] Patents were thus issued for medicinal pills, cordials, elixirs, oils, plasters, and a variety of unspecified "cures," including cures for pain, gout, rheumatism, cholera, syphilis, and cancer. Despite the wide variety of products patented, however, the small number of patents actually issued for medicinal products is striking given the early vibrancy of the market in patent medicines and other manufactured remedies.

This was probably due, at least in part, to a popular tradition of ignoring patent rights on medicines. Patent medicines had long been popular in the colonies, but the ability of British manufacturers to enforce their monopoly rights on these goods was close to nonexistent. Given the decentralized nature of the American colonies and the distance of English law, it is unlikely that most buyers and sellers actually cared much about the legal rights of manufacturers in England. As the manufacturing capabilities of the young country developed, this tradition of ignoring patent rights probably translated into a general disregard for patents on medicines. Patent medicine manufacturers occasionally threatened their competitors with legal action if they infringed on their rights, sometimes even doing so in cases where—as far as I have been able to determine—the manufacturer in question did not actually secure a patent on his good.[19] However, even those manufacturers who actually did patent their medicines rarely sought legal relief for what they considered infringement on their rights. I discuss one notable exception to this below—Samuel Thomson's efforts to enforce his patent rights over his medical system—but in general few drug manufacturers sought relief for patent infringement before 1836. Indeed, outside of Thomson's efforts I have been unable to find another example of a manufacturer working to enforce his or her monopoly rights over a patented medicine during this period. The dispute between the two Lees was vociferous, but it does not appear to have ended up in court.[20]

The reluctance to patent medicines may also have had to do with the fact that traditionally such patents were acquired as much for the benefit to the product's reputation as for the legal protections they might offer. In the eighteenth century, the granting of patents on medicines by the king of England had conveyed not just legal protections but also a degree of prestige to the manufacturer.[21] This appears to have been a motivating factor in the early decades of the new republic as well. At least some manufacturers who did acquire patents for their medicines prominently displayed this fact in their advertising.[22] Of course, they may have done so in order to discourage counterfeiting of their products, but they may also have believed that customers would be impressed by the apparent validation of the product by the national government. Indeed, some manufacturers appear to have advertised their products as patented when they probably were not. In 1809 George Rogers, for example, advertised his "Patent Vegetable Pulmonic Detergent" as "secured to the subscriber by letters patent from the President of the United States," despite the fact that I have found no records of his having actually patented his medicine.[23] Given the sometimes heated debate about patents in the early republic, however, this strategy was almost certainly less appealing than it had been in England. In a country where patents were sometimes understood as a form of class privilege, and where monopoly more broadly was seen as a threat to economic and political freedom, patents did not always seem to grant prestige.

Although patents on medicinal formulas were probably granted routinely, in the early years of the Patent Office there was also significant confusion about the scope of the patent law when it came to other types of medicinal substances. Patenting was, in English law, traditionally grounded on the idea of *manufacture*, and it was assumed that things that were not manufactured could not be patented. Following this doctrine, the 1793 act stated that patents could be obtained for "any new and useful art, machine, manufacture or composition of matter," or any "new and useful improvement" of the same. There was some ambiguity about the meaning of the term "art" in this clause, but in general it was assumed that patents only applied to things that had been manufactured or to the means of manufacture. Thus it was generally believed that patents could not be taken out on plants, minerals, or other things that were not manufactured products. There was some debate about this owing to the fact that both the Constitution and several sections of the 1793 act used the language of discovery as well as invention; as a result, commenters occasionally argued that anything that was discovered could be patented, including newly discovered species of plants.[24] Such arguments were not particularly persuasive, however, and by the 1820s it was assumed that patents could not be issued for plants or "elementary substances" such as newly discovered minerals. As Willard Phillips noted in his groundbreaking 1837 overview of patent law, "The use of the ordinary known materials cannot be monopolized by patent. We must understand this doctrine to be limited to known materials, and to such as naturally exist, whether known or not; for the discovery of a new elementary substance or material, by analysis or otherwise, does not give a right of a monopoly of it."[25]

At the same time, however, there was a tremendous amount of confusion about the issue of patenting both methods and natural principles. In the landmark 1795 case *Boulton and Watt v. Bull* an English court had found that patents could not be granted for what were called "principles of nature." The case is justifiably famous because it upheld James Watt's patent on an improvement to the steam engine and thus lay at the heart of the origins of the British industrial revolution. Watt's patent and its disputed relationship to economic growth in Britain have been covered in detail elsewhere.[26] Here I simply want to point out that previous to Watt's invention, engines had heated and cooled steam in the same chamber and were therefore relatively weak; Watt's insight was to separate the two processes into different chambers and thereby significantly increase the power of engines. As a result, his patent covered a new *method* of harnessing a familiar natural process rather than a new article of manufacture itself. The 1795 decision upholding his patent hinged on the idea that principles of nature and the processes used to harness those principles are two different things; as the magistrate in the case noted, "Undoubtedly there can be no patent for a mere principle; but for a principle so far embodied and connected with corporeal substances as to be in a condition to act and to produce effects in any art, trade, mystery or manual occupation, I think there may be a patent."[27]

Boulton and Watt v. Bull was tremendously influential in the United States. By the 1830s the doctrine that "principles of nature" cannot be patented was an important part of patent jurisprudence.[28] This included natural phenomena such as lightning or rainbows, mathematical formulas, and other principles and elements of nature that seemed beyond the domain of manufacture and invention. Elixirs, cordials, and other medicinal products that combined multiple ingredients could thus be patented, since they were understood to be new and useful compositions of matter. Alkaloids and other similar chemical preparations, however, were understood as refinements of the "active principle" of the plant in question. They were generally seen as concentrations or extensions of the essential elements of the plant itself and thus more akin to a natural principle or elementary substance than to a composition of matter. Quinine, for example, was referred to as the "active principle of the Cinchonas" in scientific and medical texts, piperine was the "active principle" of black pepper, and so on.[29] Of course, most of these substances were first isolated in Europe, but even if American pharmacists had first isolated these substances, there is no reason to think that they would have pursued patents on them. They were conceptualized as principles of nature and as such would probably have been considered beyond the domain of patentability.

Yet there was also an important ambiguity in this doctrine. In *Boulton v. Bull*, the court had found that although principles of nature could not themselves be patented, processes that embodied these principles in practical forms could be. As the court noted of Watt's patent, "Surely this is a very different thing from taking a patent for a principle; it is not for a principle but for a process."[30] Over the next several decades the idea that principles of nature can be patented in an embodied form became a well-established doctrine in American case law. However, the distinction between principles of nature and the processes used to harness them was not always clear. The issue was especially confusing for "compositions of matter" and in particular for patents on medicinal substances. Patents on preassembled remedies, elixirs, cordials, and other such goods were assumed to cover both the means of manufacturing the medicine (the formula) and the resulting medicine itself. In many respects, these patents simply covered new ways of doing familiar things—formulas for patentable medicines did not need to include novel ingredients, for example, nor did they need to produce novel effects—and distinctions between the method of producing effects, the natural principles at work, and the effects themselves were confusing and sometimes obscure. Patents could be taken out for methods of healing that do not strike modern readers as formulas at all, for example, and that resulted in no discernible product, such as the use of leeches in medicine.[31] At times, improvements in methods of manufacturing familiar products led to patents that

covered the products themselves, even when such products were probably assumed in other cases to be beyond the domain of patentability.
In 1833, for example, two men were granted a patent for improvements
in manufacturing sulfate of quinine. The patent covered not only the
method of manufacturing the substance but also the resulting "production of sulphate of quinine" and "the benefits to be derived or derivable
therefrom."[32]

The doctrine that principles of nature could not be patented but principles that were embodied in practical form could be was also intertwined
with the question of priority. As early as 1813, in the case of *Woodcock v.
Parker*, Joseph Story ruled that in order to receive a patent, an inventor
must "reduce" his invention "to practice." According to Story, "The first
inventor is entitled to the benefit of his invention, if he reduce it to practice and obtain a patent therefor, and a subsequent inventor cannot, by
obtaining a patent therefor, oust the first inventor of his right, or maintain an action against him for the use of his own invention."[33] Invention,
in other words, had to be embodied in machines, compositions of matter,
or other practically useful forms to be patentable. This reduction to practical form served as the mark for determining who first invented something and therefore who had the right to patent it. Importantly, however,
the doctrine of reduction to practice as a criterion for patentability also
meant that the question of how invention took place was not initially
relevant; patentability was not based on "reasoning upon the meta-
physical nature, or the abstract definition of an invention," as Story later
put it. "It is of no consequence, whether the thing be simple or complicated; whether it be by accident or by long, laborious thought, or by an
instantaneous flash of mind, that [invention] is first done."[34] The doctrine
did, however, mean that an inventor could only claim that which he was
responsible for reducing to practice as his own; patents on improvements
to machines and other useful things thus covered the improvement only,
and if a patent claimed "a whole machine," then it "must in substance be
a new machine; that is, it must be a new mode, method, or application of
mechanism, to produce some new effect, or to produce an old effect in a
new way."[35] Invention, in other words, must be reduced to practice and
harness principles of nature toward the production of desired effects or
ends in order to be patentable.

Story was perhaps the most preeminent legal mind of his time, and
over the course of next two decades the doctrine that patentability was
based on the reduction of invention to practice became an important
component of patent case law. Yet in his refusal to consider the "meta-
physical nature" of invention, he sidestepped an important question:

how much of an improvement was necessary to justify a new patent and, perhaps more important, what types of considerations could be used to claim that a product was improved at all? In 1813, for example, Samuel Lee of New London applied for a second patent on his "bilious pills." William Thornton, the head of the Patent Office, initially refused the application because it was for virtually the same recipe as the original patent, and he believed that granting it would be an abuse of the patent privilege. After some back and forth, however, Thornton relented and issued the second patent.[36] He really had no grounds for refusing to do so, given the fact that there was at the time no clearly established doctrine for determining how much novelty was necessary to justify a patent.

Thornton was not the only one who thought the patent law was being misused. Over the course of the 1820s and early 1830s numerous critics argued that patents were proliferating too rapidly, that they were granted for trivial improvements, that trial juries were too cumbersome a means to resolve disputes, and that the patent law hampered innovation. Pointing out that there were more than a hundred patents for manufacturing nails, sixty for pumps, fifty for churns, and an even greater number for stoves, in 1826 one judge complained that "the very great and very alarming facility with which patents are procured is producing evils of great magnitude." The result, the judge suggested, was legal "strife and collision" as "patentees are everywhere in conflict," while the public was harmed as "frivolous and useless" alterations became the basis for further patents and "made pretexts for increasing . . . prices." From this perspective, frauds and imposters exploited the public by obtaining patents on relatively minor improvements. Yet, as the judge noted sarcastically, "All are men of genius; and surely, genius, in a new and enterprising country, must be rewarded!"[37] Clearly, this would not do.

Patent Medicines and the Problem of Counterfeiting

Drug manufacturers in the early decades of the nineteenth century produced a large number of products for the developing therapeutic market. Most of these manufacturers operated small shops and produced only a handful of goods, but some had larger ambitions. Thomas Dyott is a good example. Dyott started his career in the 1790s as a young man of modest means selling shoe polish. He soon entered the drug business, and his fortunes grew rapidly, built in part through relentless advertising in the newspapers of the young country.[38] In 1820, for example, Dyott advertised a huge selection of goods in one New York paper, including over

350 botanicals, oils, tinctures, syrups, and mineral and chemical preparations. Dyott also offered almost 150 patent medicines, including a variety of products that he sold under his own name, such as "Dyott's Antibilious Pills" and "Dyott's Tooth Ache Drops."[39] He also manufactured and sold a line of products under the name of "the late celebrated Dr. Robertson of Edinburgh," whom he claimed was his grandfather, including "Dr. Robertson's Infallible Worm Destroying Lozenges" and "Dr. Roberson's Celebrated Stomachic Elixir of Health."[40] Dyott was one of the most successful drug manufacturers of his day, but he was far from unique in the types of products he sold or the means he used to sell them. A large number of manufacturers introduced patent medicines and other manufactured products to the market, advertising heavily in the growing number of newspapers that circulated the land and offering promises of health and happiness to the people of the young republic through the purchase of their goods.[41]

Like most other drug manufacturers of his time, Dyott eschewed the use of patents as a means to protect his goods. Instead, he kept the ingredients of his products secret. This is not surprising, and we should not assume that Dyott or the numerous other manufacturers who pursued this strategy did so for nefarious purposes. English patent medicine manufacturers had long kept their formulas secret, and in the turbulent markets of the young republic secrecy seemed a natural and reasonable way for manufacturers to protect their interests. By the 1820s the market in manufactured remedies had become highly competitive, and successfully introducing and marketing new products required a significant investment in time, labor, and money. Secrecy allowed manufacturers to protect their recipes from adoption by others, thereby allowing them to develop and maintain a competitive advantage based on the quality of their goods. However, competitors frequently introduced and sold similar products under the same name once a product had been popularized, thereby benefiting from the reputation for the product that the original manufacturer had worked to develop. Not surprisingly, manufacturers such as Dyott considered this to be little more than counterfeiting, a form of theft that damaged their own fortunes and duped the public into buying inauthentic goods. Patent medicine manufacturers thus regularly warned purchasers of the dangers of buying imitation products; they also designed labels, bottles, and boxes to identify their products, hoping they would be difficult to imitate.[42] In 1819, for example, Dyott accused a former employee named Peter Kerrison of manufacturing and selling "spurious trash, which he has the temerity to call by the name of Dr. Robertson's and my Family Medicines."[43]

Disputes about counterfeiting grew out of deeply held assumptions about the relationship between the names and things. This requires some explanation. Most manufactured goods were sold under general names such as "Bilious Pills" or "Essence of Spruce." Anyone could manufacture and sell his or her own version of these products under such names, and as a result different products sold under the same name varied quite widely, depending on the whims and abilities of the manufacturer, available ingredients, and other factors. Even products sold under names based on their ingredients, such as "Essence of Spruce," probably varied significantly from one another, depending on the species of plant used, different manufacturing methods, and other factors. These types of names thus pointed to a general type of product, but they did not guarantee that different goods sold under the same name would be equivalent to one another. Of course, this does not mean that manufacturers were free to make anything they wanted and sell it under any name they wanted; it is clear from archival sources that patent medicine manufacturers maintained detailed recipe books that they used to make their products, and in general they probably stuck to their recipes as best they could. Doing so was important if manufacturers hoped to attract repeat customers, since customers probably knew at least roughly what to expect in terms of taste, appearance, and effect from products that were sold under common names—if a customer bought "Essence of Spruce" and it didn't smell like spruce, he or she was probably not going to be very happy about it. Still, a certain amount of variation was undoubtedly common. Indeed, at least some variation was probably common even among different batches of the same product made by a single manufacturer as a result of changing availability of ingredients, variation in manufacturing methods from one batch to another, and other factors.

Other products were sold under names that included the name of the product's original manufacturer, such as "Bateman's drops" and "Godfrey's cordial" (both of which had been on the market since the eighteenth century). Products sold under these types of names were generally assumed to be made according to the recipe used by the original manufacturer or one close enough to it to result in the same product. The extent to which "Bateman's drops" purchased in one location were equivalent to "Bateman's drops" purchased in another is impossible to determine, but there was probably a significant amount of variation among products that were made by different manufacturers but sold under the same name. Still, these types of goods also probably needed to conform to rough ideas about taste, appearance, and effect if manufacturers and pharmacists hoped to successfully build a market. Customers probably knew what "Godfrey's cordial" looked and tasted like if they purchased it regularly—or at least they knew what the version of the product they were familiar with looked and tasted like. If a product varied significantly from what they were familiar with, they probably would have noticed and perhaps sent their business elsewhere. People were not stupid, despite what historians sometimes assume.[44]

Over the course of the early decades of the nineteenth century a growing number of American manufacturers attached their own names to their products in the way that British manufacturers long had, hoping to distinguish their goods from those of their competitors and to build markets for themselves by developing the reputation of their goods. However this practice presented a significant problem. Today we assume that the brand name of a drug and its generic name are distinct from one another, one being the exclusive property of a single manufacturer and the other being a name that anyone who manufactures the product can— and should—refer to it by. In the early decades of the nineteenth century, however, the names of manufactured products operated very differently. As long as no one held a patent on a medicine, anyone was entitled to manufacture and sell it under what was taken to be its true name, even if that name actually included the personal name of the original manufacturer. This was a well-established legal principle that dated back to the 1783 English case *Singleton v. Bolton*, if not earlier, in which Lord Mansfield had ruled that the manufacturer of "Dr. Johnson's Yellow Ointment" could not prevent a competitor from manufacturing the same medicine and selling it under that name because he had no patent on it. Mansfield ruled that if the defendant had sold a medicine of his own under the name of the plaintiff's product, then that would have been fraud; however, since the plaintiff had no patent, he could not prevent a competitor from manufacturing the same product and selling it under that product's proper name—which, in this case, was "Dr. Johnson's Yellow Ointment."[45]

Disputes about counterfeiting thus hinged on whether or not competitors knew the formula for the product in question. When Thomas Dyott accused Peter Kerrison of manufacturing and selling "spurious trash," he printed and distributed a long series of pamphlets describing Kerrison's efforts and listing the names of agents from whom his own products could be purchased.[46] He also spent what must have been a very large amount of money calling attention to the situation through advertising; as one of his announcements asserted, "All the Medicines offered for sale as Dr. Robertson's, or Dr. Dyott's, with the name of *Peter Kerrison* on the bottles, on the packages, or on the directions, are counterfeits. None are genuine but with the name of T. W. Dyott."[47] What is important here is that Kerrison attached *his own name* to the bottles of "Dr. Robertson's Family Medicines" and "Dr. Dyott's Family Medicines" that he sold. This strongly suggests that he did not actually consider himself to be counterfeiting these products but instead believed that he was manufacturing goods and selling them under their proper names. In other words, the terms "Dr. Robertson's Family Medicines" and "Dr. Dyott's Family Medicines" pointed to a set of products that were manufactured according to certain recipes. If Kerrison in fact knew the recipes for these products—which was likely, given that he was one of Dyott's former employees—then, at least from his perspective, he had every right to manufacture them and sell them under those names. After all, Dyott had no patent on them. Indeed, from Kerrison's perspective, he would have been acting dishonestly to call them anything else. Dyott, in response, argued that only his products were "genuine," strongly implying that Kerrison did not know the recipes involved and was therefore fraudulently selling goods under the wrong name.

Here we see both the importance and the dangers of secrecy to manufacturers. Under the doctrine established in *Singleton v. Bolton*, anyone could legally manufacture and sell products under names such as "Dr. Robertson's Family Medicines," just as anyone could manufacture and sell products under names such as "Godfrey's cordial" or "Essence of Spruce." Indeed, as long as manufacturers used roughly the correct recipe, they probably believed that they were in fact manufacturing the products in question and thus assumed that they *should* call them by those names—after all, if Peter Kerrison actually used the correct formula for "Dr. Robertson's Infallible Worm Destroying Lozenges," then what else could he properly call the resulting medicine? There was no other name to refer to it by, and to call it something else would have been fraudulent. Dyott's accusation that Kerrison's products were spurious imitations thus depended upon the assumption that he retained exclusive control over the recipe for the products in question. Yet as long Dyott kept his formulas secret, there was really no way to establish whether Kerrison's products were made according to his recipes or not. As long as one product appeared and acted more or less similarly to other products sold under the same name, who could really say that they were not in fact the same thing?

Secrecy was thus a double-edged sword. On the one hand, it prevented competitors from manufacturing the same product by restricting access to the formula in question, thereby protecting the name and reputation of the product that the original manufacturer had successfully built through his or her investment of time, labor, and money. At the same time, however, the use of secrecy also meant that competitors could claim to have access to the original recipe—and thus to claim a legitimate right to manufacture the product in question—in a way that was impossible to disprove without actually revealing the formula in question. It was a difficult problem. Manufacturers responded to the dilemma by loudly accusing their competitors of fraud and spending tremendous amounts of money on advertising, both to convince the public of the quality of their own products and to undermine the reputations of their competitors.[48] There was little else they could do to stop manufacturers who, they felt, stole their good name. Courts offered little recourse. As I argue in the following chapter, an emergent body of trademark law offered manufacturers little protection.

Druggists initially watched the growth of the patent medicine trade with little concern. Preassembled remedies were popular and convenient to sell, and few druggists were initially concerned about their use. Indeed, as late as the first decade of the nineteenth century, even those who explicitly rejected quackery had few qualms about selling patent medicines.[49] By the 1820s, however, a small but growing number of apothecaries had become concerned about the drug market and begun efforts to reform their trade. At the heart of this incipient effort was the belief that secrecy allowed unscrupulous manufacturers to use inferior, inert, or even dangerous ingredients. This not only seemed to threaten the health of the public; it drove the price of other goods downward and placed respectable pharmacists in the difficult position of having to choose between selling inferior products at a low price or trustworthy products at a higher one. Perhaps even worse, the use of secret ingredients meant that druggists could not compound the products themselves. This seemed the exact opposite of good pharmacy. In 1824, for example, the Philadelphia College of Pharmacy published the formula for eight well-known patent medicines in an effort to undercut the trade in preassembled remedies.[50] The goal was to allow pharmacists to compound these products themselves rather than having to rely on selling what might be inferior goods. A few years later, the college approvingly noted that the Medical Society of New York had published the formula for another popular patent medicine. "The Medical Society of New York merits the thanks of the community," noted a representative of the college, "for having stripped quackery of some of its mystery and borrowed plumes, and exposed, in naked deformity, its shallow and wicked foundation."[51]

This was an unusually strong statement for pharmacists at the time. Despite occasionally vituperative language, reform-minded druggists were relatively measured in their critiques of patent medicines, in part because they recognized that many of their peers were beginning to rely on the sale of these products for their livelihood. Reformers in the orthodox medical community were not nearly so circumspect. As the trade in patent medicines grew, orthodox physicians became increasingly concerned about medical patenting, secret ingredients, and other forms of what they took to be monopolistic quackery. In response, a small but growing number of physicians loudly denounced the growing trade in preassembled remedies, doing their best to fight what they saw as a menacing foe.

Orthodox Medicine and the Critique of Monopoly

Orthodox physicians in the first decades of the nineteenth century confronted a world in which disease was common and suffering both plentiful and difficult to treat. Although their levels of education varied, as did their practical and theoretical commitments, orthodox physicians in the late eighteenth century had generally believed that the diversity of medical problems they confronted could be understood and treated according to broad and unitary theories of disease. Deriving from the Enlightenment belief that unified principles can be used to explain the operation of the natural world, these theories understood the wide diversity of symptoms physicians treated as expressions of a single underlying pathology; as a result, physicians tended to practice in a relatively routinized way, doling out the same or similar treatments—such as bloodletting or the use of strong purgatives—for a wide variety of problems. By the 1820s, however, orthodox physicians had started to reject such broad theories and turn toward a more empirical approach to therapeutics. Diagnosis and the evaluation of treatment options was increasingly based on the experience of the physician and empirical observation of the case at hand, with treatment individualized for each case based on a wide variety of factors, including the patient's age, sex, and race, the season, environmental considerations, and other factors.[52]

This change in orientation was deeply intertwined with both the practical nature of medicine and a rapid expansion of medical knowledge. As John Harley Warner argues, one of the defining features of physician identity in the early decades of the nineteenth century was the willingness to *act*—and to do so according to the values of the orthodox medical community. Practical knowledge was essential to the identity of the physician, and this knowledge centered on the proper use of plants and other healing substances. Botanists, physicians, and others had long investigated the natural world in search of healing remedies, and as physicians began to turn toward a more empirical therapeutics, this process increased. The result was a rapid growth in knowledge about the healing properties of plants and other goods, supported by what an editorial in the *Medical Repository*—probably written by Samuel Mitchill—called the great "revolution" in medical publishing then taking place across the Atlantic world.[53] Journals and other publications from the period are filled with descriptions of new plants and compounds and new or varied uses for familiar ones. Physicians also tried to organize this expanding domain of knowledge into useful guidebooks and taxonomies, producing a growing number of texts intended to help guide physicians in the choices that they made. There was much to be learned, and human suffering was ubiquitous.

Elite members of the orthodox medical community approached this work with a rhetorical—if not always practical—commitment to medical science as a cooperative and benevolent enterprise. As self-described members of what they sometimes called the *republic of science*, elite physicians believed that their efforts to advance medical knowledge were part of a larger collaborative project dedicated to the common good, one that proceeded slowly and methodically over time.[54] They described medical science, both to themselves and to others, as a benevolent practice in which all contributed to a common fund of knowledge: "Facts when once ascertained, and experiments when once made," noted one physician in 1823, "are no longer the property of the individual but of the republic of science at large."[55] Medical science was thus rhetorically juxtaposed to the pursuit of self-interest and thus to the practice of medicine as a means toward earning wealth. While medicine itself might be practiced as a trade and pursued primarily for pecuniary gain, medical *science* was driven by nobler motives. It was difficult, elite physicians said; it required long study, the gradual accumulation of knowledge, and a selfless rejection of personal gain in favor of advancing the public good. The true physician was both benevolent and self-sacrificing—or at least that is what elite physicians told themselves. Medical science and the pursuit of self-interest were rhetorically distinct and mutually exclusive enterprises.

Notably absent from this rhetoric of benevolence, at least from today's perspective, was any significant concern about the conduct of human experimentation. Clinical experiments were understood as a normal part of the gradual process through which scientific knowledge was accumulated. It was, of course, obvious that experimenting with new drugs might have untoward effects on patients, but in general, testing new substances, or testing the use of familiar substances in new ways, was assumed to be benevolent in nature in that whatever harm occurred during the experiment would lead to greater benefits for others. There was a general sense among physicians about what constituted appropriate treatment of patients, and there were clearly limits to the risks that most physicians considered appropriate to take, but these ideas grew out of the overall frameworks that structured social life as a whole—including notions of gentility, gender roles, and racial hierarchy—rather than a clearly articulated ethical framework that regulated conduct in this area. Experiments were carried out in the combined contexts of hierarchal social relations, the therapeutic relationship, and the desire to advance scientific knowledge. Modern notions of informed consent were not at work, and, in general, physicians simply did what they believed to be appropriate as they experimented on their patients.[56]

Elite physicians thus conceptualized medical science as a benevolent, self-denying, and cooperative endeavor. Quackery, on the other hand, was characterized by its unscientific methods, its predatorial nature, and the greed of its practitioners. From the perspective of orthodox physicians, an important component of all this was the willingness to monopolize medical knowledge for selfish reasons, including through the use of patents, secrecy, and other means that seemed to interfere with the collaborative and benevolent nature of medical science. Where orthodox physicians freely shared information among themselves in order to advance medical science, the logic went, quacks monopolized and controlled information in pursuit of individual profit. Where orthodox medicine was dedicated to the public good, quackery threatened the public by restricting medical knowledge to private interests and thereby preventing other physicians from treating their patients as they thought best. Medical patenting, from this perspective, was—by definition—quackish in nature because it restricted the ability of other physicians to use the patented methods or remedies if they were deemed useful. In 1805, for example, John Kunitz of Philadelphia patented the use of leeches for bleeding. The medical community reacted with horror, calling the patent "unjust and illiberal" and "a knavish piece of monopoly."[57]

The critique of patenting had deep roots in the orthodox medical tradition. English physicians in the eighteenth century had understood themselves as distinct from midwives, astrologers, magicians, bone setters, and others who practiced various forms of healing that regular physicians loosely lumped together under the pejorative term "quackery." Of particular concern were nostrum vendors, who frequently obtained patents as a means to raise the prestige of their goods. To qualify for a patent, British vendors had to certify that their nostrums were made from an original recipe, but they did not need to demonstrate their efficacy or safety. Moreover, patent medicine manufacturers were under no obligation to reveal the ingredients of their goods to the public, a fact that struck English physicians, influenced by Enlightenment ideals, as both unscientific and dangerous.[58] By the end of the eighteenth century, English physicians had developed a strong critique of nostrum vending, understanding the selling of medicines with secret ingredients and the practice of obtaining patents on medicines in overlapping terms. Indeed, English critics typically used the terms "patent medicines," "quack medicines," "nostrums," and variations of these phrases in overlapping ways.[59] According to this tradition, patented medicines had secret ingredients, and medicines with secret ingredients might as well be patented. They were both nostrums and essentially the same thing.

Physicians in the young republic inherited this critique and combined it with a distinctly American hostility to monopoly. If a medicine was patented, it was reasoned, other physicians could not investigate it freely or prescribe it to their patients as they saw fit. Secrecy was also understood as a form of monopoly that undermined the progress of medical science and threatened the public. After all, if a remedy actually did have a new and useful therapeutic effect, perhaps as a result of using a newly discovered medicinal plant as one of its ingredients, then by keeping its ingredients secret the manufacturer unfairly limited the spread of useful knowledge that might advance medical science and benefit patients. If not, then at best the so-called remedy was little more than an effort to dupe the public into purchasing a useless good; at worst, it was an effort to conceal the use of dangerous ingredients. Whatever the case, secrecy was understood as an effort to restrict the free circulation of information about the product, thereby interfering with medical science and threatening the health of the public. From the perspective of therapeutic reformers in the orthodox medical community, patents and secrecy were thus overlapping categories. As one critic noted, "He who advertises a secret or patent medicine is aiming for the money of the credulous and ignorant, and when he has obtained it cares no more for them."[60] Monopoly had no place in the practice of a truly scientific medicine.

Elite physicians responded to the problem of quackery by working to reform their profession along what they considered both ethical and scientific lines. The critique of monopoly, and of patent medicines specifically, was an important part of this effort. The English physician Thomas Percival's 1803 text, *Medical Ethics; or, A Code of Institutes and Precepts, Adapted to the Professional Interests of Physicians and Surgeons*, thus denounced the use of patent medicines as "disgraceful to the profession, injurious to health, and often destructive even of life." "No physician or surgeon should dispense a secret *nostrum*," wrote Percival, "whether it be his invention, or exclusive property."[61] Percival's denunciation of secret nostrums, even if the "exclusive property" of the physician, points to the fact that concerns about patenting and secrecy were deeply intertwined in the English medical framework. As medical reformers in the United States thought about how to improve the practice of medicine in their own country, they drew heavily on Percival's system. The articulation and enforcement of ethical codes seemed central to the process of reform, and as Percival's work made clear, reform demanded the suppression of secret and patent medicines. Decrying the "pestilential touch" of "nostrum mongers and venders of infallible cures," one critic in 1812 thus noted that "a system of medical ethics must be taught and enforced, otherwise there is no security against those mean artifices to which some men resort to obtain professional business."[62] The goal was to use the "moral power" of the physician, exerted through the enactment of formal codes of ethics and the institutional apparatus of medical societies, to "relieve sufferings produced by patent medicines and the use of nostrums."[63]

One of the earliest formal codes of medical ethics adopted in the United States was the "Code of Medical Police" adopted by the Association of Boston Physicians in 1808. The association adopted Percival's language closely, including his prohibition on quack medicines as "disgraceful to the profession, injurious to health, and often destructive even of life."[64] In the 1820s and 1830s a handful of other medical societies followed suit and adopted formal codes of ethics to regulate the behavior of their members. These codes often drew liberally on Percival's system, combining it with their members' own views on these topics.[65] They also typically included bans on associating with quack doctors and recommending or dealing in nostrums and sometimes included explicit bans on holding patents. In 1823, for example, the Medical Society of the County of New York instituted a formal system of medical ethics. Among other provisions, it declared that "the right of a patent medicine being incompatible with the duty and obligation enjoined upon physicians to advance the knowledge of curing diseases, it constitutes quackery and cannot be professionally countenanced." The society also declared associating and consulting with quacks to be a form of quackery, and promised expulsion for those who violated the new code.[66]

By the 1830s the critique of patent medicines was widespread among therapeutic reformers in the orthodox medical community. The image of the patent medicine vendor as a monopolistic quack, driven by the desire for profit and uncaring about the health of the public, was rhetorically juxtaposed to the image of the orthodox physician as fundamentally benevolent and self-sacrificing. It was also intertwined with a developing critique of the means by which patent medicine manufacturers advertised their wares. From the perspective of reformers in the orthodox medical community, manufacturers like Thomas Dyott made claims for their products that defied reason; they mystified a gullible public and led people away from truly benevolent and scientific care. As one physician put it in 1808, "It is the lower classes of society that are more especially liable to be taken in by the false assertions of these infamous venders of poison, and these almost uniformly prefer the use of a patent medicine to the advice of a regular practitioner."[67] Clearly, there was a self-interested component to such arguments, which should be understood in the context of the fact that most physicians at the time earned only a modest livelihood at best from their trade. Yet there was a real concern for the good of the public at work as well. From the perspective of such critics, advertising was used to sell dangerous and ineffective goods to the ignorant; it exploited people's suffering and distorted the therapeutic market away from how it should properly operate. Patents, secret ingredients, and unethical advertising to the public were all part of the selfish efforts of quacks to exploit human suffering for personal gain.

There was also an important contradiction at the heart of this rhetoric. Although patents and secrecy might both have been means of restricting medical knowledge for personal gain and thus forms of monopoly, they were also quite different because of the basic fact that patents were temporary in nature and intended, at least according to supporters of patenting in other areas, to promote the public good. The Patent Office thus required descriptions of inventions, with the assumption that these descriptions would be available to the public and, after the patent expired, that the inventions would be put to use by other parties. Patents were thus, in an important sense, the exact opposite of secrecy: although they operated as a form of monopoly, they did so for only a short amount of time and were intended to promote the circulation of information about inventions in ways that secrecy did not. This irony was occasionally noted in the medical community; as one astute observer put it in 1836,

No patent is or can be granted for a secret process. The very meaning of the word "patent" is "open," public, not private or secret; and the very first prerequisite of a patent is that the inventor shall furnish the government with a written description of his discovery . . . so that at the expiration of the term during which the law secures the exclusive right of the discovery under a patent to the inventor, the public at large may be at full liberty to make and use such discovery.[68]

Such arguments were made only very rarely in the early decades of the nineteenth century. Far more common was the attack on patents and secrecy as overlapping and pernicious forms of monopoly.

The Conflict with Unorthodox Medicine

In 1825, Alexander Coventry, the president of the Medical Society of the State of New York, gave a lecture on endemic fever to the society. It was a scientific discourse on the nature of fever and a discussion of what treatments might be used to combat what was, at the time, a serious disease. It was also a biting attack on quackery. Coventry began his lecture by calling on his colleagues to suppress "the innumerable patent medicines, whose virtues are blazoned forth and fill the columns of every newspaper." Without referring to them by name, Coventry also denounced the followers of Samuel Thomson, practitioners of an alternate medical system that was gaining tremendous popularity. Coventry derided Thomsonians as "a set of impostors, whose impudence is only equaled by their ignorance" and bemoaned the fact that they "are allowed to rob and murder the good citizens, under the pretence of using only herbs and roots."[69] Like his colleagues, Coventry considered both patent medicines and Thomsonism to be among the worst forms of quackery.

By the early nineteenth century orthodox physicians had begun to distinguish themselves from other healers through efforts to consolidate professional authority and establish institutional structures such as medical schools, journals, and professional societies. In 1800, for example, there were just four functioning medical schools in the country; by 1825, there were eighteen, and the number more than doubled in the next two decades. At the same time, physicians worked to pass licensing laws and to suppress the practice of other forms of healing. State governments had imposed various restrictions on medical practice during the colonial period, but such efforts were sporadic and largely ineffective. In the early decades of the new republic this type of authority was increasingly shifted to professional societies that pushed for the authority to regulate the practice of their trade and suppress what they saw as irrational and dangerous forms of quackery. In 1800 just six states had state medical societies. By 1830, nearly every state in the country had one, each clamoring for the authority to license physicians.[70]

The conflict with Thomsonian medicine was deeply intertwined with this process. Samuel Thomson grew up in rural New Hampshire in the late eighteenth century. Between about 1806 and 1808 he developed a theory of health based on the idea that all animals are made up of the elements of earth and water and that these elements are kept in motion by the elements of air and fire; illness develops when the ability of the fire element to generate "vital energy" from food is reduced by some external force, such as cold or damp weather. Over the next few years, Thomson developed a six-part medical system that he claimed cured all illness through the use of induced sweating and various emetics, most notably a powerful plant called lobelia. In 1812, he wrote a pamphlet describing his six-part system and outlining the rules for a "friendly botanic society" of people who purchased the right to use it. The following year he patented his method, and in 1822 he expanded his pamphlet into a book, *New Guide to Health; or, Botanic Family Physician.* Thomson traveled the country selling rights to his system, establishing societies of license holders, and proselytizing about the value of his system and the supposed evils of orthodox medicine, including the high prices orthodox physicians charged. Thomson did not coin the term "botanic"—it was used occasionally before he popularized his system to refer to medicines made from plants, as opposed to metallic or chemical remedies—but as the popularity of Thomsonian medicine rapidly spread across the country, the term became closely associated with his system and its various offshoots. As Thomson's system was popularized, it was both institutionalized and modified in many ways, leading to numerous factions and sometimes bitter debate about the proper practice of botanic medicine. By the late 1830s the botanic medical movement was a highly complex phenomenon that often had only a tenuous relationship to Thomson's original system.[71]

An important part of this movement was the development of so-called botanic medicines. Thomson's 1813 patent had six different components, each of which included descriptions of how to prepare and use various healing plants. The first two parts described how to use lobelia and cayenne pepper to cleanse the stomach and induce sweating. Part 3 involved a tea made from a combination of rosemary, the bark of bayberry or candleberry, and a choice of several other ingredients to "scour the stomach, promote perspiration, and repel the cold." Part 4 was a recipe for "bitters for correcting the bile," while part 5 involved a syrup made from peach kernel or cherry stones, gum myrrh, water, sugar, brandy, and wine intended to "strengthen the stomach" and "restore the digestive powers." Finally, Thomson included a formula for "rheumatic drops," which were made from gum myrrh, wine, camphor, cayenne pepper, and spirits of turpentine.[72] Thomson appears to have manufactured remedies based on his system and sold them through his growing network of license holders.

In 1825, for example, a man named John Locke of Boston advertised that he had acquired a license from Thomson to practice medicine according to his system. He also advertised that he carried "all kinds of Botanic Medicine prepared by Dr. Thomson, and . . . warranted genuine; several kinds of which will be found very convenient for families."[73]

Thomson faced a significant amount of competition in this trade. As Thomson's system and its variants were popularized, druggists began to advertise that they carried botanicals such as bayberry bark, cayenne pepper, and lobelia to the developing market. Enterprising druggists also began to manufacturer and sell medicines that they claimed were made according to Thomson's principles, selling them under names such as "Thomsonian medicines" and "Botanical Drops."[74] Thomson considered these types of products to be counterfeit goods. He regularly warned people against distributing his system without his authorization, and he attacked the flourishing trade in these products as a violation of his patent rights. In 1821, for example, Thomson accused a man named Elias Smith of selling counterfeit botanic medicines.[75] Smith manufactured his products according to the rough outlines of Thomson's system, but he also varied some ingredients and changed the amounts of others. Still, Thomson arranged to have Smith arrested for the "trespass" on his patent, and Smith was held on an extraordinarily high bail of $3,500. The case came to trial in early 1822. According to one account, after the judge in the case read a copy of Thomson's patent, he "observed that there was no *patent*, though there might be good medicine described" and dismissed the case. "After about nine years of worry, threatening and advertising to individuals," noted the account, "it comes out that other men have as much right to prepare and use medicine, as *the Doctor*."[76]

Following the trial, Smith advertised the fact that Thomson had lost his patent and that anyone was free to use his system or products made according to the precepts of it.[77] Thomson, in turn, secured a second patent on his system in 1823.[78] Thomson continued to threaten other manufacturers, but his efforts had little success in stopping the flourishing trade in botanic medicines. Smith continued to manufacture his products, as did numerous other small manufacturers.[79] In doing so, these competitors undercut Thomson's market by offering lower-priced and—according to Thomson—inferior goods. Referring to Smith's products in the early 1830s, for example, Thomson argued that "these cheap rights, and cheap medicine, will produce cheat practice." Pointing to one family in Boston that took Smith's cures with bad results, Thomson argued that "they have had a pilfered right, a counterfeit practitioner, poisoned medicine, neglect of steam, and no cure. . . . If the people want Thomsonian cures, they must employ Thomsonian doctors, and Thomsonian medicine, and pay Thomsonian prices; then they will not only have Thomsonian cures, but also health, at low prices."[80]

Debates about monopoly were also central to the fracturing of the Thomsonian movement. By the 1830s Thomson had lost the trust of much of the botanic medical community, which increasingly saw his control of the movement as an odious monopoly. "If he had confined himself to obtaining a patent, for any particular medical compound . . . there can be no doubt of his having a right to the exclusive sale of it," one botanic physician argued. "But to claim a patent right for his whole system of medical practice, is as we conceive the height of absurdity; for it would be an injurious and unjust interference with the rights of the whole community."[81] Other medical sects criticized Thomson and his followers as well. Eclectic medicine, for example, was established in 1827 by a physician named Wooster Beach as an offshoot of the Thomsonian movement. Beach focused on the use of plant remedies, but as the eclectic school developed, it also incorporated a variety of techniques from orthodox medicine and rejected the dominance of Thomson's system in the practice of botanic medicine. Eclectics also rejected Thomson's patents as an unethical form of monopoly. "The tendency and aim of the Thomsonian system," noted one eclectic medical journal in 1838, "is a total subversion of all medical science, and a substitution of a limited patent system of practice, founded upon the ignorance, prejudices, and dogmas of a single individual."[82] Beach himself cautioned against the dangers of "nostrums and patent medicines," and he denounced the "pure" Thomsonians as "rigid followers of . . . an illiterate, conceited, arbitrary, and selfish individual who obtained a patent for curing all diseases."[83]

Of course, from the perspective of orthodox physicians, Thomsonian medicine—and botanic medicine more broadly—was little more than a dangerous form of monopolistic and profit-driven quackery. In 1808 critics accused Thomson of sweating two children to death; in 1809 he was charged in court with killing one of his patients with lobelia. Thomson was acquitted on the grounds that, under New Hampshire law, anyone could administer medicine with the intent to heal and that there was no legal way to determine sound treatment from poor.[84] Thomson was harshly criticized not just for promulgating what orthodox physicians considered a nonsensical and dangerous system but also for keeping his system a secret and monopolizing it through a patent; both indicated his predatorial intent and the quackish nature of his supposed cure. Thomsonians more generally were derided as uneducated "patent doctors" who acquired their right to practice not through learning and knowledge but simply by purchasing a license.[85] A significant number of botanic physicians were legally charged with endangering the health of their patients, with some trials attracting a substantial amount of publicity in the popular press, and orthodox physicians bitterly attacked the idea that "any ignoramus" might be "transformed into a doctor," "Minerva-like, full grown and completely armed," from "the brain of this legislative Jupiter."[86]

The conflict between orthodox and botanic medicine lay at the heart of the vitriolic debates about licensing laws in the first half of the nineteenth century. Over the course of the first half of the century, a significant number of medical societies established formal codes of ethics, many of which included prohibitions on the use or promotion of patent and secret medicines. These codes were used to enforce norms of behavior and to distinguish orthodox medicine from Thomsonism and other forms of quackery. Violation could and did lead to expulsion from medical societies and the ruining of one's reputation, the destruction of social networks, and even the loss of the right to practice. In 1823, for example, the Medical Society of New York established a code of ethics that rejected medical patents as "being incompatible with the duty and obligation enjoined upon physicians to advance the knowledge of curing diseases."[87] Six years later, the state legislature prohibited the practice of medicine without a license or diploma from an incorporated medical society or medical school. The new law also required that physicians be members in good standing of a county medical society in the state in order to be licensed; as the county medical societies modeled themselves after and were closely aligned with the state society, this effectively meant that violations of the state society's code of ethics could lead to the loss of the right to practice. As a group of leading physicians in the state noted, the code gave the state society the power to "control, correct and punish all irregular acts or immoral habits of individuals." The "Medical Police," they noted, had sufficient power to "inspect or regulate" all licensed physicians in the state.[88]

From the perspective of botanic physicians, medical licensing laws were little more than an effort by orthodox physicians to unfairly restrict the practice of medicine to their own kind. Botanic physicians loudly denounced licensing laws as a "mockery of freedom" and "an unrighteous and oppressive monopoly," and they flooded legislative bodies with petitions demanding the repeal of what they considered oppressive licensing laws.[89] "We do not ask for the license of the Medical Society," noted a petition written to the South Carolina legislature demanding the repeal of that state's 1817 licensing law. "But we demand freedom from their tyranny."[90] From this perspective, Thomson's system was highly democratic in nature because anyone could purchase rights to it. It was not controlled by an elite group, and it was not used to extract unfair prices from a suffering population. Botanic physicians thus proudly adopted the name "patent doctor" as a sign of their populist orientation, the effectiveness of their system, and the implied critique of those who condemned patent medicines. Indeed, Thomson himself adopted the term.[91] Political debate about licensing laws was thus conducted, in part, through the lens of antimonopoly reform. Both sides saw themselves as battling against monopolistic forces that sought to bend the therapeutic market toward their own selfish ends.

One of the consequences of this struggle was the clarification of the relationship between federal patent law and state licensing laws. As a part of their legal strategy, botanic physicians sometimes argued that a license from Thomson constituted a legal right to practice medicine. Under this theory, Thomson's patent rights gave him the legal authority to use and license his system as he saw fit, federal patent law trumped state licensing laws, and the right to practice was transferred to those who purchased a license from him. Indeed, Samuel Thomson argued that this had been his primary reason for securing a patent; as he noted in the 1825 edition of his *New Guide to Health*, "In obtaining a patent, it was my principal object to get the protection of the government against the machinations of my enemies, more than to take advantage of a monopoly."[92] Botanic physicians across the country pursued this strategy when they got into legal trouble under the new wave of licensing laws, but they were unsuccessful. In 1836, for example, the president of the Albany Medical Society sued John Thomson, Samuel Thomson's son, for practicing medicine contrary to the laws of New York. In his defense, Thomson produced a license from his father and claimed that he had a right to practice medicine under its authority. The court thought differently and ruled that he had no right to practice medicine for a fee unless he had a diploma from some regularly incorporated medical school or society, as the law required.[93] Other courts came to similar conclusions, and by 1840 it was a well-established legal doctrine that patents on medicines did not confer the right to use those medicines if doing so violated the licensing laws of the state in question.[94]

The battle between orthodox and botanic medicine left a profound and lasting impression on the contours of orthodox medical thought. The critique of Thomson and his followers overlapped significantly with the growing concern about patent medicines, blending together into a generalized attack on quackery as monopolistic, unscientific, and dangerous to the health of the public. Orthodox physicians would remain hostile to unorthodox medical systems for many years to come. They would also continue to see drugs and related products that were manufactured according to different theoretical frameworks than their own as fraudulent and dangerous impositions on the therapeutic market. And they would continued to see the willingness of ordinary people to purchase such products as an irrational form of behavior driven by the duplicitous efforts of a predatorial industry. Such forms of self-help had no place in the developing therapeutic framework.

The Evils of Irregularity: Standardization and Early Therapeutic Reform

Concerns about the therapeutic market among the orthodox medical community extended beyond the critique of monopoly and the attack on what its members considered quackery. In the decentralized environment of early America, assumptions about the proper use of plants and other healing materials varied from place to place and practitioner to practitioner, depending on tradition, level and place of education, access to reference books, personal preferences, and other variables. What systemization existed across this diversity of practice tended to be the product of similar educational backgrounds, both formal and informal social networks, and the limited circulation of medical texts, many of which were either imported or reprinted texts from Europe. These texts were themselves quite diverse in their recommendations, were often of only limited availability, and frequently did not meet local needs because they detailed plants and remedies that were not readily available, while ignoring many of the medicinal plants that grew abundantly in the forests and fields of the young country. With the diversity of opinions about proper use, dosage, and other factors, worried observers in the medical community faced an unruly situation that they considered both dangerous and unseemly.

In response, early therapeutic reformers worked to develop a thoroughly American, and normative, pharmacopeia. As early as 1790, John Redman Coxe suggested to the Medical Society of New Haven County that an "American Pharmacopeia" be established. Members of the society found the idea intriguing and promised to help by submitting lists of native medicinal plants, providing both "their botanical & vernacular names, and virtues." The need for such a text was clear. As the society noted, "Nothing can be more obvious than the necessity of some standard amongst ourselves to prevent that uncertainty & irregularity which in our present situation attends the composition of the Apothecary, & the prescriptions of the Physicians."[95] Coxe appears to have abandoned the effort, but over the next several decades a small number of indigenous pharmacopeias were produced, including the *Pharmacopoeia of the Massachusetts Medical Society* (1807) and the *Pharmacopoeia of the New-York Hospital* (1811).[96] The latter was intended to serve as a standardized guide not just for doctors at the hospital but also for physicians in both New York and other parts of the country who were isolated from one another and faced a confusing diversity of formulas and substances in whatever texts happened to be available to them.[97]

The first edition of the *Pharmacopoeia of the United States of America* (*USP*) appeared in 1820.[98] In 1817 a young physician named Lyman Spalding had submitted a plan to the New York County Medical Society to establish a national pharmacopeia, and later that year the society formed a committee and sent out circulars across the country calling for the establishment of a convention dedicated to that goal.[99] Work proceeded rapidly, although the resulting document was not really national in scope—only marginal representation from areas outside the Northeast appeared at the supposedly national convention. Still, imagining themselves representative of the nation as a whole, this small group of physicians created a text that they declared to be good for the entire nation. It included 217 of the "most fully established and best understood" remedies and a variety of formulas for compounded preparations. It was also an explicitly normative work, designed to combat "the evil of irregularity and uncertainty in the preparation of medicines" by replacing existing texts and resolving conflicts about the proper compounding of remedies.[100]

One of the chief goals of the *USP* was to standardize not just medical practice but language itself. As the introduction noted, "[the *Pharmacopeia*'s] usefulness depends upon the sanction it receives from the medical community and the public; and the extent to which it governs the language and practice of those from whose use it is intended."[101] Establishing standardized names for the remedies that were included was crucial to this goal. The authors of the first USP considered the use of standard names in medical practice to be vitally important in order to avoid confusion and errors in both medical prescribing and dispensing of drugs; after all, the same plant might be referred to by a variety of different vernacular names, and the same vernacular name might refer to different plants. At the same time, names in the *Pharmacopeia* needed to be practically useful if they were to be adopted by physicians in their daily practice. The authors of the first *USP* thus made it a principle to use simple, single terms for what they called the "officinal names" of medicines wherever possible. For example, whereas the Edinburgh Pharmacopoeia used *Convolvulus Jalapa* and the London Pharmacopoeia used *Jalapa Radix,* the first *USP* simply used *Jalapa*. As the authors noted, "The advantages of this mode are, that the name stands in the nominative case; that it expresses the medicine, and nothing else; that it is short and explicit, and does not require to be mutilated in practical use, as long names will inevitably be." However, the authors of the *USP* were also conscious that they were using names that did not strictly apply to the plants in question from a scientific perspective; as they put it, "The words *Jalapa* . . . and other [names] of the same kind, are not, strictly speaking, the names of any plants, but the names of drugs and medicines." In order to preserve scientific accuracy, the first edition of the *USP* therefore listed both the officinal name of the substance in question and the "scientific term, or the systematic name of the plant, animal or mineral, from which each substance is derived." For chemical substances, the authors followed the "modern language of chemistry" where appropriate, although "a few names of inconvenient length have been superseded by shorter terms, on previous pharmaceutical authority."[102]

The first *USP* was not received particularly well. Critics denounced it as confusing and filled with errors and other problems.[103] Competing revisions were issued in New York in 1830 and Philadelphia in 1831, and after some controversy, the 1831 revision served as the basis for a subsequent revision in 1840, after which revisions were issued once a decade. In order to help ensure accuracy and practical utility, pharmacists were brought into the revision process beginning with the 1840 edition.[104] In 1833, Franklin Bache and George B. Wood issued the first edition of the *Dispensatory of the United States of America*. While the *USP* listed drugs that members of the convention considered the most important and well understood and provided recipes for important preparations, the *Dispensatory* offered detailed discussions of their history, chemical properties, taxonomical questions, and other issues. Beginning with the second edition, the *Dispensatory* also carried a section detailing the use of "unofficial" remedies. These were drugs that had been demonstrated to be useful by respectable physicians that Bache and Wood deemed to be in wide enough use to merit description but had not yet made it into the *USP*. The *Dispensatory* was a thoroughly practical text, and it quickly went through multiple editions as Bache and Wood issued regular revisions to keep up with the introduction of new remedies. It quickly became, as Gregory Higby puts it, "the de facto guide of pharmacy practice during the middle half of the 1800s."[105]

During this period the physicians involved in the revision process of the *USP* faced an increasingly complex semantic environment. Scientific nomenclature in the early nineteenth century was surprisingly fluid: names changed rapidly as new terms were invented and categorizations debated, and multiple scientific terms for the same thing—or what might (or might not) be the same thing—were common, much to the dismay of many naturalists.[106] At the same time, the discovery of new plants, advances in chemistry, and other scientific and technological innovations generated a constant flow of new things that needed to be named and categorized. By the early 1830s, for example, botanists and others had identified a large number of different types of cinchona bark and debated endlessly about how to categorize them, whether different varieties were or were not the same thing, what was a true variety of cinchona and what belonged to other genera, and other issues. As Wood and Bache noted in the second edition of the *Dispensatory*, "To form a correct and lucid system of classification is the most difficult part of the subject of bark, which is throughout full of perplexities."[107] At the same time, chemists had isolated what they believed to be four major alkaloids of the bark and a variety of secondary alkaloids, which were manufactured by the chemical treatment of the four "primary principles." There was a significant amount of debate about the relationship of these substances to one another, how these substances should be derived, and other complex issues.[108]

From a practical standpoint, such taxonomical complexity could not be captured in the *Pharmacopeia*. Simple, straightforward names were crucial to the effort to establish a common language on which to base prescription and dispensing practices, and the "classifications of naturalists," as one observer put it in reference to the 1831 edition, "would be but a clumsy and inconvenient substitute for that more simple nomenclature which general use has sanctioned, and which the convention has adopted."[109] Subsequent revisions of the *USP* thus included short officinal names for the remedies listed as well as one commonly used vernacular name and one scientific name for the remedy in question. In the 1831 edition, for example, the officinal name "Cinchona" was paired with "Peruvian bark" and *Cinchona lancifolia*.[110] At the same time, however, the growing taxonomical complexity of botanical and chemical knowledge could not be fully avoided. After all, different varieties of cinchona bark might have different therapeutic uses; they might be of greater or lesser quality, cost more or less, and otherwise be different from one another from both scientific and practical perspectives. There was thus an important tension between the need to use short, practical names that referred to general types of goods and the increasing complexity of scientific knowledge and commercial practices.

Much of this tension was resolved through a division of labor between the *USP* and the *Dispensatory*. The *USP* served as compendium of the most important treatments available and worked to standardize the language of medical and pharmaceutical practice by promulgating officinal names with relatively little regard for the complexities of taxonomy. The *Dispensatory*, on the other hand, served as a guide to the practical conduct of pharmacy itself and included a wealth of information about the chemistry, natural history, and taxonomy of each remedy, including detailed discussions of the different species of the plant included under officinal names. For example, the 1831 *USP* listed three subtypes under the officinal name "Cinchona"—"Cinchona Flava," "Cinchona Pallida," and "Cinchona Rubra," with three corresponding vernacular names: Yellow Bark, Pale Bark, and Red Bark. The *USP* left it at that. The second edition of the *Dispensatory* (1834), however, devoted a significant amount of text to explaining the complexities of each of these categories, offering detailed descriptions of the various species of plant included in each, taxonomical debate about the relationship of these species to one another, and other issues. Wood and Bache also included a description of a fourth category, "Carthagena Barks," as well as a discussion of both "false barks" and detailed discussions of the various alkaloids, their chemical properties, and their relationship to the various species of the plant.[111]

The shortened names in the first *USP* were not generic in the sense of the term that we mean today. They did not, in other words, refer to the common name of a substance in distinction to technical scientific names on the one hand and proprietary names controlled by specific manufacturers on the other. Instead, these were generic names in the sense that they operated at a higher taxonomical level than the names of the different subtypes of the product in question, a naming formulation that drew heavily on the taxonomical system developed by the Swedish natural scientist Carl Linnaeus during the eighteenth century.[112] In Linnaean taxonomy, plants and animals were categorized into both genus and species, among other categories, and were given Latin names that combined both. By the early nineteenth century this system was widely used in the classification of plants, with the name for the genus of a plant sometimes being referred to as its *generic* name and the name of its species sometimes being referred to as its *specific* name.[113] Other sciences sometimes used the terms in a similar way. In nosology, for example, the *generic* name of a disease—such as "fever"—referred to a general type of ailment, while *specific* names referred to particular types of diseases within the genus, such as "nervous fever" or "putrid fever."[114] Occasionally, words such as "opium" were also understood to be generic in the sense that they pointed to a general type of good that included different subtypes. Opium could thus be divided into at least three different species named after the place of their manufacture, including *Smyrna, Constantinople,* and *Egyptian.*[115]

The shortened names included in the *USP* were generic in this sense. They referred to general types of goods that, in many cases, could be divided into a variety of subtypes. However, with a few exceptions, they were listed in the *USP* with no further enumerated subclassifications and thus masked the rich debates among botanists, chemists, and others about taxonomy and nomenclature that frequently operated at the level of subtype. Rhubarb, for example, was included in the 1831 *USP* under officinal name "Rheum" and the scientific name *Rheum palmatum.* However, taxonomical debate about the varieties of rhubarb were highly complex; as one text from 1834 put it, "The precise species of rheum, which afford the rhubarb of commerce, is still a subject of doubt and uncertainty."[116] Even in cases in which subtypes were listed—cinchona, for example—these subtypes themselves masked significant taxonomical complexity. The same can be said of manufactured products and the formulas used to describe them: the formula for "Vinegar of Opium" in the 1830 *USP,* for example, included both "opium" and "nutmeg," but there were no statements of what *type* of opium or nutmeg should be used (botanists recognized at least three distinct species of nutmeg at the time).[117]

Perhaps more important, these names were not juxtaposed to proprietary names that could be monopolized by manufacturers. This is not particularly surprising: although competition among those who sold botanicals, opium, and other medicinal products was often stiff, trademark law was only just beginning to coalesce in the United States, and it simply did not occur to wholesalers or manufacturers of chemical products or other prepared remedies to try to protect their interests through the legal control of a thing's name. Even in the case of newly discovered alkaloids or other chemical substances—cases where the things themselves had never before existed as isolated therapeutic and commercial goods—the names given to these products seemed to be the names of the things themselves. As such, they were, like the name of the sun or the moon, simply a part of language itself. They were the common property of all, and the possibility that other names might adhere to them did not yet exist in any significant sense. The distinction between *generic* and *brand* names had not yet been born.

Monopoly and Ethics in the Antebellum Years

By the 1840s the drug trade in the United States had become a diverse and complex enterprise, one part of the chaotic economic growth of the time. The antebellum drug market was a vibrant place, filled with an almost endless number of products to satisfy the appetites and meet the needs of the public. Distributors worked at the local, regional, and national levels and combined the importing, manufacturing, and processing of drugs with the rapidly growing trade in both indigenous and imported botanicals. A growing number of domestic manufacturers produced tinctures, elixirs, pills, and other manufactured products, including what sometimes seemed like a tremendous flood of preassembled remedies made with secret ingredients. A small but growing number of manufacturers also produced chemical preparations intended for the therapeutic market. Drugs were popular and widely consumed; as one physician noted, "We swallow [them] as greedily as the catfish swallow's the schoolboy's bait."[1]

Orthodox physicians and their allies among reform-minded pharmacists looked at the drug market with deep concern. Adulteration, widespread secrecy in ingredients, sensationalist advertising, and other problems deeply disturbed therapeutic reformers in the antebellum period. Among orthodox physicians, the rejection of quackery was intertwined with both a belief in their own benevolence and the rhetorical distinction between medical science and private interest; as physician Robley Dunglison put it in

1844, "In science there is but *one* republic," to which "we freely furnish" and from which "we ourselves select from all that which is good."[2] Articulated through a framework that positioned science and monopoly as oppositional categories, reformers in the orthodox medical community were highly critical of both patenting and secrecy in drug manufacturing, as well as related problems such as adulteration, duplicitous advertising, and association with quackery. Among reform-minded pharmacists, the critique was similar but less biting, in part because pharmacy was recognized as a commercial enterprise. In both cases, however, reformers sought to rationalize the therapeutic market by denouncing quackery and working to suppress the use of patent medicines.

Therapeutic reformers in the antebellum period pursued their goals haltingly and with mixed results, confident in the ethics of their cause but with few real successes; by the outbreak of the Civil War, most laws restricting the practice of medicine to orthodox physicians had either been repealed or went unenforced, and what new laws reformers managed to pass were largely ineffective. Despite a small handful of victories, the drug market remained, from the perspective of early therapeutic reformers, a troubling place. Yet even as efforts at reform were largely unsuccessful, the critique of monopoly had a profoundly important effect. It led to the emergence of a segment of the drug industry that self-consciously adopted the ethical norms of the orthodox medical community and embraced the prospect of science as a cooperative, ethical, and benevolent enterprise. So-called ethical manufacturers rejected the use of patents, secrecy, and other monopolistic practices in an effort to distinguish themselves from patent medicine manufacturers and other forms of what they took to be quackery. At the same time, however, changes in both patent and trademark law began to shape the drug market in important ways. In the years following the Civil War, these changes would underlie ethical manufacturers' gradual embrace of monopoly.

The 1836 Patent Act and the Question of Novelty

In 1836 Congress passed a major revision of the patent law. As Steven Lubar has suggested, the new law was a Jacksonian answer to the contradictory attitudes toward patenting at the time. By the early 1830s robust patent rights were understood by many people to be an integral part of economic growth that promoted the general welfare. Yet monopoly— whether in banks, corporations, or numerous other forms—was also the target of significant public critique, and patents were widely understood as one form of monopoly. The 1836 law thus reproduced much of the basic framework of the 1793 Patent Act, including the essential fact that patents could be taken out for any "new and useful art, machine, manufacture, or composition of matter" or any new and useful improvement on the same. Yet there were also extremely important changes in the revised law, perhaps the most important of which was the reestablishment of an examination system for applications. By establishing a process of professional review, the new law promised to restrict patents to true inventions that would, supposedly, benefit the public. It also appeared to treat patentees without favoritism, reflecting the democratic ethos of the time.[3]

Despite the significantly more rigorous application process, what seemed to be a tremendous number of patents were issued under the 1836 patent law. Between 1790 and 1836 about ten thousand patents had been issued; three decades later, more than eighty thousand had been.[4] This enthusiasm for patenting resulted in part from the shift to equity courts to resolve patent disputes. Originally established to bring judicial flexibility and mercy to the common law tradition, equity courts operated under more flexible rules than other courts, verdicts were made by judges rather than juries, and they were intended to promote general principles of justice rather than the strict application of legal doctrine. By the 1830s and 1840s the equity system was being used to untangle increasingly complicated business transactions, including patent disputes, through judicial verdicts rather than jury trials. While some critics believed that equity courts gave judges "despotic" powers, most saw their efficiency as good for business and judicial verdicts as protecting inventors against the passions and ignorance of the common man.[5] Equity courts allowed supposedly sober, gentlemanly judges to make highly technical decisions and to have flexibility in the rules of evidence. As Lubar writes, "The growth of equity—its triumph over the common law—allowed judges to take patent law into their own hands. A hearing before a judge, followed by an injunction, became the general rule in patent cases." This helped transform patent law into a reliable basis for economic growth, encouraging businesses to increasingly rely on patents.[6]

However, despite the rapid growth of patenting, the shift to an increasingly friendly legal environment for resolving patent disputes, and a patent system that clearly allowed medicines to be patented as "compositions of matter," drug manufacturers did not share the enthusiasm for patenting. Between 1836 and the outbreak of the Civil War only a tiny number of patents were granted for medicinal goods. One reason for this may have been the public antagonism toward patenting in general. In the early years of the republic, patents had been acquired on medicines at least in part to enhance the reputation of the product, a function they had long served in England. The widespread critique of monopoly during the Jacksonian period probably meant that patents lost whatever luster they may once have had in this regard. Manufacturers may also have believed that patenting actually threatened the commercial viability of their products. Because of the importance of secrecy to the business strategies of many companies and the fact that obtaining a patent required that the applicant disclose the recipe to the Patent Office, patenting threatened manufacturers with the possibility that their formulas would become public knowledge. The extent to which manufacturers recognized this possibility is not clear, but it was probably another reason for the small number of patents issued for medicines.

At the same time, skepticism at the Patent Office toward patents on medicines also played an important role. As Kara Swanson has pointed out, the establishment of an examination system meant that a new group of patent experts were placed in the position of evaluating applications. These examiners rejected a significant number of the applications that crossed their desks: in the first year following the passage of the act, the first patent examiner rejected about 75 percent of all applications, compared with a rejection rate of between 25 percent and 67 percent over the rest of the antebellum period.[7] These rejections were based in part on the failure of applicants to demonstrate sufficient utility for their supposed inventions. Although details of the process remain scarce, the utility requirement appears to have led patent examiners in the antebellum period to reject many of the applications for medicines that they examined. It also appears that patent examiners relied on both testimonials and, in at least some cases, experiments to establish the effectiveness of supposed inventions. In 1843, for example, the commissioner of the Patent Office noted that that although "some very valuable discoveries have been made in the healing art," in order to "prevent injury and imposition, it becomes necessary, to a certain extent, in cases of patent medicines, to call for tests of their efficacy. In several cases their value has been proved by direct experiments; and in others, testimonials of the most creditable character have been received, vouching for their genuineness."[8] The following year, noted the commissioner, "Many unsuccessful applications have been made for patents for medicines professing to be infallible cures for various diseases."[9] The exact number of patent applications for medicines is probably impossible to know, so it is not really possible to determine the ratio of medicinal products that were rejected. Whatever the actual number of applications, however, few patents were approved for medicinal products.

This included medicinal substances made by chemical manufacturers. American chemical manufacturing grew significantly in the three decades before the Civil War, particularly in Philadelphia and other urban areas in the North. American chemistry was practically oriented in the antebellum period, and the arts of chemistry and pharmacy maintained close ties. Manufacturing chemists, for example, had helped establish the Philadelphia College of Pharmacy in 1821 as a means to improve chemical science in the bustling city, and the college maintained close relations with the city's manufacturing chemists throughout the antebellum period. Leading members of the effort to standardize the practice of pharmacy who were associated with the college, such as Franklin Bache and George B. Wood, worked closely with affiliated chemists such as James Curtis Booth.[10] At the same time, a small but growing number of manufacturers specialized in producing chemical products specifically for the therapeutic market. Some of these manufacturers were part of a developing segment of the drug industry that self-consciously conformed to the norms of the orthodox medical community. In doing so, they contributed to the emergence of what historians have called the "ethical" wing of the pharmaceutical industry (a term I use in a nominal rather than normative sense). As I explain below, these manufacturers generally refrained from securing patents on either their products or the methods that they used to manufacture them. In addition to paints, varnishes, and numerous other products, chemical manufacturers thus produced a variety of substances that were used medicinally but almost never patented.

However, patenting grew increasingly important in other segments of the chemical industry during the antebellum period—as the patent officer pointed out in 1843, "The application of chemistry to the arts presents a vast variety of patentable subjects, and a most fruitful field for inventors."[11] Vulcanized rubber provides an important example and is worth examining in some detail. By the late 1830s, the rubber manufacturing industry had become highly competitive, but it also faced significant problems because rubber products tended to melt in hot weather. In 1844 Charles Goodyear obtained his first patent on the process of vulcanization, and Goodyear then licensed the rights to make specific products using his method to individual manufacturers.[12] Goodyear's process had significant drawbacks—he had rushed to secure the patent for commercial reasons—and it took another four years of work by Goodyear and a number of other inventors in the field before high-quality vulcanized rubber could be reliably produced. At the same time, Goodyear and his licensees instituted numerous lawsuits against competitors for violating his patent, and by 1848 cartels had begun to form among licensees that used the patent to keep competitors out of the field. In what Cai Guise-Richardson calls "a brilliant maneuver," in 1849 Goodyear's backers convinced him to get his patent reissued in a way that covered innovations in the field that had been made over the past five years; as one critic at the time asked, "How is it then, that a man who claimed so little in 1844, should pretend, in 1849, to so much?" Goodyear and his backers then used the new patent to consolidate control over the rubber market over the course of the next decade.[13]

Goodyear's ability to control the rubber market depended on the fact that the reissued 1849 patent was, as Guise-Richardson argues, "less focused on a specific process and more conceptual" than his original 1844 patent.[14] In the reissued patent, Goodyear described how a conceptual advance—the application of heat to caoutchouc—was at the heart of his invention rather than the development of any specific manufacturing techniques. In doing so, Goodyear dramatically broadened his claim to include numerous innovations in the field that had been made over the past five years by other inventors, such as the use of steam in vulcanization.[15] Goodyear also asserted that the reissued patent covered not just the general process used to make it but also vulcanized rubber itself. This was a controversial assertion. By the 1840s it was clearly understood that previously existing things could not be patented, whether previously known or not.[16] At the same time, novelty was increasingly understood through the doctrine of reduction to practice, which courts took to mean that the inventor of something was the first person to sufficiently perfect it to make it useable.[17] Principles of nature, of course, could not be patented in the abstract, but methods that harnessed those principles in practical form could be.[18] However, the distinction between methods and the results of methods remained confusing. In composition patents, the method for producing a product and the product itself were often thought to be the same thing, or at least so closely related that, in most cases, patents on one covered the other—were patents on medicinal formulas, for example, patents on methods or on things? However, in other cases the article produced by a new method might not itself be new; in these cases the method and the result could obviously not be the same. This raised numerous difficult questions: Were different methods of producing the same thing infringements on each other? If a patented process was used in a new way to produce a new thing, was that an infringement? Similar difficulties characterized the relationship between methods and effects: Were methods of producing effects distinct from the effects them-selves? Could effects themselves be patented, and if so were different means of producing the same effect infringements on one another? These and other questions swirled around the question of how methods and the results of methods were related to one another.[19]

Goodyear's 1849 reissued patent raised these types of difficult ques-tions. Was his invention a method, or was it an attempt to patent a natural principle? Was vulcanization an effect that applied to a previously known substance, or was vulcanized rubber a new thing? These issues were exceedingly complex and took many years to resolve, but the question of the patentability of vulcanized rubber itself, as opposed to the processes used to make it, was settled in a series of cases that took place between 1850 and 1853 in New Jersey. In these cases, the circuit court ruled that since "we know the substance only by its qualities," Goodyear "may be said to have discovered a new substance" and therefore the substance itself was covered by the patent.[20] In the 1853 case *Goodyear v. Central Rail-road*, the court ruled that Goodyear's patent covered not just the method used to make vulcanized rubber but also the substance itself. Ironically, the decision hinged in part on a distinction the court drew between pro-cesses used to manufacture things and things themselves. Noting that "there is not only a distinction, but a wide difference between one who merely invents a new method or process . . . and the discovery of a new compound, substance, or manufacture, having qualities never found to exist together in any other material," the court ruled that "in the first case the inventor can patent nothing but his process, and not his composition of matter. In the latter, both are new and original, and both patentable; not severally, but as one discovery or invention." Patents on new methods that simply produced better versions of familiar things, therefore, did not cover the resulting thing itself. When the result of a new method, how-ever, was a new "compound, substance, or manufacture" that had "quali-ties never found to exist together in any other material," a patent might cover not just the process used to manufacture it but also the thing itself. As the court noted, the composition "now known as 'vulcanized rubber'" had "certain qualities not possessed by caoutchouc in its natural state, or any other known substance" and as a result was itself covered by the patent.[21]

The controversy over Goodyear's patent was only one of numerous decisions in the 1840s and 1850s that clarified the relationship between patents on methods and patents on the results of those methods. In 1851, for example, the Supreme Court ruled that in order to be patentable an invention must show a degree of ingenuity or inventiveness that went beyond that ordinarily employed by one skilled in the relevant art.[22] This decision settled many of the questions related to the degree to which something needed to be novel in order to be patentable; no longer could trivial improvements be patented. The following year, the court decisively ruled that although "the discovery of a new principle" is not patentable, methods or processes in which those principles are "embodied and brought into operation" are. The court also found that although the method used to harness a natural principle can be patented, the *effect* of the harnessed principle cannot be. "A patent is not good for an effect, or the result of a certain process, as that would prohibit all other persons from making the same thing by any means whatsoever," the court noted. "This, by creating monopolies, would discourage arts and manufactures, against the avowed policy of the patent laws."[23]

These types of decisions did several important things. Perhaps most important, they confirmed that in order to be patentable, a substance must be both new and the result of inventive activity; previously known things, compositions, or processes could not be patented, nor could natural principles or substances in their "natural state" such as caoutchouc. Even though natural principles could not be patented, however, inventions that reduced these principles to practical form could be, even if the invention was to a certain degree conceptual rather than embodying any one specific manufacturing process. Modifications of natural substances might also be patentable, as long as they had qualities not possessed by the substance in its natural state. These decisions also made it clear that, at least in some cases, there was a distinction between the processes used to achieve practical ends and the ends themselves. In the years that followed, this distinction would play an increasingly important role in chemical patenting as manufacturers realized that by separating their claims into distinct process and product patents, they would be able to retain one even if the other was invalidated.

At first these types of deliberations were of little interest to drug manufacturers. By the time of these decisions, "patent medicine" manufacturers rarely acquired patents for their products—indeed, by the early 1850s virtually no patents were being issued for simple formulas at all. At the same time, manufacturing pharmacists and chemical manufacturers that were closely allied with the pharmacy community also shied away from patenting. These manufacturers probably assumed that alkaloids and other chemical products were not patentable, in part because they seemed to be naturally occurring "elementary principles" and in part because drug patenting had long been associated with products made from secret nostrums, many of which were thought to be little more than frauds. More important, manufacturers with close ties to the pharmacy community self-consciously adopted the norms of the orthodox medical community and became cautious about patenting. In doing so, they constituted themselves into a distinct segment of the drug manufacturing industry, defining themselves and their goods as "ethical" in contradistinction to the supposed quackery of patent medicines and those who produced them. Yet despite the distance at the time between changes in patent law and the concerns of drug manufacturers, in the coming years the two would become increasingly intertwined.

Trademark Law and the Problem of Equivalence

I now turn to a discussion of early trademark law and the difficulties that many manufacturers found in protecting their interests in the highly competitive antebellum drug market. By the 1840s the patent medicine industry was intensely competitive and keenly focused on advertising as a basic component of its business strategy.[24] Manufacturers of pre-assembled remedies increasingly realized that sales were not based simply on informing the public of their goods, which they had long done, but on actively creating markets for their products through advertising that enhanced and promoted the reputations of their companies and products. Patent medicine manufacturers began to invest significant amounts of money producing long pamphlets, almanacs, and other vehicles for printing interesting narratives designed to capture the attention of readers. Some of these stories had little to do with the products in question, while others told stories about the remarkable value of their products and personal stories of self-sacrifice, creative discovery, and financial struggles to spread the word about their discoveries.[25] At the same time, manufacturers of successful products faced intense competition from other firms that claimed to be making the same products—and perhaps were. Not surprisingly, successful manufacturers described this type of competition as unethical counterfeiting that exploited the public. Although they did not always articulate their critiques in these terms, it is clear that they saw the efforts of their competitors as undermining their investment in time, labor, and money that they had made in developing their products. They also saw the efforts of their competitors as fundamentally unfair and as distorting the natural operation of a just market. Their competitors, understandably, had a different view of the situation.

John Sappington's remarkably successful pill business is a good example. Born in 1776, Sappington grew up in Tennessee, where he studied with his father to be a physician. In 1819 Sappington moved to central Missouri, where he practiced medicine, established a series of businesses, maintained a large plantation, and bought and sold slaves. In 1832 he began manufacturing and selling pills for the treatment of fever. Sold under the name "Dr. Sappington's Anti-Fever Pills" or some variation thereof, Sappington's pills became immensely popular, in part thanks to his business acumen and in part because they were made with quinine and were therefore unusually effective against fevers in the South, where, we now know, malaria was rampant.[26] Sappington does not appear to have patented his pills, which is not surprising, but he did conceal their ingredients as a means to protect himself from the fierce competition of the regional drug market. In doing so, however, he also alienated the local medical community. In 1837 physicians in Saint Louis organized the state's first medical society. Among other rules for membership, the society prohibited members from holding patents or promoting secret remedies. Members of the medical society denounced Sappington as a quack and appear to have refused him admittance to the society.[27]

As the popularity of his pills grew, Sappington faced numerous competitors who sold their own products under the name "Sappington Pills" or some similar formulation. In the early 1840s, for example, a pharmacist named Green Hill in Columbus, Missouri, began manufacturing pills in his store and selling them under this name. In 1842 a controversy over his pills broke out after several people in the area died after taking them. After examining the pills, a local physician suggested that they contained arsenic and should not be used.[28] The physician also accused Hill of manufacturing the pills himself, using an incorrect recipe, and selling them under a false name. This was a serious charge, since it amounted to an accusation of adulteration. Hill disputed the claim that the pills contained arsenic, and while he acknowledged that he manufactured them himself, he also argued that he was within his rights to do so because he had purchased Sappington's recipe from a company in Philadelphia and Sappington had no patent on them. "I have had considerable experience in the business of a Druggist and apothecary," he noted, and "know it to be their custom to prepare and sell any medicine not patented, or where the patent has expired, by limitation, when they can obtain a recipe which they consider as genuine." Given that there was no patent, Hill argued, "why should not any person who has the 'recipe' be at liberty to make them?"[29]

Hill also asserted that he had every right to call his products "Sappington's Pills" because the Philadelphia company that he bought the recipe from was "of the highest respectability" and he had no reason to question its genuineness.[30] From Hill's perspective, he had simply marketed and sold the pills under their proper name. As he pointed out, this was common practice among pharmacists: "Where is the Druggist in the United States, who does not prepare and vend 'Lee's Pills,' 'Bateman's Drops,' 'British Oil,' 'Godfrey's Cordial' . . . and numerous other medicines, some of which never have been patented and others of which the patents have expired?" Indeed, from Hill's perspective there was really no other name available for him to use if he wanted to market his pills honestly. Of course, this assumes that Hill was being honest and that the recipe he used was in fact Sappington's—if it was not, then he would have been properly accused of producing an imitation product, whether he did so unwittingly or not. Given the fact that Sappington kept his recipe secret, however, there was no way to know if Hill was manufacturing his pills according to the true recipe or not. Chemical analysis was one possibility for sorting all this out, as Hill's accuser suggested, but chemical techniques in the antebellum period were generally not sophisticated enough to make this type of reverse engineering persuasive. It is not clear from the available sources if anyone at the time used chemical analysis to compare the two sets of pills, but even if they had done so, there is little likelihood that the results would have settled the controversy.

I have been unable to find any evidence that Sappington ever pursued legal action against competitors such as Hill. As they confronted this type of competition, however, numerous other manufacturers began to seek legal protection for the names of their goods. There was a long legal tradition of seeking redress against unfair business practices through the use of torts, and as economic productivity increased and markets expanded in the young country, numerous manufacturers in multiple industries sought legal relief against what they considered to be unfair poaching on their reputation. By the 1840s, if not earlier, a small body of common law had been established protecting trade names and trademarks.[31] Even as they largely rejected the use of patents, patent medicine manufacturers began to draw on—and contribute to—this emergent legal tradition. Patent medicine manufacturers wanted to protect their investments in the reputation of their companies and products: as they increasingly invested resources in advertising and other promotional activities, they began to see their distinctive reputation as something that was valuable and needed to be protected.

Yet the developing body of trademark law offered little relief to patent medicine manufacturers. This requires a bit of explanation. Trademark law was grounded in the fundamental principle that no one has the right to pass off his own goods as those of another, since doing so was assumed to unfairly harm the interests of the original manufacturer.[32] As trademark case law developed, the essential characteristic that allowed a mark to be appropriated was thus taken to be what courts and legal observers called its "designating" character: in order to be eligible to be claimed, a name or word had to designate the manufacturer or businessperson who had an interest in the product with which it was associated. In other words, the key function of a trademark was to distinguish the product from other products by associating it with a manufacturer or business. One way to do this was for a manufacturer to make his own name or the name of his company a part of the name of the product. This type of designation clearly indicated the source of the product to the purchaser and clearly distinguished it from similar products. Outside of unusual situations, such as when two manufacturers shared the same name, this type of mark could be adopted by a manufacturer because it was both distinctive and designating.

On the other hand, names or marks that did not clearly designate a specific manufacturer or company could not be appropriated. In juxtaposition to marks that had a designating quality, names and marks that denoted "nothing more than the kind, character, or quality of the article" and did not point to "the origin or ownership of the article" could not be appropriated. Letters, for example, could not be adopted as marks unless they were accompanied by some sort of modification that made them distinctive, such as a circle with a color background. Nor could the common names of items, such as the word "tree." Instead, words of this type could be used "with equal truth and propriety" by all manufacturers.[33] As the Supreme Court put it in 1849, "The owner of an original trademark, has an undoubted right to be protected in the exclusive use of all the marks, forms or symbols, that were appropriated, *as designating the true origin or ownership of the article or fabric to which they are affixed*—but he has *no right* to an exclusive use of any words, letters, figures or symbols, which have no relation to the origin or ownership of the goods, but are only meant to indicate their name or quality."[34]

The idea that only words or marks with a designating character could be adopted grew out of the basic assumption that the role of trademarks was to convey information about the reputation of the company in question to the buyer and distinguish the manufacturer's goods from those of his competitor. This meant that trademark infringement was understood fundamentally through the lens of misrepresentation and the passing off of one's own product for what was made by another: the "essence of the wrong" in all cases of infringement, as one observer put it, "consists in the sale of the goods of one manufacturer or vendor as those of another."[35]

It also meant that competitors could adopt the name of a good for their own use as long as they did not deceive the public in doing so. The doctrine first established in *Singleton v. Bolton* that the control of a product's name depends upon having a patent on the product was thus an important part of early trademark jurisprudence.[36] From the perspective of the courts, having an exclusive right to the good in question was fundamentally linked to having an exclusive right to its name. This meant that if the original manufacturer had no patent, in cases of infringement the legal question at hand was whether or not the competitor deceived the public into thinking that his good was manufactured by the original party. If a competitor manufactured or sold the same product under the same name but presented it in a way that misled the purchaser into thinking it was produced by the original manufacturer, that was considered fraud and actionable. However, if the original manufacturer had no patent on the medicine, as long as the competitor did not imply that his version of the product was made by the original manufacturer, then the competitor had every right to sell it under the original name.

One of the most important antebellum decisions about this issue was the 1837 case *Thomson v. Winchester*. Samuel Thomson had sued competitor Hosea Winchester for selling medicines under the name "Thomsonian medicines" without a license. Thomson's second patent had expired the previous year. The court found that

> where certain medicines are designated by the name of the inventor, as a generic term, descriptive of a kind or class, the inventor is not entitled to the exclusive right of compounding or vending them, unless he has obtained a patent therefor; and if another person prepares such medicines of an inferior quality, and sells them, and by this means all medicines of that class are brought into disrepute, such inventor can maintain no action for any loss sustained by him in consequence thereof, unless they are sold as and for medicines prepared by him.[37]

Thomson v. Winchester was a groundbreaking decision. It is sometimes taken as the first case to clearly condemn the false passing off of one product for another.[38] Whether or not this is the case, for my purposes the significance of the decision lay more in the fact that it clearly asserted the doctrine that manufacturers did not have an exclusive control of the name of their products unless they also had a patent on those products. The decision made it clear that even products sold under names that *included personal names* operated as what the court called "generic" names—meaning that they covered all instantiations of the same product, no matter who manufactured them. This type of name thus operated at something like the level of the genus, in that such names applied to all examples of the product, but different manufacturers that produced the same good were not yet understood as manufacturing distinct "species" of the item in question. That would come later. For the time being, products that were considered equivalent to one another and were therefore properly called by the same name were thought to belong to a single and undifferentiated class of goods, and as long as there was no patent involved, everyone had a right to manufacture and sell them under what was taken to be their true names. As Francis Upton put it, commenting on *Singleton v. Bolton*, the court had found that "the plaintiff had no *exclusive right* in the medicine itself—and *therefore*, the defendant had an equal right to prepare and sell it under the same name that the plaintiff used— that, having become the proper name of the thing." This was, as Upton noted, "unquestionably, the established doctrine."[39]

During the antebellum period, manufacturers thus had little ability to control the names of their products unless they also obtained a patent on them. Even the most highly designating names, including names that incorporated the personal name of the manufacturer, could legally be adopted by competitors if no patent rights were involved because of the deeply held assumptions that trademark infringement was fundamentally a problem of misrepresentation and that the names of goods were available for all to use. As long as the company manufactured the good according to roughly the same recipe as the original manufacturer and made it clear that it was the company's own product, the original manufacturer really had no legal recourse. As the New York Superior Court put it in 1857, "A name may, in some cases rightfully be used and protected as a trademark; but this is only when the name is used as indicating the true origin or ownership of the article offered for sale; never when it is used to designate the article itself, and has become, by adoption and use, its proper appellation. . . . All who have an equal right to manufacture and sell the article have an equal right to designate and sell it by its appropriate name, provided such person is careful to sell the article as prepared and manufactured by himself, and not by another."[40]

In the two decades before the Civil War manufacturers of patent medicines thus began to turn to the courts for relief from what they considered unfair competition, but the courts, in general, did not provide a robust common law tradition of trademark rights that served their interests. Occasionally manufacturers tried to assert property rights in their names in other ways, such as by copyrighting their labels, but these efforts were generally frustrated as well.[41] Instead, patent medicine manufacturers kept the ingredients of their products secret and relied on aggressive marketing to maintain and expand sales, call the attention of the public to what they saw as the duplicitous behavior of their competitors, and protect their own good name. Of course, all this deeply offended the orthodox medical community, and throughout the antebellum period orthodox physicians leveled an unrelenting torrent of criticisms against "the host of secret and patent medicines" that, as one physician put it, "swarm[ed] like locusts."[42] Most patent medicine manufacturers do not seem to have cared. Their products were profitable, at least in many cases, and the critiques of physicians were not particularly damaging to their reputations among the public at large. Benjamin Brandreth, for example, was one of the most successful patent medicine manufacturers in the antebellum period, and despite the hostility of the orthodox medical community to his widely used "Brandreth Pills," in 1849 he was elected to the state senate of New York. By this point he was purportedly spending $100,000 a year on newspaper advertising for his pills. Clearly, the vociferous criticism of patent medicines had only a limited impact on public attitudes toward those who manufactured them.[43]

Still, some manufacturers began to carve out a different approach to the market, one in which they conformed to the ethical framework of the orthodox medical community and self-consciously rejected the use of patents, secrecy, and other supposed hallmarks of quackery. In 1844, as the popularity of his pills spawned endless imitators, Sappington published a medical text titled *The Theory and Treatment of Fevers*, in which he revealed that his pills were made from quinine, gum myrrh, and licorice.[44] The exact reasons for his doing so are unclear. Perhaps Sappington recognized that the widespread use of secret ingredients was dangerous to the health of the public and wanted to encourage other manufacturers to use the proper ingredients when they manufactured his pills. Perhaps he wanted to be accepted by the local medical community as a reputable physician. Whatever the case, it was certainly obvious to Sappington that trying to protect his financial interests by keeping his recipe secret and his acceptance as an ethical physician among his peers were mutually opposed possibilities. In Sappington's rejection of secrecy, then, we see the origins of a new approach to drug manufacturing, one that sought to reconcile the pursuit of individual profit with membership in the republic of science through the rejection of monopoly.

Yet conforming to the norms of orthodox medicine, at least to the extent that Sappington did, brought its own risks. Revealing the secret of his recipe may have helped his reputation with his peers, but it also meant that his competitors could now freely copy his recipe and make equally effective products. Correspondence from his sales agents during the late 1840s and 1850s are thus filled with complaints about his competitors encroaching on his markets, undercutting his prices, and selling high-quality imitations of his pills. As George F. Bicknell, one of Sappington's agents, noted, "None [had] been counterfeited to so great an extent" as Sappington's pills. Bicknell recommended a variety of strategies to counteract this, such as designing a special wooden box with Sappington's signature imprinted on it.[45] He also recommended a variety of means to improve sales, such as distributing a large handbill engraved with a picture of Hercules destroying the Hydra. Sappington never fully embraced the ethical norms of the orthodox medical community, and he continued to advertise and sell his products directly to the public. Yet in his limited efforts we see a dilemma that manufacturers in the emergent ethical wing of the industry would face for many years to come: without some form of monopoly, therapeutic innovation dispersed rapidly and was quickly adopted by one's competitors. The only response, it seemed, was to expand markets as rapidly as possible and thereby push the boundaries of what the orthodox medical community considered respectable behavior. "Humbug is the order of the day," Bicknell wrote. "If you do not keep up to the efforts of others, sales will be small. It is these efforts that make the medicines sell, not so much their intrinsic value. Too much economy here is no economy at all."[46]

The American Medical Association and the Formalization of Medical Ethics

In 1831 the physician and medical reformer Charles Caldwell called for a national organization made up of state medical societies organized under state law that would include "all educated and respectable physicians that reside within [the states'] limits." The goal would be threefold: to "throw light" on the diseases of each state, to "give to the members a more correct and intimate knowledge of each other," and "to regulate the practice by a code of ethics."[47] Caldwell's dream of a national medical organization would take more than fifteen years to become a reality. Yet in his call for the organization we see how the dreams of reformist physicians combined the desire to heal with a belief in the importance of codes of ethics in regulating medical practice. At the same time, running deeply through this vision was a profound critique of quackery, which described both patents and secrecy as antiscientific, unethical, and impositions on the public. Caldwell thus denounced "professional charlatanry," including "secret nostrums," "patent remedies," and the "numerous panaceas and catholicons . . . which constitute, in part, the disgrace of the age."[48]

In 1847 a national medical association was established that, a few years later, came to be known as the American Medical Association (AMA). The concern about patented and secret remedies was important enough that it was incorporated as a central part of the Code of Ethics adopted by the organization at its founding. The code stated that

equally derogatory to professional character is it, for a physician to hold a patent for any surgical instrument, or medicine; or to dispense a secret nostrum, whether it be the composition or exclusive property of himself, or of others. For, if such nostrum be of real efficacy, any concealment regarding it is inconsistent with beneficence and professional liberality; and if mystery alone give it value and importance, such craft implies either disgraceful ignorance, or fraudulent avarice. It is also reprehensible for physicians to give certificates attesting the efficacy of patent or secret medicines, or in any way to promote the use of them.[49]

This was a clear and forceful statement that condemned both the holding of medical patents and the promotion and use of nostrums, whether patented or made with secret ingredients. It was also widely influential among orthodox physicians, many of whom celebrated the code as the dawning of a new age in the profession. Medical societies across the country adopted it as their own: some adopted the code word for word, others modified it slightly, and some used it for inspiration to write their own codes of ethics, perhaps blending in some of the national organization's language. In 1849, for example, the Medical Society of North Carolina adopted the AMA's Code of Ethics to regulate its affairs. It also adopted a new constitution that, among other things, barred membership for "any physician who shall procure a patent for a remedy, or instrument of surgery, or who shall hereafter give a certificate in favor of a patent remedy, or instrument."[50]

The promulgation of the AMA's Code of Ethics and the institutionalization of formal prohibitions on medical patenting through parallel codes in state and local medical societies was not a trivial matter. The AMA quickly gained a significant amount of authority over the daily practice of medicine by the linking of its code to membership in the national organization and the parallel enforcement of ethics codes at the state and local levels. In 1852, the AMA adopted a resolution stating that "no medical society shall have the privilege of representation which does not require of its members an observance of the code of ethics of this

Association."[51] This meant that for a state or local medical society to participate in the new association, it had to prohibit physicians both from holding medical patents and from recommending patent medicines to their patients or the public. This was a significant escalation of the threat of formal sanction to physicians who dealt in patented goods, since membership in a medical society was, at least in some states, a condition for licensing. Indeed, the promulgation of the code and its subsequent enforcement should be understood as central means through which the orthodox medical community—under the leadership of some of its most elite members—organized itself into a body with a relatively unified ethical framework. This process was halting and highly contested, but its impact should not be ignored. It played out in heated debates about the expulsion of members from state and local societies, in bitter denunciations of quackery, and in ongoing debates about how to further institute and modify codes of ethics.[52]

The prohibition on patents and other forms of monopoly was thus a central part of what Robert Baker has called "the American medical ethics revolution" that took place as a result of the adoption and promulgation of the AMA's Code of Ethics. This was not so much a transformation in ideas as a practical effort to improve therapeutics by enforcing the commitment to an antimonopoly stance. Many physicians did prescribe patent medicines because they were convenient and increasingly popular. Others lent their names and reputations to manufacturers by penning testimonials that were used for advertising purposes. Still others entered the business themselves, manufacturing and selling products made with secret ingredients and perhaps even acquiring a patent on their goods. All this struck reformers as both deeply unethical and contrary to medical science. Medical science was a gradual process, they believed, one built on the renunciation of profit and the free circulation of knowledge within a community of respectable peers. Individual character, medical practice, and the advancement of science were deeply intertwined. This framework had little room for dramatic scientific advances. Nor did it recognize that the pursuit of individual interest might spur inventive activity, at least not when it came to medicines.[53] Patents and other forms of monopoly were thus clearly and unequivocally juxtaposed to "scientific medicine," in which, "in the spirit of true science," every physician should "feel himself bound to contribute to the general stock of knowledge as much as may be in his power."[54] The Code of Ethics was an effort to enforce this framework by simultaneously raising the character of the profession, suppressing quackery, and maintaining a strict boundary between medical science and monopoly.

Occasionally, however, physicians were forced to confront challenges to their ethical and therapeutic framework. Indeed, the founding of the AMA and the promulgation of the Code of Ethics took place at the same time that a major controversy over a medical patent was roiling the orthodox medical community. In 1846, the year before the national organization was established, a dentist named William T. G. Morton demonstrated the ability of sulfuric ether to induce general anesthesia at the Massachusetts General Hospital. Morton administered ether to a "thin, spare man" named Gilbert Abbott, who, according to a later account, was "suffering from a tumor on the jaw, composed of a knot of enlarged and tortuous veins." Before the introduction of general anesthesia, surgery was a grim, brutal affair. Patients suffered excruciating pain under the operating knife, writhing in agony as the surgeon cut open their flesh. "Under the first influence of the agent," however, Abbot "became flushed and exhilarated, but soon its more powerful effects became manifest, and in four or five minutes he lay as quietly and soundly asleep as any child." A well-known surgeon named John C. Warren then began the operation and, "seizing the bunch of veins in his hand, made the first incision through the skin." Remarkably, "the patient made no sound nor moved one muscle of his body; as the operation progressed, all eyes were riveted on this novel scene in eager expectancy and amazement."[55] It was a truly remarkable discovery. The ability of ether to make surgery painless electrified the medical community, provoking widespread enthusiasm, heated debate, and numerous efforts to find new anesthetic agents. Pain, it seemed, had been conquered. "I regard Etherization as one of the greatest meta-physical discoveries of the age," noted one physician. "What extraordinary phenomena it presents. The understanding awake and conscious—the body impassive. The soul becoming almost a stranger to this body, even in this life."[56]

To the endless dismay of the medical community, however, Morton also sought to monopolize his discovery by securing a patent on his "new and useful Improvement in Surgical Operations on Animals."[57] Morton also initially sought to conceal the fact that the gas he administered was sulfuric ether by referring to it as "letheon"; ether was well known to physicians at the time as a treatment for respiratory ailments, and Morton hoped to conceal the identity of the gas as a means of protecting his ability to sell rights to its use in surgery. This particular gambit did not last long (ether's distinctive smell soon gave its identity away), but Morton's successful effort to patent the surgical use of the drug caused a tremendous amount of consternation in the medical community. Immediately following the discovery, some physicians criticized the surgical use of ether as a form of quackery because of the patent. "We are persuaded that the surgeons of Philadelphia will not be seduced from the high professional path of duty into the quagmire of quackery by this will-o-the-wisp," declared one critic, before denouncing the use of "this new 'patent medicine.' "[58] Others recommended that Morton's patent simply be ignored. "Why, if I wish to avail myself of any of the possible effects of an article of our materia medica," wrote one physician, "why must I now purchase the right to use it, and use it as a patent medicine? I doubt the validity of such letters patent. It would seem to me like *patent sun-light* or *patent moon-shine*."[59] The surgical use of ether was either a quackish and monopolized imposition or it was a valuable therapeutic advance that belonged to all. For most physicians there were no other possibilities.

Morton's patent was profoundly troubling to the orthodox medical community in part because it did not fit their understanding of how medical science progressed. Although some physicians initially dismissed it as quackish, it rapidly became clear that the discovery was far too important to simply be ignored. Yet the sudden nature of the discovery and Morton's subsequent behavior did not fit the image of scientific progress as a slow, gradual affair built upon the open sharing of information within a community of peers. As Stephanie P. Browner has suggested, this contradiction was eventually resolved by the characterization of Morton as a lowly dentist not bound by the ethics of the medical profession.[60] As one physician at the Massachusetts General Hospital later noted, "We all thought it very strange that any regular physician would, even for a moment, consent to apply for a patent for such a boon to humanity as this promised to be. But Dr. Morton was the only person known as the administrator of the article and he was at that time a dentist only, and therefore not subject to the medical ethics contained in the unwritten law of the profession."[61]

From this perspective Morton himself was something of a barrier to scientific progress. In 1855, for example, the AMA appointed a committee to study the issue of how the national government could best support the development of the medical profession. The committee spent a substantial amount of time discussing patents and, not surprisingly, strongly denounced their influence on medical science. The committee members saw science as a gradual process built upon the cooperative efforts of myriad investigators, and therefore inventions or discoveries "in the healing art" could not be regarded by physicians as a form of private property and monopolized through patenting. Even the discovery of general anesthesia was the result of the slow progress of medical science, not the efforts of a single individual. As the committee noted, "Not until it had been stripped of its secrecy, and Letheon had become sulphuric ether, under the demands of the profession; not until the principles of medical science had been applied to the administration of its vapor by inhalation, by the profession . . . did anesthetic etherization become a boon to humanity, or anything else than a seductive and dangerous nostrum." Morton's patent was thus an "arrant piece of quackery," medical science gradual and benevolent, the Code of Ethics proper and just. Suffering humanity owed a debt of gratitude, but not to the efforts of a misguided dentist. "We submit," asserted the committee, "that whatsoever debt of gratitude the world has incurred in this behalf was due to the medical profession, and not to Dr. Morton."[62] Within a few years Morton's discovery had thus been successfully incorporated into the dominant narrative that positioned scientific progress and monopoly as mutually opposed categories. Morton might have been a charlatan, or he might have simply been a lowly dentist, but either way, his patent did not conform to orthodox medical ethics and did not merit respect.[63] It was not really his discovery in any meaningful sense. It was the product of a community of like-minded, benevolent, and self-sacrificing peers.

A small number of physicians dissented from this view. The most notable was Henry J. Bigelow, who wrote and published the initial article describing the surgical use of the drug about a month after Morton's demonstration. In it, Bigelow not only described the effects of the gas but also made a point to address the patent status of the discovery. Bigelow argued that the patent was ethically legitimate because inducing anesthesia was highly dangerous, and patenting the discovery and licensing the right to use it were the only practical ways to ensure that only respectable and qualified physicians used it for this purpose. Equally important, he argued, the fact that Morton intended to be "liberal" in granting licenses indicated that he did not really intend to monopolize its use at all.[64] Two years later, Bigelow argued that the initial skepticism toward ether and the continued concerns about its safety were the result of an irrational prejudice against patents. Bigelow believed that opposition to the use of ether based on its patent status was profoundly short sighted and needlessly subjected patients to unnecessary suffering. As he noted, "Those who stood between this agent of mercy and the world, those whose duty it was to deal out to mankind this inestimable blessing, have seen fit to refuse it to the unhappy victims of surgical art, and have condemned them to severe suffering which might easily have been avoided." Bigelow had little use for an ethics that prevented patients from receiving the benefits of this wondrous discovery. In his opinion, the patent status of the drug had nothing to do with its effectiveness and thus nothing to do with the question of whether it should be used or not. "A want of ability has been displayed in confounding the questions of ether patent and ether inhalation," he noted. "Those who have declaimed against the ether patent . . . have found it very difficult to give a candid hearing to the separate question of ether insensibility."[65]

Bigelow's analysis was the rare exception to the assumption that medical patenting was, by definition, an unscientific form of quackery. The orthodox medical community was able to maintain this position because there were few innovations in the drug market that truly forced its members to confront the fact that even patented medicines could be obviously and unequivocally useful. By the outbreak of the Civil War the primary case that might have forced such a reevaluation had been success- fully contained within the dominant ethical and scientific framework. Despite occasional defenses, Morton's behavior was either denounced or ascribed to his lowly status as a dentist, his patent was widely ignored, and his efforts to gain compensation for his discovery frustrated.[66] By the 1850s the controversy had largely died out within the medical community, although disputes about who should be credited with the discovery con- tinued to play out in other areas for many years. In 1862, Morton's patent was overturned by a court in New York, which ruled that the use of ether in surgery was a "discovery" rather than an "invention" and was therefore not suitable for a patent.[67] The ruling both invalidated Morton's patent and upheld the ability of the orthodox medical community to believe that medical science and monopoly were oppositional to one another. Yet in the coming years this belief would be increasingly challenged.

The Development of Ethical Pharmaceutical Manufacturing

During the 1840s and 1850s reform-minded druggists worked to trans- form the practice of pharmacy along what they considered both scientific and ethical lines. William Procter Jr., for example, played a central role in the organization of professional pharmacy in the middle decades of the nineteenth century. Born in Baltimore in 1817, Procter began working in a drugstore at a young age. He graduated from the Philadelphia College of Pharmacy in 1837, was elected to the college in 1840, and in 1843 opened a drugstore in Philadelphia, which he operated until his death. Procter maintained his pharmacy along what he considered ethical lines, and he worked assiduously over the course of his professional life to reform the pharmacy trade as a whole. He was an important figure in the organi- zation of the American Pharmaceutical Association (APhA), which was established in 1852, and for many years edited the *American Journal of Pharmacy*, the most important pharmaceutical journal in the nineteenth century. He believed deeply in the need for a truly scientific pharmacy and worked tirelessly to uplift the practice of his trade.

Efforts to suppress unethical practices were an important part of the broader effort to reform pharmacy. In 1848, the Philadelphia College of Pharmacy promulgated a set of pharmaceutical ethics that required that "any discovery which is useful in alleviating human suffering . . . be made public for the good of humanity and the general advancement of the healing art." It also required that "no member of this College should originate or prepare a medicine, the composition of which is concealed from other members, or from regular physicians."[68] The language of the Philadelphia College of Pharmacy's code was closely mirrored in the first code of ethics established by the APhA, which prohibited pharmacists from using "secret formulae" and discountenanced other forms of "quackery and dishonorable competition in their business."[69] Unlike reformers in the medical community, however, reformers in the pharmaceutical community confronted the basic fact that a growing number of druggists depended upon the sale of patent medicines to keep their businesses afloat. The original constitution of the APhA thus restricted membership to pharmacists who were willing to subscribe to its code of ethics. As Edward Parrish pointed out in 1854, however, by that time many ordinary pharmacists depended upon the sale of patent medicines for their livelihood. Establishing this litmus test for membership would therefore disqualify many ordinary pharmacists who "desire[d] a reform in their business, and would be glad to co-operate in the laudable objects of the association."[70] The following year, facing the realities of the situation, the association voted to drop the original code of ethics from its constitution. A new constitution was adopted in 1856, which instead declared it the obligation of pharmacists to "improve and regulate the drug market, by preventing the importation of inferior, adulterated, or deteriorated drugs"; "to improve the science and the art of pharmacy by diffusing scientific knowledge among apothecaries and druggists"; and "as much as possible to restrict the dispensing and sale of medicines to regularly educated druggists and apothecaries"—all of which implicated the widespread trade in patent medicines without explicitly attacking it or establishing a litmus test for membership.[71]

Reformers in the pharmaceutical community also succeeded in passing a small number of laws regulating the practice of their trade in the years before the outbreak of the Civil War. These laws included educational requirements for the practice of pharmacy, prohibitions on selling dangerous drugs without a prescription, and labeling requirements that mandated that dangerous products be sold under their "true name" and bear labels with the word "poison" on them.[72] In 1847, for example, the state of Maine passed a labeling law that required that patent medicines bear labels listing their formula, under penalties "that would make a common pedlar [sic] wince."[73] The law never seems to have been seriously enforced, but its passage illustrates the growing concern about secret ingredients, adulteration, and related issues among pharmacists and their allies in the medical community and state governments. Yet reformers were only able to push these types of efforts so far. Public opposition was, at times, strong, and patent medicine manufacturers were strongly opposed to such laws. "The tendency to make medicine *ad libitum* is a feature of the Anglo-Saxon race," noted one critic of such efforts, "[and is] duly inherited by the American people, which, whatever may be its faults, is as much their nature as is the love of political and personal freedom."[74] Freedom, in this view, was the freedom to engage the therapeutic market as one thought best and without the tyrannical hand of the government interfering in the choices that one made.

Despite their modest successes at passing restrictive pharmacy laws during the antebellum period, efforts by reformers to professionalize pharmacy and improve the drug market had a tremendously important effect. The critique of secrecy, adulteration, and other unethical practices overlapped with the critiques being leveled by the orthodox medical community, and taken together, the two created a niche market for supposedly ethically manufactured products. The patent medicine industry continued to flourish, of course, but a smaller group of manufacturers began to explicitly conform their business practices to both the norms of the orthodox medical community and the developing ethos among pharmacists. So-called ethical firms manufactured alkaloids, tinctures, fluid extracts, and other products made from ingredients and formulas listed in the *USP* or other highly respected texts. They manufactured, or claimed to manufacture, these goods according to officinal or other respected standards, and they refrained from making products with secret ingredients. They advertised only to physicians or pharmacists and generally refrained from making therapeutic claims about their products, since doing so was believed to undermine the therapeutic privileges of physicians. What little advertising ethical firms used tended to be sedate, listing the products offered by the company and avoiding the use of fancy fonts or graphics.[75]

Ethical manufacturers were also cautious about patenting. In general, pharmacists who thought about such things assumed that medicines themselves should not be patented, both because doing so would be an unethical form of monopoly and because such patents would be a misapplication of the patent law. "The compounding of drugs cannot be regarded as an invention," noted an 1849 editorial published in the *American Journal of Pharmacy* that was probably written by William Procter. "We have little hope that nostrum selling will be interrupted, but the sanction of the government ought not to be given to medicines as crude and incongruous as those which are patented."[76] However, pharmacists were more open-minded about patents on manufacturing processes. Leaders in the pharmaceutical community occasionally argued that patents on manufacturing processes were legitimate, that they acted as a stimulus to innovation, and that the ethical prohibition against patents in the medical community did not apply to manufacturing methods because pharmacy was, by definition, a form of commerce.[77] Still, despite such arguments, manufacturers in the emergent ethical wing of the drug industry generally refrained from obtaining patents on their manufacturing processes. Physicians did not have a clear understanding of the developing distinction between patents on methods and patents on goods themselves, and they bitterly attacked manufacturers who employed patents with little regard to such distinctions.

Ethical manufacturers (a term I use in a nominal rather than normative sense) thus avoided the use of patents and other forms of monopoly to protect their interests. Instead, these manufacturers—some of whom came from the developing chemical industry, some of whom came from roots in pharmacy, and some of whom were trained as physicians—based their business strategies on producing high-quality goods that conformed to the scientific and ethical norms of the medical and pharmaceutical communities. They also understood themselves as filling preexisting demand rather than creating new markets for their goods. Since medical science was assumed to progress through the circulation and gradual accumulation of knowledge, respectable manufacturers who discovered new medicinal plants or made other therapeutic innovations were expected to share their discoveries with the scientific community as a whole through publication in medical and pharmaceutical texts before commercially bringing them to market. Introducing a new remedy to market before it had been thoroughly investigated by the medical community would be a form of imposition: it would impose the manufacturer's own selfish interests on the cooperative and benevolent process of medical science and thus be seen as a form of quackery. If a manufacturer refused to disseminate information about its discovery, introduced it to market before it had been thoroughly investigated by the medical community, or resorted to using secrecy, patents, or other means to protect its commercial interests, it violated both the ethical and scientific norms of its target markets and thus risked damaging its reputation. Ethical manufacturers therefore largely refrained from introducing new products at all and instead generally limited themselves to selling remedies listed in the *USP* or the *Dispensatory* and to manufacturing derivatives of well-known basic ingredients. In this way, they conformed to the understanding of medical science as a cooperative and gradual endeavor within a community of peers. They also protected their reputations for ethical conduct and thus their ability to sell their products to their target markets.

As a result, ethical manufactures had to be quite cautious about how they introduced therapeutic advances to the medical community. Henry Tilden's efforts to introduce two different products serve as a useful illustration of this dynamic. Beginning in the mid-1840s a handful of manufacturers that served the eclectic market had begun to develop new methods for making liquid derivatives of botanicals. William S. Merrell of Cincinnati, for example, began marketing what he called "concentrated" eclectic remedies around this time.[78] In 1849, Tilden began manufacturing extracts using an innovative vacuum technique. The process led to high-quality products that were relatively standardized in strength and to the popularization of fluid extracts among orthodox physicians. By 1855, Tilden's manufacturing plant in New Lebanon, New York, employed thirty-five people and produced thirty thousand pounds of extracts per year, consuming nearly a million pounds of raw materials.[79] The development of Tilden's vacuum manufacturing process, in other words, was an important therapeutic advance and had a significant impact on orthodox medical practice. Clearly, patenting the fluid extracts themselves would have been considered unethical, even if it had been legally possible. Nor did Tilden patent his manufacturing process. Instead, he relied on the reputation of his company and his products to create and maintain his market. Indeed, Edward Parrish praised the firm in his influential *An Introduction to Practical Pharmacy* (1859), pointing to the "fine quality" of Tilden's products and noting that his company's "enterprise in this department of pharmacy [is] a great improvement in the quality of medicinal extracts."[80]

Tilden was careful to promote his firm in ways that conformed to the norms of the ethical medical community. Fluid extracts were considered modifications of the original substance from which they were derived and therefore, as the *Medical and Surgical Reporter* put it, "an elegant form of administering medicines" rather than something truly new.[81] As a result, Tilden's extracts were well within the boundaries of what was acceptable for an ethical manufacturer to produce and sell. In 1858, however, the company faced criticism on a number of fronts. Some of its extracts were discovered to have fermented, changing their composition and rendering them useless. Even worse, the company had introduced a sugar-coated "improved" compound cathartic pill "without calomel." This was clearly going too far. The assertion that the product was "improved" by removing calomel was a therapeutic claim that an ethical manufacturer had no business making. Such a claim "tends to cripple the practitioner who knows how to use medicine," noted the *Boston Medical and Surgical Journal*, "and it is unbecoming, in any pharmaceutical house, to put forth a pill of the style of Tilden & co.'s."[82] Thus, while fluid extracts were considered improved versions of the underlying botanicals and therefore conformed to medical orthodoxy, making therapeutic claims for products that exceeded or violated medical consensus threatened to damage the reputation of the firm. In the view of orthodox medicine, a commercial house had no right to "improve" the practice of medicine by innovating new remedies, a practice that came dangerously close to quackery. "They have insulted us," wrote the *New Orleans Medical News and Hospital Gazette*. "[They] have extensively preyed upon the credulity of the profession and inflicted infinite injury on suffering humanity."[83]

The same dynamic applied to introducing new botanicals. For example, take the efforts of George W. Carpenter to introduce "wahoo" to medical practice. Born in 1802, Carpenter began apprenticing in a drug store in Philadelphia in 1820. In 1828 he opened his own store and by the 1840s had built a successful manufacturing and wholesale business.[84] In 1842, Carpenter introduced the fluid extract of the bark of a plant he called "wahoo" as a cure for dropsy. Carpenter published a lengthy announcement of his new product in which he described how he had learned about the plant on a tour in the West. He also described the appearance of the plant, what conditions it grew under, and some of its medicinal properties, but he did not provide a detailed account of its botanical character.[85] He then began to sell his extract, despite the fact that the medical community as a whole had not yet investigated it—or, for that matter, even really knew what it was. The product began to attract both scientific and medical interest, and within five years it had developed a modest reputation as a useful product.

Not surprisingly, critics in the medical community attacked Carpenter for undermining scientific medicine and described the product as a "quack preparation."[86] The problem was that a variety of different plants were referred to by the common name "wahoo," Carpenter had not clearly identified the plant that he was using by a scientific name, and as a result he appeared to be monopolizing information about his remedy. What's more, he had commercially introduced his extract before the medical community had come to a consensus about its therapeutic utility; Carpenter had reversed the proper relationship between commerce and science, and his behavior thus bordered on quackery. It is not clear whether Carpenter was intentionally hiding the scientific name of the plant as a way of monopolizing its use, but sometime in the early 1850s he appears to have given either George Wood or Franklin Bache a sample of the plant for analysis, and as a result the 1854 edition of the *Dispensatory of the United States* identified the plant as *Euonymus atropurpureus* and included a summary of scientific knowledge about its medicinal properties.[87] This rescued Carpenter from the charge of quackery, but it also meant that other firms were able to exploit the market in the drug that he had developed. The following year, Tilden & Company began to sell a fluid extract of the same plant.[88]

Carpenter's experiences introducing wahoo to the therapeutic market point to an important dynamic. Ethical manufacturers were supposed to refrain from introducing new products to market lest they undermine the practice of scientific medicine and be labeled with the charge of quackery. Yet as the introduction of wahoo demonstrated, it was also possible for manufacturers to introduce new remedies to market before they had been thoroughly investigated, for these products to be adopted into medical practice, and *then* to have them incorporated into the *USP* or some other highly respected text. Ethical firms, in other words, had the ability to introduce useful products to market before the medical community as a whole had signed off on their utility. Yet doing so had little benefit and brought significant risk. As Carpenter learned, it was certainly possible to introduce a new product to market commercially and protect one's interest in that product by keeping its true nature secret or by some other means. Doing so, however, threatened to bring down the wrath of the medical community and thereby undermine the very markets that a manufacturer such as Carpenter hoped to cultivate. On the other hand, it was also possible to introduce a new remedy in an ethical way and, perhaps, enhance one's reputation by contributing to the progress of medical science, but this made it difficult to commercially benefit from the introduction of the new remedy. When manufacturers were unwilling or unable to protect their goods with secrecy or patents, innovations could rapidly be adopted by competitors. For ethical manufacturers, scientific innovation and the pursuit of profit were thus distinct domains. The two could not be joined without risking the charge of quackery and the potential destruction of one's reputation and markets.

Making Medicines "Officinal"

During the antebellum period the *Pharmacopeia of the United States* became an increasingly influential document. Following the dispute over the first revision in the early 1830s, additional revisions were issued in 1840 and 1851, with a fourth revision issued in 1863 during the middle of the Civil War. By the 1850s the *USP* had acquired a significant amount of rhetorical authority among therapeutic reformers, who increasingly believed that officinal drugs should be used whenever possible. From this perspective, physicians should prescribe officinal drugs using officinal names, and pharmacists should compound remedies using officinal ingredients and formulas. Pharmacists could also ethically dispense a manufactured remedy, such as a tincture or extract, that had been prepared by an ethical firm and that had been made according to the standards of the *USP*. This was increasingly considered both the basis of good medical and pharmaceutical practice and an ethical obligation toward one's patients and customers. As one committee of the American Pharmaceutical Association put it in 1854, "It is the duty as well as interest of apothecaries and druggists to advocate the use of the officinal medicines in lieu of the quackery of the day."[89]

Concerns about the relationship between names and things were at the heart of the effort to standardize the use of drugs. Prescribing and dispensing pharmaceuticals according to their officinal names seemed imperative to reformers in order to avoid confusion and standardize practice. Colloquial names for botanicals varied from place to place, of course, but even scientific names sometimes changed quickly. Chemical and botanical taxonomy was a matter of much debate, and multiple names for the same thing—or what might be the same thing—often seemed to be a source of confusion rather than precision. Indeed, names used in commerce were generally understood to be more stable and reliable than scientific names for precisely these reasons. In 1840, for example, the president of the Massachusetts College of Pharmacy urged the *USP* revision committee to stick to "ancient & well known names" wherever possible. He pointed to the substance "commonly known for a long period by the name of Corrosive sublimate" as an example of the "confusion & trouble" introduced by the use of scientific names in prescribing. Corrosive sublimate, he noted, had been given at least six different scientific names over the years, leading to many "disagreeable if not fatal results." Products that had long been known, he thus suggested, should be included under their common names—which were also those names used in commerce—while products new to science should be given names "such as are likely to be permanent and as scientific as they can be made at the time of their adoption."[90]

Nomenclature thus remained a difficult issue. In most cases, the physicians and pharmacists involved in the revision process decided on relatively simple names to refer to plants and other substances that botanists, chemists, and others frequently divided into multiple types. In the fourth revision of the *USP* (1863), for example, rhubarb was listed simply as "Rheum," with a brief note indicating that the term covered "the root of Rheum palmatum, and of other species of Rheum."[91] Such names and descriptions not only masked a tremendous amount of taxonomical complexity. They also subsumed multiple species into a single category and thus drew boundaries around the definition of the thing that the name applied to. There was therefore a significant amount of tension around the question of how precisely names should be drawn: overly general names might include versions of the substance that should not be included, while overly precise names might exclude versions that should be. For example, the fourth revision of the *USP* discarded the term "Cinchona" as a "generic" term (i.e., as a term that included all subtypes within it) but kept the three "varieties" of the plant that had formerly been subtypes of the broader term. Of course, these three types themselves masked significant taxonomical complexity (the identity of *Cinchona rubra*, for example, was a matter of much debate among botanists). Yet by eliminating the generic term "Cinchona" in favor of a higher degree of specificity, a number of types of cinchona that had previously been included under the generic name were now excluded from official status. This had the positive effect of excluding inferior types of the plant that had formerly been included under the general term, but as one observer noted, it also excluded "other kinds which are of great importance, which are by this course not recognized as officinal, and which formerly were embraced under a general head as Cinchona."[92]

Beginning in the third revision (1851), brief descriptions of included drugs were added to the *USP* in order to specify with greater detail what was included in the meaning of the officinal name. Botanicals were generally defined in relatively simple terms. Chamomile, for example, was listed under the official name of "Anthemis" and described as "the flowers of Anthemis nobilis." Formulas for tinctures, syrups, and other products that a pharmacist might prepare, however, were given in relatively detailed terms, while chemical substances and other manufactured products were given descriptions that included their color, melting points, the types of substances that they could be dissolved in, and other characteristics. These descriptions became more precise in the fourth revision, with additional descriptive characteristics used to define various products. Opium, for example, was defined in the 1863 version for the first time according to its morphine content, which was set at 7 percent.[93] Officinal names thus began to be linked to standardized characteristics. A tremendous amount of variation was thus subsumed under increasingly precise definitions and linked to presumably stable names in an effort to render different products made by different people in different times and places equivalent to one another. In other words, it was increasingly assumed that remedies made according to officinal standards and sold under officinal names were in fact the same thing from a scientific perspective.

There is another important issue that should be pointed out. Prescribing and dispensing drugs according to officinal names increasingly seemed like a solution to the distortions of the market caused by patent medicine manufacturers and other unethical actors in the drug industry. Secret ingredients, adulteration, and other forms of quackery not only undermined the practice of medical science, they seemed to distort the natural and fair operation of the market. "It is the duty as well as interest of the apothecaries and druggists to advocate the use of officinal medicines in lieu of the quackery of the day," noted a committee appointed by the American Pharmaceutical Association to study patent medicines in 1852. "It is the rightful interest of regular pharmaceutists to divert, in this manner, the thousands which now annually flow into the coffers of quacks, into their own limited stores, where of right it belongs."[94] Quackery may have been profitable, but it was in no sense fair. In a properly functioning market, free from the impositions of predatorial adulterers, patent medicine vendors, and other quacks, legitimate profit would naturally flow to those who rightfully deserved it. Therapeutic reformers, like many other people at the time, were deeply concerned about what seemed to be the growing power of monopolistic practices to distort the market and shape it toward their own selfish interests. And like numerous other people at the time, they began to seek legislative solutions to their concerns. The power of monopoly might be counteracted by the power of an activist state.

The most important early effort to regulate the drug market was the successful passage of legislation banning the importation of adulterated drugs. Beginning around 1845 the Philadelphia College of Pharmacy, the Medical Society of New Jersey, and a number of other organizations had begun to call for a federal law to deal with the problem of the large amounts of putrid drugs entering the country.[95] This caught the attention of Thomas Owen Edwards, who was both a physician and a congressional representative from Ohio, and he spearheaded a successful effort to pass a federal law banning the importation of adulterated drugs. Passed in 1848, the law mandated that imported drugs conform to the standards of strength and purity established by the *USP*—unless the product was manufactured in either England, Scotland, France, or Germany, in which case the standards in use in those countries would apply. Within a few months the federal government was actively enforcing the law. In New York, for example, inspectors rejected the entry of thousands of pounds of putrid goods, including almost four thousand pounds of jalap and more than six thousand pounds of rhubarb root. Yet the law also quickly exposed the limits of antebellum federal power. There were not nearly enough inspectors, inspections aroused the hostility of importers, the law was widely ignored, and numerous other problems plagued the effort. By the outbreak of the Civil War, the 1848 law was barely enforced.[96]

A few other laws were passed in the antebellum period that linked the *USP* to the practice of pharmacy. In 1849, for example, physicians in Philadelphia successfully lobbied for a tax increase on the sale of patent medicines in an effort to suppress their sale.[97] The law defined a patent medicine as any drug that was not sold under an officinal name and that did not conform to the standards of the *USP*. Reformers in the pharmacy community had mixed feelings about the effort. Although they were generally opposed to the sale of patent medicines, both Edward Parrish and William Procter criticized the law for lumping all medicines not in the *USP* with overtly quackish nostrums. Even reputable pharmacists dispensed many goods that were not officinal, and Parrish and Procter argued that the resulting increase in tax burden would make it very tempting for even the best pharmacists to "pay himself by increasing his assortment of the objectionable articles."[98] Despite criticisms of the tax, however, the basic idea of somehow legally enforcing the standards of the *USP* seemed reasonable.

The *USP* was intended as a normative text, but even outside the failure to enforce its standards using legal means, the document carried only very limited practical authority. In 1859, for example, the New York delegates to the 1860 revision convention sent out a questionnaire to medical societies in the state asking for suggestions about what new remedies should be included in the next revision. One physician replied that his society had nothing to contribute, as "country practitioners in this region are so much in the habit of preparing their own medicines and making extemporaneous prescriptions that they make but little use of the pharmacopeia."[99] Even those pharmacists and physicians who tried their best to conform to the officinal standards faced numerous challenges. The various editions of the *USP* were riddled with what critics took to be errors, and at times officinal nomenclature was difficult to use, appeared outdated, or was otherwise considered inappropriate for daily use. New remedies were also regularly introduced, and because the *USP* was only reissued once a decade even the strictest physicians and pharmacists sometimes found themselves recommending and dispensing goods that were not yet sanctioned. As a result, the effort to standardize practice led to a deep tension both within the orthodox medical community and among reputable pharmacists. A general fidelity to professional norms was important, of course, but conformity to what constituted ethical practice—at least according to therapeutic reformers—could not be enforced as much as its proponents would have liked.

One result of this was the beginning of a problem that has long characterized the practice of medicine: a disjuncture between medical science and formally articulated ethical norms on the one hand and the actual daily practice of physicians on the other. As long as orthodox physicians stayed within the general domain of what was considered reputable behavior they could freely use whatever drugs or remedies they considered efficacious for their patients, whether or not those drugs were included in the *USP*. In some ways this violated the developing norms of the profession, which increasingly suggested that the practice of medicine should be guided by the investigations and decisions of the delegates to the pharmacopoeial conventions. This idea significantly overlapped with the ethical prohibition against prescribing patent or secret nostrums, since it was grounded on the principles that the republic of science was cooperative and benevolent in nature, that the experts who made the decisions about what to include in the *USP* were drawing on a common fund of knowledge generated by all, and that the promotion of individual interest had no place in the practice of truly scientific medicine. Yet as therapeutic reformers began to clearly articulate and enforce their vision of what ethical medicine should be, there was also a way in which quackery began to be an almost inescapable reality. Virtually no one could fully conform to the increasingly rigid code of behavior.

Still, the *USP* held a tremendous amount of potential power to reform the therapeutic market. In 1860, for example, E. R. Squibb described the new pharmacopeia that had recently been adopted in Belgium. Squibb was a young physician who had recently opened a manufacturing business in Brooklyn.[100] Deeply committed to the ethical norms of his own profession, Squibb approached the market from the perspective of both an ethical manufacturer and a dedicated therapeutic reformer. He thus noted approvingly that according to royal mandate, under the law of Belgium all physicians must prescribe according to the weights, formulas, and nomenclature of the new pharmacopeia and "if they desire a remedy to be otherwise prepared, they must give the formula for it in their prescription, or at least indicate the Pharmacopeia in which it may be found." Moreover, he noted, all containers that held medical substances must bear, in plain language, "the names of the substances contained in them, these names to be in conformity with those used in the official Pharmacopoeia." The royal act also mandated that the offices, stores, warehouses, and laboratories of pharmacists be inspected at least once a year, at indeterminate periods and with no advance notice, and that all adulterated medicines be immediately confiscated. Squibb strongly approved of such laws. Indeed, he seemed a bit envious of the unilateral ability of the Belgian throne to force its citizens to conform to that nation's pharmacopeia. "Some such regulations as these," he suggested, "would very soon improve the materia medica, and revolutionize the pharmacy of this country, and would yield an element of certainty and uniformity in our practice of medicine hitherto unknown."[101]

In the Shadow of War

In 1866 Thomas Antisell and David Prince prepared a report on medical patenting for the American Medical Association. Antisell had been trained as a physician in his native Ireland before emigrating to the United States in the late 1840s; he was also a former chief examination officer of the Patent Office, had a long interest in chemistry, and had recently returned to Washington after serving in the Civil War. Prince was a surgeon from Illinois who developed important techniques in plastic surgery. It is not completely clear what prompted the report, but Antisell and Prince provided a strong and reasoned defense of patenting by physicians. While acknowledging that the "aversion" among the medical community to patents had a long history, they also noted that patenting in general was widespread and argued that there was no compelling reason why one set of inventors should be more "liberal" than any other. They argued that allowing medical patents would stimulate invention in the field, that so-called patent medicines were generally not patented, and even suggested that requiring patents on new medicines would suppress the use of nostrums because of the requirement to disclose ingredients in patent applications.[1] Predictably, the report was widely panned by the medical community. The *Medical and Surgical Reporter*, for example, provided a vociferous rebuttal, calling it "utterly subversive" to professional ideals and "part of that materialism which despises philanthropy, and sneers at the teaching of Christ as impracticable dreams."[2] The secretary of the American Medical Association even distributed a letter apologizing for the report being printed in the transactions of the association. Clearly, patenting had no place in a truly scientific medicine.[3]

The negative reaction to Antisell and Prince's report points to the continued assumption within the orthodox medical community that the republic of science had no place for monopolistic practices. It also points to the belief that therapeutic innovation and the commercial pursuit of profit should be kept distinct from one another and that new remedies were supposed to be thoroughly investigated by the medical community before being commercially marketed. To patent a therapeutic innovation was to reverse the proper order of things: it was to place private interest before the advancement of science. Yet during the 1870s this framework came under increasing strain as numerous new remedies were introduced to the medical profession. In addition to a large number of new plants and other remedies introduced by physicians and pharmacists, in some cases manufacturers who did not feel constrained by orthodox medical ethics commercially introduced effective new products before they had been thoroughly investigated by the medical community. They also sometimes patented and trademarked their products, prompting significant debate among physicians, pharmacists, and manufacturers in the ethical wing of the industry about their use.

Manufacturers in the ethical wing of the industry also increasingly chafed against the ethical prohibition on introducing new remedies. Of particular importance is the work of the pharmacist and physician Francis E. Stewart. Stewart worked closely with Parke, Davis & Company, which became one of the leading pharmaceutical manufacturers of the time. Stewart formulated a means for Parke-Davis to introduce new remedies to market in a way that did not antagonize the medical profession. Stewart discovered that Parke-Davis was able to commercially introduce new products by circulating information about new remedies to the medical community and publishing the results. However, this also prompted Stewart to argue for the importance of patents—and to a lesser extent trademarks—to the ability of firms to engage in this type of work. Central to this project was Stewart's reinterpretation of the ethics of intellectual property rights in pharmaceutical manufacturing. Like other reformers of his time, Stewart was deeply concerned about the relationship between commerce and science. Yet he also believed that the two could be productively, and ethically, intertwined.

Patenting and Pharmaceutical Manufacturing

Following the Civil War the United States underwent a massive and chaotic transformation as the country shifted toward increased manu-facturing, a nationally integrated market, and other characteristics of the coming industrial and corporate order. The patent system played an important part in this turbulent and sometimes violent process. In the two years following the end of the war, applications to the Patent Office increased dramatically, and the number of patents issued doubled to about thirteen thousand a year. The increased volume overwhelmed the Patent Office and, along with other perceived problems, prompted calls for reform. In 1870 a number of statutes related to patenting that had been enacted over the past three decades were consolidated into a single law, which, despite some modifications, maintained the basic system that had been in place since the 1836 revision. Among the more significant changes, the law extended the life of patents to seventeen years and estab-lished a special examiner to handle priority disputes in so-called interfer-ence cases. Over the course of the next decade, the Patent Office issued between roughly 13,000 and 15,500 patents per year.[4]

The relationship among patenting, economic growth, and innovation is a matter of much debate among historians and other scholars, but it was clear to observers at the time that patenting and other forms of mo-nopoly had immense economic consequences. During the 1870s contro-versy swirled around the ability of newly formed industrial corporations to dominate the economic landscape, the way in which various levels of government seemed to privilege certain economic actors over others, the ability of technological innovation to reshape social relations, and other complex issues. Debates about patenting were deeply intertwined with much of this controversy. These debates boiled down to a series of inter-related questions about the proper relationship between the individual right to a limited monopoly granted by patents and the operation of the market: Did patent-based monopolistic practices distort and thereby undermine the market by artificially inflating the prices of goods? What was the appropriate role of the federal government in promoting or restricting monopoly, and how did that role affect the market? Should the patent law be changed, and if so to whose benefit? The railroad and agricultural industries were particularly concerned about these types of questions, but virulent debate about patents, monopoly, and the market occurred across numerous sectors of the rapidly changing economy.[5]

The drug industry was one part of the rapidly changing economic order. The trade in botanicals expanded quickly as domestic agricultural producers increased production and, equally important, as the number of imported goods increased.[6] This expansion was at least in part a result of integrating both the South and the western border states into a national market. At the same time, these regions also provided an abundant supply of natural resources to be exploited. Locally grown plants, noted one observer, "have been more than usually abundant, as parties at the South have largely engaged in gathering them since the cessation of the war."[7] The emergence of a national and increasingly integrated market meant that even as local resources were extracted from rural areas, particularly in the South and western border states, the distribution of botanicals was increasingly concentrated in wholesale firms that were located in large cities, most of which were located in the Northeast. The growth of this trade is probably impossible to document, but it is clear that in their journey from plant to commodity, therapeutic substances increasingly traveled long distances before being consumed. One observer from Kentucky thus noted the "remarkable fact" that "our Louisville wholesale druggists depend upon the New York markets for their supplies of indigenous drugs, many of which abound and frequently are collected in our immediate neighborhood." While local retailers were still being supplied in "limited quantities" by "small gatherers" of local plants, this trade was quickly being supplanted by botanicals grown and gathered in various parts of the country that were then shipped to New York or some other large city before being brought to Kentucky. Indeed, the same observer noted, "When first making inquiries regarding the collection of indigenous drugs, I met with the invariable response, 'Inquire in New York.'"[8]

The manufacturing segment of the drug industry also experienced rapid growth in the years following the war. Union spending had provided a huge market for northern manufacturers during the conflict, and small firms such as Frederick Stearns & Company in Detroit (founded 1855), E. R. Squibb & Sons in Brooklyn (1858), and Sharp & Dohme in Baltimore (1860) grew from tiny manufacturing operations into large companies partially as a result of military spending; as Squibb noted in 1863, the demand from the military was such "that it has been difficult, with a small building, to keep up the supply," leading him to build "a large and moderately complete laboratory."[9] Numerous other firms were formed in the decade following the war, including Parke, Davis & Company in Detroit (founded 1866); Schering & Glatz in New York (1867); the Lydia E. Pinkham Company (1875) in Lynn, Massachusetts; and Eli Lilly & Com-

pany in Indianapolis (1876). These and other companies produced a tremendous diversity and volume of manufactured products for the expanding drug market.[10] In 1874, for example, Parke-Davis offered 254 types of
fluid extracts, 300 types of pills, 74 solid extracts, 46 medicinal elixirs, 53
concentrations, 23 medical syrups, 15 medicinal wines, 8 alkaloids, and
chloroform.[11] In 1875, Parke-Davis sold about $87,000 worth of products.
Five years later, sales had jumped to more than $450,000.[12]

Parke-Davis, like many other drug manufacturers at the time, advertised itself as "manufacturing chemists" because most of the goods it
produced were considered chemical products. Unlike companies that
manufactured fertilizers, paints, and other products that were not used
medicinally, however, the company hewed closely to the norms of the
orthodox medical community as it developed and marketed its goods.
Parke-Davis was not alone in this practice. Following the Civil War, the
manufacturing wing of the drug industry developed along the lines that
had been established over the past three decades, with a basic division
between so-called patent medicine manufacturers and so-called ethical
manufacturers as one of the fundamental components of how the production and trade of pharmaceuticals was organized. Patent medicine
manufacturers such as the Lydia E. Pinkham Company frequently introduced new remedies to market, kept the ingredients of their products
secret, advertised directly to the public, and at times made extravagant
claims for their goods—all of which aroused the wrath of the orthodox
medical community. Ethical firms, on the other hand, continued to
focus on manufacturing familiar goods from well-known ingredients,
largely refrained from making therapeutic claims for their products, and
refrained from advertising to the public. During the 1870s the differences
in business practices between these two segments of the manufacturing
industry were thus shaped in large part by how they positioned themselves vis-à-vis the orthodox medical community.[13]

Patenting played only a minor role in this process. In 1875, just fourteen patents were issued for products explicitly identified as medical
compounds or cures out of almost fifteen thousand total patents issued;
in 1880 just twenty-five patents for medicines were issued out of more
than thirteen thousand total patents.[14] From the perspective of patent medicine manufacturers, with some exceptions, the cost of acquiring patents, the potential risk that came with revealing their formula to
the Patent Office, and other issues meant that patenting was not generally a desirable means of protecting their interests. Secrecy, on the other
hand, allowed patent medicine manufacturers to monopolize effective
formulas, innovate freely with the ingredients of their goods, and adver-

tise their products in ways that may or may not have conformed to the generally held ideas at the time about the effects of the ingredients that they used. Secrecy also allowed manufacturers to distinguish their own products from what very well may have been similar products made by other manufacturers: as long as one's ingredients were kept secret, it was possible to make claims for the unique nature of one's product, even if it might in fact be a relatively ordinary preparation. Large manufacturers in particular benefited from this strategy because of their ability to marshal significant resources in advertising campaigns. Advertising was important not only for promotional purposes. It was also critically important because it was the only vehicle available to spread accusations of counterfeiting against one's competitors, a practice that also depended upon keeping the ingredients of one's products secret. This may explain why the handful of manufacturers who did obtain patents for their remedies all appear to have been small and relatively unsuccessful businesses, presumably unable to afford expensive advertising to promote or defend their products.

Ethical firms also refrained from patenting their goods but for different reasons. Patents, of course, were considered an unethical form of monopoly within the medical community, and ethical manufacturers recognized that they would face significant damage to their reputations if they patented their products. This was certainly an important consideration—possibly the most important—but ethical manufacturers also generally believed that patents on medicinal substances, as opposed to patents on manufacturing processes, contradicted the norms of scientific practice and probably were not valid anyway. Raw botanicals could not be patented, it was assumed, but neither could alkaloids, extracts, tinctures, or other products that were still considered more concentrated forms, or in some cases the "principle," of botanicals. It probably did not occur to most pharmacists to try to patent these types of products, but even if it had, they would almost certainly have considered it unethical to do so.

In general, ethical manufacturers were also careful not to introduce new remedies to the therapeutic market during the 1870s. The orthodox framework for the ethical introduction of new drugs that had developed over the past three decades clearly held that new drugs were to be thoroughly investigated by the medical community before being commercially distributed. According to this framework, physicians and pharmacists experimented with new remedies and with the use of familiar remedies in new ways, and the results of these experiments were then published in the medical and pharmaceutical presses. If a remedy seemed promising, then its experimental use spread and eventually a consensus emerged about its appropriate therapeutic use, at which point it was incorporated into the *USP* or some other highly authoritative text such as the *United States Dispensatory*. At that point, and not before, manufacturers could ethically distribute the remedy for commercial purposes. Manufacturers might therefore introduce new drugs to the medical community for testing, but they were not supposed to offer these drugs for sale or actively promote their use until sufficient investigation had demonstrated their utility and, ideally, they had been adopted into the *USP*. Doing so was viewed with deep suspicion by the medical community because it seemed to subordinate the cooperative methods and noble goals of science to the selfish pursuit of individual interest. Manufacturers thus occasionally introduced new drug plants that one of their representatives had acquired from some distant land, but for the most part they were quite careful to do so through the scientific literature and to refrain from promoting their use until the plant in question had been thoroughly investigated. The same can be said of new alkaloids or other chemical products. It was certainly acceptable for manufacturing pharmacists to investigate new remedies and to publish the results of their studies—indeed, doing so benefited their reputations—but they had to be cautious about how they introduced their discoveries to the medical profession lest they be accused of exploiting science for private gain. Most ethical firms thus commercially manufactured only familiar goods made with well-known ingredients.

Instead, ethical manufactures worked to improve their manufacturing processes and to develop competitive advantages by increasing productivity, reducing costs, and offering improved versions of familiar goods. Whether or not they secured patents on these improvements is difficult to determine, but it appears that they generally refrained from doing so. The pharmaceutical community was generally opposed to patents on medicines themselves, understanding such patents as a barrier to scientific progress, the right of physicians to prescribe whatever they thought best for their patients, and their own rights as pharmacists to compound remedies.[15] Patents on manufacturing processes, however, were not considered particularly disreputable as long as they did not lead to a monopoly on the resulting product itself; anyone was free to make pills, after all, and free to invent a better machine for doing so. During the 1870s inventors thus took out a variety of patents on improved methods to coat pills, make lozenges and extracts, and other manufacturing processes.[16] Discussions in the pharmaceutical press sometimes mentioned these patents, but generally without comment or concern about their propriety unless they were thought to be unfairly broad, to monopolize the resulting product, or to otherwise hinder scientific progress.[17] In 1874, for example, there was a minor debate about whether patented machinery could be "advertised" by reading papers describing them at the annual meeting of the American Pharmaceutical Association and, if so, whether it was appropriate for these papers to be published in the association's *Proceedings*. One critic argued that they should not be read at the meeting at all, while others argued that reading them would "do no harm" but they should not be printed. The debate was, at heart, about whether a scientific meeting, and the subsequent scientific literature that resulted from it, could be used to promote the commercial interests of a patent holder. There was no suggestion, however, that obtaining patents on machinery was itself unethical. As one observer noted, "If a gentleman gets up an apparatus for pharmaceutical uses which is an improvement, he deserves credit for it, and the thanks of all pharmaceuticists, and he is perfectly right in having his invention protected by a patent."[18]

The extent to which ethical manufacturers used patented machinery is difficult to determine, since patents related to drug manufacturing were taken out under the names of the inventors and typically did not include licensing information.[19] Certainly, a fair number of such patents were taken out—during the 1870s, for example, at least twenty-seven patents were secured for pill-making machines or improvements thereof. As far as I have been able to determine, however, the largest and most successful ethical manufacturers generally avoided the use of patented machines.[20] Despite their general acceptance, patents on manufacturing processes retained a whiff of commercialism that some found unbecoming.[21] More important, physicians had little understanding of the difference between patents on manufacturing processes and patents on products themselves. From the perspective of the orthodox medical community, all patents were unethical forms of monopoly, and any medicine protected by a patent—whether on the machinery used to make it or on the product itself—was, by definition, both unscientific and unethical. As a result, most successful ethical manufacturers appear to have either refrained from using patented manufacturing processes or, if they did, kept it very quiet. Acquiring such patents might tarnish their reputation somewhat among their peers, but it had the potential to seriously damage it among the orthodox medical community and thereby threaten their ability to market their products.

Manufacturers thus avoided introducing new remedies to the therapeutic market, worked to improve their manufacturing processes, and, it appears, generally avoided the use of patents on these improved manufacturing techniques. However, there was an important ambiguity in all this. Improved methods of manufacturing pills, extracts, and other products were typically understood to result in better versions of already known things, and the introduction of improved products was not generally thought to impinge on the therapeutic privileges of physicians. Yet new manufacturing methods also sometimes led to products that were substantially more powerful than previous versions of the same good, or that had different effects, or that appeared to some observers to be completely new products. These ambiguities meant that manufacturers who invented new manufacturing techniques sometimes affected medical practice in ways that went beyond what was assumed to be their proper role. Fluid extracts were an important example of this process: first introduced into the eclectic community, by the Civil War they had become popular among orthodox physicians. By the 1870s the orthodox medical community had fully accepted their use and considered them little more than a convenient way to administer the underlying botanical. Some critics, however, raised concerns about the fact that fluid extracts tended to be more powerful than the underlying botanicals taken in raw form, while others were concerned about extracts made from new combinations of ingredients. In each case, critics suggested that manufacturers were themselves introducing changes into medical practice and thereby subordinating the methodical and rational progress of scientific medicine to the pursuit of profit.[22] Were fluid extracts simply more convenient forms of already known remedies, or were they something truly new? The answer to this and many similar questions would play an important role in the future shape of medicine.

The Commercial Possibilities of the New Therapeutics

During the late 1860s and 1870s a large number of new remedies were introduced to medical science. Physicians, of course, continued to investigate new drug plants and other substances and to experiment with using familiar remedies in new ways. Pharmacists were also an important part of this process: they experimented with new compounds, modified familiar substances in new ways, and otherwise worked to advance the science of therapeutics. "The demand for new remedies is constantly agitated by our industrious co-laborers in the healing arts, the pharmacists," noted one physician in 1869. "During the last year they have added a host of new compounds to the already innumerable list of elixirs, pills, wines, syrups, powders, & c., heralding them through the columns of our medical journals as specifics for particular diseases."[23] Reflecting this expanding world of therapeutic possibilities, in 1872 a new quarterly journal was established, edited by Horatio C. Wood, titled *New Remedies*. It published descriptions of new compounds and formulas, means of preparing familiar remedies in new ways, descriptions of clinical experiments, and other such information. The first volume, for example, included formulas for treating infantile eczema and calculus nephritis, a description of a new elixir made from calisaya bark, and the results of an experiment on dogs in which a physician determined that "the injection of chloroform into arteries or veins . . . does not cause anesthesia, unless a sufficient quantity is injected to produce coma or death."[24]

The rapidly changing therapeutic landscape presented ethical manufacturers with an important dilemma. On the one hand, they were committed to maintaining an ethical stance toward manufacturing, in part because their markets in the orthodox medical community depended upon their reputations, but also because of their own self-image as promoters of pharmaceutical science. Their reputations mattered to them, and not just for financial reasons. Yet they were confronted with two problems: first, intense competition from manufacturers they considered less reputable than themselves meant that ethical manufacturers faced an increasingly difficult market for their products; the difficulty in distinguishing their own products from those of their competitors in the ethical field did not help matters. Second, ethical manufacturers recognized that new drugs were sometimes adopted by the medical profession outside of the strictly ethical framework in which scientific drug development supposedly operated. Ethical manufacturers recognized that therapeutic innovation had a tendency to outpace the traditional ethical framework that had been established for the investigation and incorporation of new drugs into orthodox medical practice. Deeply concerned about damage to their reputation, during the 1870s most firms in the ethical segment of the industry stuck to manufacturing well-known goods, selling them only to physicians, and largely avoiding therapeutic advertising. In doing so, they understood themselves to be filling markets that already existed. Yet they also confronted the basic fact that they were unable to enter these markets until they had already been firmly established. Nor were they able to commercially introduce new products, or new uses for familiar products, without running afoul of the ethical framework of the orthodox medical community. As a result, they began to see themselves as operating under a competitive disadvantage vis-à-vis firms they considered less ethical than themselves.

An important example is the introduction of salicylic acid. In 1874 Herman Kolbe, a professor of chemistry at Leipzig University, announced

that he had successfully synthesized the chemical, which occurs naturally in willow trees. Salicylic acid quickly attracted a tremendous amount
of medical interest because of its ability to lower temperature, reduce
swelling, and ease pain in cases of rheumatic fever. The chemical was
commercially introduced by a German firm that Kolbe helped establish,
the Chemische Fabrik von Heyden. Kolbe acquired a patent on the manufacturing process, which caused some controversy within the medical
community, but the patent was not very protective, and competitors
rapidly entered the market using other production methods. Since Kolbe
rejected other trappings of the patent medicine industry, most orthodox physicians appear to have decided that his violation of ethics was
not severe enough to warrant the dismissal of such an important therapeutic advance; although the chemical caused severe irritation to the
stomach, many physicians considered it, as *Scientific American* put it, "the
most important . . . antipyretic ever discovered." Salicylic acid was rapidly
incorporated into medial practice and became a common treatment for
rheumatic fever. Indeed, it was acceptable enough to the medical community that it was incorporated into the 1880 revision of the *USP*.[25] There
was some discussion within the medical community about whether this
was appropriate in light of Kolbe's patent, but defenders of its inclusion
pointed out that anyone could manufacture the product using different
methods.[26] "Neither the *name* nor the *product* . . . are proprietary, but,
on the contrary, open and free to the use of all mankind," noted one
observer. "In our opinion, therefore, there can be no objection made to
the reception of salicylic acid into the pharmacopeia on any grounds."[27]

The willingness of the medical community to accept salicylic acid
points to the way in which the scientific introduction of new remedies
sometimes outpaced the ethical framework in which drug development
supposedly took place. Ethical manufacturers were quite aware of this
dynamic and as a result increasingly believed that the system in which
they operated was too rigid, that it did not allow them to manufacture
new remedies in a timely manner, and that it held back scientific progress. As early as 1860, for example, E. R. Squibb had begun to suggest that
the *USP* needed to be revised more frequently in order to accommodate
the rapid pace of therapeutic innovation. Over the course of the next fifteen years he repeated the suggestion, and in 1877 he made a concerted
effort to convince the medical community that the *USP* should be issued
annually. Squibb framed his arguments largely in terms of promoting
science, suggesting that a revision once every ten years might have been
good enough in years past but that "in order to keep pace with the more
rapid progress of general medical science the revisions should be more

frequent."[28] Squibb thus proposed that the American Medical Association take over publishing the volume so as to expedite the approval of new drugs. However, his efforts were opposed by physicians involved in the revision process who, among other concerns, objected to the idea that the AMA had the "legal or moral right" to monopolize what should be a cooperative endeavor of interested parties. Unable to convince his colleagues about the need for reform, Squibb limited himself to manufacturing well-known and officinal goods.[29]

Squibb was not the only one dissatisfied with the state of affairs. Even the most reputable manufacturers sometimes crossed the boundary of ethical behavior and attracted criticism, both from orthodox physicians and from their colleagues in the pharmaceutical industry, as they struggled to stay profitable in a highly competitive environment. One important example was Frederick Stearns's efforts to commercially introduce "sweet quinine." Stearns had founded a small manufacturing laboratory in 1855 in Detroit. Stimulated by war spending, the company grew rapidly and soon developed a reputation for both ethical behavior and quality products. Stearns became a respected leader in the pharmaceutical community and served as president of the American Pharmaceutical Association in 1867.[30] The following year, however, he began selling a substance under the name "sweet quinine." This was a new term, and the pharmaceutical community initially assumed that it referred to a preparation of quinine. However, when William Procter investigated the substance, he discovered that it did not contain quinine at all but instead contained the alkaloid cinchonia and that Stearns had simply prepared it in a way to make it more palatable (physicians tended to prefer quinine because cinchonia was extremely bitter and patients had trouble tolerating its taste).[31] Stearns was quickly accused of adulteration, and Squibb introduced a resolution to expel him from the APhA. In his defense, Stearns argued that cinchonia and quinine were therapeutically identical—a claim that went against the consensus view of cinchonia at the time because of its taste—and that the use of quinine was nothing more than "fashion." He stated that his goal was simply to market a useful product to the medical community and that he felt it legitimate to "adopt as much of the reputation of quinine as [he] possibly could to make it a saleable commodity." The membership of the APhA was not pleased and voted to strip his membership. Stearns left the organization in disgrace.[32]

It would be easy to interpret Stearns's expulsion from the APhA as simply a result of his violating the prohibition on adulteration. Yet the issues involved were more complicated. The controversy over "sweet quinine" was really a dispute about the proper relationship between the name of a thing and its underlying object and thus about the relation-ship between commerce and science. In Stearns's opinion, he was not mislabeling his product; he had simply marketed cinchonia under the new name "sweet quinine" as a way to illustrate its therapeutic proper-ties while gaining a competitive advantage over other cinchonia prepara-tions. Yet in the eyes of his critics it was not legitimate to simply rename an already known therapeutic substance for commercial purposes; as Procter noted, "When physicians want cinchonia they can get it by pre-scription, and it is not in accordance with our ideas of fair dealing to serve it up as a new substance."[33] Stearns had also referred to sweet quinine as an "invention" and a "valuable discovery," which to his accusers made it seem like he was claiming that it was a new substance, but which Stearns meant to refer to his means of making the substance more palatable. Stearns's problem was not so much that he had substituted one good for another and been dishonest about it. It was that he had marketed a previ-ously known substance under an unfamiliar name and, more important, made claims for it that exceeded the consensus opinion about the drug. By claiming that his product had the same therapeutic effect as quinine and basing both the name of his preparation and his marketing strategy on this claim, Stearns was effectively selling an old drug for a new pur-pose and under what his critics took to be a false name. This reversed the proper order of things—in which ethical manufacturers followed the lead of the medical community—and made claims for a drug that exceeded the consensus medical opinion about its effects. Even worse, it seemed to do so under a deceptive name. As his critics saw it, Stearns had thus engaged in the sin of adulteration.

Despite his humiliating experience, Stearns continued to manufacture pharmaceuticals, and in 1876 he introduced what he called a "new idea" to the trade. Stearns began to manufacture preassembled remedies made from popular formulas, but only to the extent that the formulas con-formed to accepted medical use; he refrained from making therapeutic claims about the products; and he vowed that he would not monopolize them in any way, including through the use of patents and trademarks. As a part of this strategy, Stearns also promised to list all the ingredients in his products on their labels. This was an extremely important innovation: it allowed Stearns to manufacture preassembled remedies drawn from popular formulas without facing the charge of quackery. It also allowed him to expand the reach of his company into a field that had previously been dominated by patent medicine manufacturers and, at the same time, legitimized the use of preassembled remedies among the orthodox medical community. Not incidentally, the promotional campaign for his "new idea" also worked to rehabilitate his reputation among the pharmaceutical community. It worked spectacularly well. By the time Stearns retired from active management of the company in 1887, his reputation had been rebuilt, the company was considered fully within the domain of ethical manufacturing, and sales reportedly exceeded $1 million annually.[34] Stearns's promotion of his "new idea"—and the rapid growth of the company—also inspired a number of imitators to copy his methods.[35] Indeed, the phrase "non-secret medicines" proliferated through manufacturing circles, becoming a frequently used synonym for preassembled remedies that explicitly rejected secrecy.[36]

Stearns was not the only one to realize that introducing new products to the medical community, if done properly, had the potential to generate tremendous profits. Another important example was Parke, Davis, & Company, Stearns's chief rival in Detroit. Parke-Davis was first established in 1866 by Detroit pharmacist Samuel Duffield and businessman Hervey C. Parke. In 1867 a manufacturer named George Davis was brought into the company, and in 1875 the firm was incorporated under the name Parke, Davis & Company after Duffield retired because of ailing health.[37] Under Davis's leadership the company conformed to many of the norms of the ethical segment of the industry, marketing its products only to the medical community and rejecting the use of patents, trademarks, and secrecy to protect itself against competition.[38] However, unlike other ethical manufacturers—who understood themselves as supplying preexisting demand—Davis believed that commercially introducing new remedies could be profitable, would promote the goals of medical and pharmaceutical science, and might be worth the risk of angering the orthodox medical community.[39] Davis thus sponsored botanical expeditions to distant parts of the country as early as 1869 and over the next decade sent agents to California and the American Southwest in search of new drug plants.[40] The company then advertised these new remedies in the medical literature, sending free samples and promotional material to interested physicians. The company also made therapeutic claims for these products in its advertising, a practice that many of its critics found disturbing but that, in Davis's view, was necessary both to spread information about the new drugs to the medical community and to generate demand for what were otherwise unknown remedies. This did not strike Davis as contrary to the goals of science. Quite the contrary. He saw no real difference between promoting medical and pharmaceutical science and promoting the financial interests of his firm. The two were deeply intertwined projects.

Davis's willingness to commercially introduce new drug plants before

they had been thoroughly investigated by the medical community provoked substantial controversy. Between 1877 and 1878, for example,
the firm introduced a number of new medicinal plants from California,
including cascara sagrada, which was sold as a remedy for constipation.[41]
Critics denounced the effort as a form of "mercantile exploitation of
the profession and their patients."[42] Horatio Bigelow, for example, writing in the *New England Medical Monthly*, declared the efforts of the firm
"an immense Quackery," and argued that "their arguments are unsound,
their methods improper, and the extensive advertising is reprehensible,
as is the very reprehensible practice of the patent medicine quackery."[43]
The stakes in the struggle were high: Parke-Davis's competitors funded
the distribution of thousands of copies of editorials attacking the firm in
an attempt to destroy the company's reputation.[44] Violating the norms
of the orthodox medical community clearly carried grave risks for a firm
that marketed its products exclusively to physicians. Yet despite the initial
controversy, many of these products began to attract significant attention
from the medical community as useful new remedies. Cascara sagrada, for
example, was quickly recognized as a valuable treatment for constipation
and other digestive problems. By the early 1880s it was being reported on
favorably in the pharmaceutical and medical presses, and over the course
of the next decade Parke-Davis actively worked to develop a market in the
drug.[45]

The willingness of the orthodox medical community to give these
new drugs a chance grew out of an important change in attitudes toward
therapeutics. By the 1870s many physicians had grown disillusioned with
the established body of officinal remedies and had begun to actively seek
alternatives; at the same time, advances in chemistry, physiology, and
other areas suggested that drug actions produced in the laboratory setting should serve as the basis for clinical practice.[46] The turn toward what
John Harley Warner has called "physiological therapeutics" was based on
an emerging consensus that new remedies should be thoroughly tested
in the laboratory, with uniformity in laboratory results ensuring uniformity in therapeutic effect. As Warner puts it, proponents of this view saw
"reductionist knowledge of physiological processes and drug action produced in the laboratory as the chief starting point for scientific reasoning
in the clinic."[47] This was a dramatic reconceptualization of therapeutics,
and although this was still a minority position, by the late 1870s it was
increasingly persuasive to large sections of the medical community. The
popularity of the California remedies thus grew, at least in part, out of
the substantial number of articles published in the medical press about
the drugs detailing the results of "chemical and microscopical analysis,"

as one article from 1879 put it.[48] Laboratory studies suggested a scientific basis for the use of cascara sagrada and other new botanicals that had little to do with the ethical concerns of orthodox physicians. Indeed, in this respect Parke-Davis had done the medical community a favor by bringing useful new remedies to its attention.

Davis's great insight was to recognize that the shifting basis of therapeutics offered a tremendous opportunity to a firm willing to challenge the traditional ethical framework of orthodox medicine. Davis embraced the challenge fully and invested heavily in developing and introducing new remedies to the therapeutic market. Between 1879 and 1881, for example, Davis established a scientific laboratory for the company and hired chemist Albert B. Lyons to work on the standardization of fluid extracts. Lyons developed new assay methods that allowed standardization of drug strength in a variety of extracts, and in 1883 the company announced a groundbreaking line of twenty standardized fluid extracts, which it termed "Normal Liquids." The company also continued to sponsor botanical expeditions, sending its agents across the globe in search of new drug plants to exploit. One of the agents employed by the company, for example, was a young botanist named Henry Hurd Rusby, who had developed an interest in botany at a young age and begun collecting specimens around his home state of New Jersey. Sometime during the late 1870s he began collecting specimens for the company, and in the mid-1880s he undertook a major botanical expedition to South America on behalf of the firm.[49] Finally, Davis embarked on medical publishing and established a series of scientific journals, the most prominent of which was the *Therapeutic Gazette*, first published as *New Preparations* in 1877.[50] These journals were clearly intended to promote the interests of his firm by publishing information about the company's new products, but they were also intended to promote the cause of medical and pharmaceutical science more broadly. Davis saw no real difference between the two goals. As he saw it, the promotion of medical and pharmaceutical science and the creation of profitable markets were intertwined goals.

As part of this effort Davis began working with a young pharmacist and physician named Francis Stewart. After graduating from the Philadelphia College of Pharmacy in 1876, Stewart had earned a medical degree from Jefferson Medical College and moved to New York City, where he opened a private practice and began to develop a name for himself as a dedicated young doctor. In 1879 he developed a remedy for wasting diseases made from desiccated bullock's blood. When a representative from Parke-Davis approached Stewart, who was short on financial resources, and offered to market his new remedy, he readily agreed. Davis also suggested that

Stewart publish a medical article describing the usefulness of the product, which appeared in 1880.[51] Unfortunately, Stewart's colleagues were not amused by his collaboration with the firm: as he later recalled, "The New York physicians took exception to it as a serious breach of medical ethics on my part" and accused him of "the worst form of quackery." Their objection was to the commercial introduction of the drug before it had been thoroughly investigated by the medical community. As E. R. Squibb explained to Stewart, the only way to ethically introduce a new therapeutic agent into medical practice was to openly disseminate information about it through the medical press and other preestablished educational channels of the profession. If the remedy proved sufficiently useful, it would then be adopted into the *USP*; at that point and not before, manufacturers could ethically bring it to market. To do otherwise was to enter the realm of quackery.[52]

Stewart considered this prohibition stifling. He also considered it deeply unscientific. As part of its initial promotional efforts, Parke-Davis had sent samples of his new remedy to physicians across the country, who had then experimented with the drug in their own practice and reported on its utility in the medical press. According to Stewart, this type of testing had demonstrated the therapeutic value of his invention and therefore justified both its widespread use and its commercial sale. As Stewart put it in 1880,

I speak in the name of scientific medicine. . . . To protect the medical profession, it is proposed to discountenance the employment of any new remedy until it is introduced into the Pharmacopoeia. Such an introduction can be effected but once every ten years, at which time the Pharmacopoeia is revised. I have introduced a new remedy. It has been carefully tested clinically in a number of hospitals, both in New York and Philadelphia, with favorable reception. And now, by the above proposition, its use is to be discouraged until it is admitted to the Pharmacopoeia, ten years hence. Against this I most strongly protest.[53]

From this perspective, the ethical norms that prohibited the use of remedies until they had been adopted into the *USP* slowed the progress of medical science. A more rational approach to investigating new remedies was needed; as Stewart later noted, "It is a matter of regret that the ultra-conservatism of the medical profession has been from the outset a serious, but gradually yielding, obstacle to the progress of [a] truly scientific method of investigation."[54]

In either 1880 or 1881 Stewart proposed a plan for systematizing the experimental investigation of new remedies that he thought would assuage the concern of the medical establishment. In what he dubbed the "hospital plan," Stewart suggested that scientists at Parke-Davis investigate new remedies and then publish their results as a series of "working bulletins." Samples of the remedies, along with the working bulletins, would then be sent to various hospitals for clinical use and investigation. Once at least twenty-five reports about a particular remedy had been returned to the firm, the reports would then be provided to the medical press to be published, "whether good, bad, or indifferent." The point of the plan was to establish the therapeutic value of new remedies through a research system dedicated to the norms of cooperative investigation, transparency, and fidelity to both laboratory and clinical experiment. The plan also worked to rearrange the traditional relationship between the scientific investigation and commercial introduction of new drugs, justifying the introduction of new products to the therapeutic market before they had been incorporated into the *USP*. The goal was, as Stewart later put it, to "harmonize the interests of science and commerce, so that one may aid the other without jeopardizing the interests of either."[55] Davis was pleased by the idea, and the company quickly adopted the plan and began issuing working bulletins.[56]

From Principle to Product: Patenting and the Chemical Industry

In the chaotic years immediately following the Civil War, the American chemical industry grew rapidly and became an important sector of the emergent industrial and corporate order. As part of this broader transformation, manufacturers that did not primarily focus on manufacturing pharmaceuticals occasionally developed products that had therapeutic properties. Most American chemical manufacturers were not constrained by the ethical norms of either the orthodox medical community or the pharmaceutical community, and as a result these companies were more than willing to patent not just the processes that they used to manufacture their goods but the goods themselves. They were also willing to market their products to the public, to make therapeutic claims about their goods, and to otherwise engage in both promotional and monopolistic practices that firms that more closely hewed to the ethical framework of the orthodox medical community were generally unwilling to engage in. As a result, their efforts to develop and market these products were shaped by interactions with the Patent Office, patent litigation, and other dynamics that did not yet directly affect ethical drug manufacturers.

The development of acid phosphate of lime by the Rumford Chemical Works is a good example. In 1855 Eben N. Horsford, a professor of chemistry at Harvard, and a manufacturer named George F. Wilson established the Rumford Chemical Works in Rhode Island to manufacture baking powder, fertilizer, and other chemicals. Within a decade the firm had become one of the largest chemical manufacturing companies in the country, fueled in part by military spending during the Civil War. By 1872 the company employed more than 120 people and produced a huge quantity of goods, including a yearly output of over 3 million pounds of bone coal, 1.2 million pounds of cream of tartar, and 400,000 pounds of muriatic acid.[57] As a part of the company's overall business strategy, Horsford took out a number of patents, including an 1868 patent on a pulverulent acid product that he named "acid phosphate of lime."[58] Rumford Chemical Works initially sold acid phosphate of lime as an ingredient for bread, but Horsford also discovered that it acted as a healthy and stimulating tonic and began to advertise it in medical and pharmaceutical journals.[59] Over the course of the next decade he also began to market the product to the public, selling his acid phosphate preparation for dyspepsia, urinary problems, mental exhaustion, and other ailments.[60]

Acid phosphate of lime was not particularly difficult to manufacture, and as the popularity of the product grew, other manufacturers entered the developing market.[61] Not surprisingly, Horsford tried to defend his patent rights in an effort to suppress this competition. In 1872, for example, he sued a manufacturer named John E. Lauer for manufacturing an acid phosphate product that Horsford claimed infringed on his 1868 patent. Lauer responded by conceding that he did indeed make an "improved acid compound for use in baking and cooking" but that he did so using a process that he had patented in 1867. He further argued that Horsford was not the original inventor acid phosphate of lime and was not the first to recommend its use in baking. Lauer's patent attorneys solicited a number of well-known experts in chemistry who testified that by following an "old formula" they had produced a substance identical to Horsford's.[62] The court agreed and declared Horsford's product patent to be void for lack of novelty.[63]

As this case shows, chemical manufacturers that patented their goods or the means to manufacturer them sometimes became involved in complex legal disputes about their patents. By the 1870s a clear distinction between patents on products and patents on processes had become a central assumption in patent jurisprudence. The separation of the two meant that different processes for arriving at the same effect or product might be patented without violating one another as long as the product

itself was not patented.[64] The distinction between the two was also codi-
fied through the act of separating process and product claims for a single
invention into two different patents or series of patents, as Horsford had
done. This appears to have been in part a defensive strategy on the part of
patent holders; by separating the two types of claims, one could be maintained even if the other was declared void. In Horsford's case, the strategy
was not successful: in addition to declaring his product patent void, the
court also ruled that Lauer had not infringed on Horsford's process patent, since he used a different method to arrive at the same result.[65]

The doctrine of "reduction to practice" also continued to have significant implications for chemical manufacturers and, by extension, for the
practice of medicine. The case of Vaseline serves as an important example.
In 1859, a young chemist from Brooklyn named Robert Chesebrough
traveled to Pennsylvania to investigate the oil industry. While there, he
discovered a gooey substance that sometimes clogged refinery machines
and that, to his surprise, workers applied to their cuts and burns. Chesebrough then developed a method of manufacturing the substance and
in 1865 acquired patents on two related manufacturing processes.[66] He
continued to improve his methods and by 1867 had manufactured a version of the substance that was free from impurities. In the spring of 1869
Chesebrough met a leather manufacturer named Charles Toppan. Toppan used a black petroleum jelly as part of his manufacturing process, as
was common in his industry, and he asked Chesebrough if he knew how
to produce a purified version of the substance. Chesebrough responded
affirmatively and showed him his process for doing so. Toppan then asked
him to refine several barrels of the substance for him. Although the chro-
nology is not exactly clear, at some point around this time Chesebrough
named his substance Vaseline and began to sell it as a treatment for burns
and other skin injuries.

Toppan also recognized the commercial value of Vaseline, and in early
1870 he was granted a patent on the substance.[67] Around the same time,
Chesebrough applied for a product patent as well, although his application was rejected after it was deemed to have been anticipated by Toppan's patent. Chesebrough then filed to have Toppan's patent invalidated
based on a claim to having been the original inventor. Chesebrough's case
rested on the assertion that Toppan had obtained the knowledge of how
to produce Vaseline from him. Toppan's case, on the other hand, rested
on the claim that Chesebrough had first made the product for practical
uses under his direction and that as a result he had "reduced the invention to practice by the hands and machinery of Mr. Chesebrough." The
argument here was that although Chesebrough had discovered a process

for producing a new substance, it was only through Toppan's initiative and inventive activity that the scientific principles involved had become embodied in an actual product with economic value. Chesebrough responded by showing that he had made small amounts of the substance before meeting Toppan and that therefore he had been the first to reduce the process he had discovered to a practical result. In 1872, the commissioner of the Patent Office ruled that the fact that Chesebrough had made only a small amount of the product could be accounted for by the fact that there was, at the time, no demand for it. He then ruled in favor of Chesebrough and invalidated Toppan's patent. A patent on Vaseline was granted to Chesebrough soon after.[68]

Toppan's argument that his patent should be upheld because it was his initiative that brought Vaseline into commercial use for the first time pointed to the developing legal doctrine that commercial utility could itself justify a claim to novelty. The most important case in this regard was the 1874 Supreme Court decision in *American Wood-Paper Co. v. The Fibre Disintegrating Co.* The court ruled that cellulose made from purified wood pulp was not patentable because it was "an extract obtained by the decomposition or disintegration of material substance." As the court noted, extracts were not patentable because they lacked novelty:

There are many things well known and valuable in medicine or in the arts which may be extracted from diverse substances. But the extract is the same, no matter from what it has been taken. A process to obtain it from a subject from which it has never been taken may be the creature of invention, but the thing itself when obtained cannot be called a new manufacture. . . . Thus, if one should discover a mode or contrive a process by which prussic acid could be obtained from a subject in which it is not now known to exist, he might have a patent for his process, but not for prussic acid.

The patent holders in this case had argued that their patent should be ruled valid because the paper pulp from which they produced their extract was not pure cellulose and therefore "the pure article obtained from wood by their process is a different and new product, or manufacture." The court declined to rule on the issue of whether purity alone could be used to justify a claim to novelty, although it did express skepticism of the idea, noting that "a slight difference in the degree of purity of an article produced by several processes justifies denominating the products different manufactures, so that different patents may be obtained for each, may well be doubted, and it is not necessary to decide." However, the court did hold out the possibility that the novelty requirement might have been fulfilled based on commercial utility. The court speculated that if previous to the patent in question cellulose "had been only a chemical preparation in the laboratory or museum of scientific men, and had not been introduced to the public," then the product "might have been patented as a new manufacture." Quoting the 1866 British case *Young v. Fernie*, the court noted that "what the law looks to . . . is the inventor and discoverer who finds out and introduces a manufacture which supplies the market for useful and economical purposes with an article which was previously little more than the ornament of a museum." In this case, however, cellulose was in commercial use before the original patent was granted, so such speculation did not apply to the patent at hand.[69]

In the coming years, the Supreme Court's musings on these issues would develop into two important legal doctrines. First, the court's skepticism that purification itself was enough to justify a patent would be validated through a series of decisions that held that modifications that result in differences of degree rather than kind are not adequately novel to justify a patent. In another case from 1874, for example, the court held that "a mere carrying forward or new or more extended application of the original thought, a change only in form, proportions, or degree, doing substantially the same thing in the same way, by substantially the same means, with better results," is not adequate grounds for a patent. "It is the invention of what is new, and not the arrival at comparative superiority or greater excellence in that which was already known, which the law protects as exclusive property and which it secures by patent."[70]

However, even as purification itself was held to be an inadequate ground for patentability, a related doctrine emerged that held that significantly improved utility—including commercial utility—might be an adequate justification for a claim to novelty. The substitution of hard rubber in place of metal in the manufacture of dentures, for example, was ruled patentable by the Supreme Court in 1877 based on the idea that the substitution revolutionized dental practice.[71] New uses for familiar things therefore might be patentable if the adaptation of the thing to its new use required inventive activity and made a substantive difference in the utility of the thing under consideration. From this perspective, familiar things that gained substantive new uses through new manufacturing methods, through their combination with other ingredients, or by other means might be patentable. In 1877, for example, Horace Bowker of Boston secured a patent on a combination of saponins extracted from vegetable material that was used to create and maintain a foamy head on beverages.[72] The following year, the Circuit Court of Massachusetts upheld his patent, despite the fact that a competitor argued that there was nothing novel in his discovery. The court ruled that Bowker had made use of saponins in a new way and had produced a useful invention as a result that was patentable. "It has further been argued to us," the court noted, "that there is nothing patentable in the discovery that the foam in beverages can be increased by the use of saponine; but we are of opinion that it is clearly a case of a patentable discovery of a new use, in a combination, to produce a better result than was known before."[73]

The degree to which the practices of the Patent Office were shaped by the evolving legal environment is difficult to determine, but during the 1870s a variety of patents were issued for extracts of hops, tobacco, wheat flour, beer, fish, meat, and other substances.[74] Presumably the utility of these products was part of the reason they were also considered to be novel enough to be patentable. Some of these products also had medicinal properties and began to be sold on the therapeutic market. Over the course of the decade, a large number of companies were established with the purpose of manufacturing these and other types of new medicinal products. Like chemical manufacturers such as Horsford and Chesebrough, these manufacturers sometimes had no real affiliation with the pharmaceutical or medical communities before they began to manufacture their products, and as a result they did not feel constrained by the ethical norms of orthodox medicine. Other companies, however, were founded by pharmacists or even physicians who had made what they considered important therapeutic advances. As a result, these manufacturers took a variety of perspectives on the propriety of advertising to the public, making therapeutic claims for their products, and other issues. They sometimes patented their products, and they sometimes kept their ingredients secret. Others did neither, in an effort to conform to orthodox medical ethics. Whatever their differences, however, these manufacturers typically sold their products under scientific-sounding names that were frequently trademarked. These manufacturers thus occupied something of a middle ground between ethical manufacturers that marketed exclusively to the orthodox medical community on one hand and manufacturers of traditional patent medicines on the other. As a result, they should be considered a new segment of the drug manufacturing industry, one that produced what were increasingly called "proprietary" goods.

Trademarks and the Thingification of Names

The first federal trademark law was passed in 1870 in response to the rapid growth in commerce following the Civil War. The law grew primarily out of domestic concerns about international trade. During the late 1860s, the United States had entered into a series of treatises concerning trademark protection with Russia, Belgium, and France in order to suppress the sale of goods under counterfeit marks. These treaties required a central registration system, which did not yet exist in the United States; they also granted access to federal courts for foreign citizens of signatory countries, despite there not yet being statutory federal law in the United States on the topic. After a number of failed efforts to pass a law fulfilling these treaty obligations, a trademark clause was inserted into the 1870 bill that consolidated the patent laws.[75] The fact that the new trademark law was primarily directed toward international trade was well known at the time; as William Henry Brown noted in 1873, the law "had a stimulating effect upon our own people; although the principal object that it had in view, it must be confessed, was the matter of reciprocity."[76]

The 1870 trademark law offered only modest protections by today's standards. The law established a central registry that could be used to document ownership of a mark in case of litigation; it also ensured access to federal courts in disputes involving registered marks. However, the law conferred no positive rights to a mark and did not supplant either state laws or the common law tradition with regard to the scope of marks or what constituted infringement. Registration did not mean that the marks were actually valid, and trademark holders were not required to register marks in order to claim infringement by other parties. Many critics saw the law as inadequate to meet the needs of a rapidly expanding economy. As a result, the 1870 law was supplemented in a variety of ways over the next decade. In 1874, for example, an earlier law that had allowed seven-year patents to be granted for prints on fabrics was expanded to allow patents on other types of designs, including labels, in order to grant positive rights to certain types of identifying symbols. Two years later the federal government passed a law adding criminal penalties for trademark infringement, which in turn spurred growing concern about the law's constitutionality.[77]

Ethical companies generally refrained from making use of the new law. In part, this was simply because they refrained from anything but very modest promotional activities at the time, generally limiting themselves to sedate advertisements in the medical and pharmaceutical press that listed their goods. Ethical manufacturers did seek to distinguish themselves from their competitors by emphasizing the supposedly high quality of their products, their manufacturing skills, and the quality of their ingredients, and in doing so they actively promoted what was sometimes referred to as their "brand" as a means of promoting the fortunes of their firm vis-à-vis other firms that made very similar products. Parke-Davis, for example, emphasized the skills of its employees, its investment in expensive machinery, and the quality of its botanicals in their advertisements.[78] Yet the company did not make use of trademarks to protect its interests. There seemed to be little reason to do so. The ability of ethical companies to sell their goods fundamentally depended upon their reputation, the quality of their products, and—for Parke-Davis, at least—the firm's embrace and promotion of both clinical and laboratory science. Counterfeiting was not really a problem, because the firm did not claim an exclusive right to anything, and advertising played only a small role in its overall business strategy. Thus, even as ethical firms such as Parke-Davis worked to distinguish themselves from other manufacturers through the cultivation of a distinctive brand identity, they generally avoided the use of trademarks during the 1870s.

Patent medicine manufacturers, on the other hand, registered a significant number of trademarks. These trademarks were typically taken out on phrases that conformed to the traditional naming formulation in the industry, in which the manufacturer's name modified the type of substance being sold. Trademarks were also were frequently written in highly stylized shapes or letters and often combined with images or pictures of recognizable people or things, such as the founder of the company or a picture of an Indian. These marks were part of a broader advertising strategy that increasingly relied on creating dramatic narratives about the history of the company and, frequently, the supposedly exotic origins of its products. These companies worked to create advertising narratives that invested value in their reputation and linked the purchasing of specific products to these narratives. Stories of company founders being captured or saved by Indians, for example, were particularly popular and served as a common selling theme for patent medicines in the decades following the Civil War. To take just one example, in 1874 a physician from New Jersey named Clark Johnson published what he claimed to be the autobiography of a man named Edwin Eastman. The narrative tells the supposed story of Eastman being captured by the Comanche Indians in 1860, his rescue from being burned at the stake, and his learning the secrets of "Indian medicine." Clark sold a product he called "Indian Blood Syrup" that was supposedly based on this information, and he included portions of the autobiography in his trade catalogs as a way to emphasize the supposed origins and effectiveness of the product. Clark also trademarked both the image of an Indian that he used in his advertisements and the term "Dr. Clark Johnson's Indian Blood Syrup." He pointedly warned the public that "any and all Indian Blood Syrups" that did not have his trademarked label were "spurious and should be avoided."[79]

Clark's efforts to monopolize all uses of the phrase "Indian Blood Syrup" points to an important issue. During the 1870s names that could be appropriated as trademarks continued to be understood as fundamentally transparent in nature, in that they designated the origins of the good in question. At the same time, courts continued to rule that descriptive words, including geographical names, colors, letters, and the common names of things themselves, could not be appropriated because they did *not* indicate the origin of the thing in question and could therefore be used by all.[80] However, during this period we also see the beginning of what would become a fundamental transformation in trademark law: in a process that legal scholars have called "thingification," marks started to acquire the legal characteristics of property as courts began to recognize that the reputations of companies had value and that trademarks acted as a sort of congealed form of this value.[81] As a result, a legal doctrine began to emerge that allowed the appropriation of another's trademark to be restrained even if there was no deception involved. In other words, courts began to treat trademarks as an intangible form of property that was linked to the accumulation of value in the reputation of the firm in question. The result was that courts began to uphold trademark rights on names that had lost their designating character.

The 1876 Circuit Court of Connecticut case *Filkins v. Blackman* serves as a good example. In 1840 Jonas Blackman invented a remedy that he called "Dr. J. Blackman's Genuine Healing Balsam." Blackman sold his remedy from door to door and over the next two decades built up a good market for his product. In 1865 he entered a contract granting his son-in-law, Morgan Filkins, the "exclusive right" to manufacture and sell the product under that name for ten years for a sum of $365 per year. At the end of the decade, if the payments had all been made, Filkins would then acquire the same right for fifty more years for no additional cost. Over the course of the decade, however, Blackman's son, Newton Blackman, had also begun to manufacture and sell what the court called "the same medicine" under the same name and in bottles with very similar labels. In court, Newton Blackman claimed that his father had also sold him both the formula and the right to manufacture the product. The question at hand was thus whether or not Filkins's contract with the elder Blackman gave him the exclusive right to manufacture and sell the product under its original name for the entire sixty years, as stipulated in the original contract.

The court decided that it did. Finding in favor of Filkins, the court determined that the right to the mark was transferred from the elder Blackman to Filkins as a result of the contractual relationship between the two and that the contract between the elder and younger Blackman was therefore invalid. Significantly, this ruling was based on the idea that the reputation of the product had value that had been built up by the elder Blackman over time. "It is obvious that the plaintiffs have expended a good deal of money in advertising and in bringing this medicine into public use," the court noted. "They have made its manufacture profitable, and have invested their property in the business." This meant that the trademarked name itself had value, since it conveyed the reputation of the company to the purchaser, and because it had value, it was covered by the contractual relations of the original agreement. As the court put it, the younger Blackman was "seeking to take advantage of the reputation which the efforts of others [had] given to the article."[82]

The importance of this new perspective on marks is noteworthy. Under antebellum trademark law it would have been acceptable for the younger Blackman to produce the medicine in question under the name "Dr. J. Blackman's Genuine Healing Balsam" because the product was not under patent. If the younger Blackman had indicated the true origin of his goods, he would have had every right to manufacture the remedy and call it by what all parties involved would have considered its proper name. By the 1870s, however, things had begun to change as a result of the realization that the names of things themselves had economic value. In this case, the court found that the value of the mark justified its transfer from the elder Blackman to Filkins and thus that Filkins had an exclusive right to use it. In other words, the accumulation of value in the reputation of the product—value that was encapsulated in the mark itself—gave the phrase "Dr. J. Blackman's Genuine Healing Balsam" status as a form of intangible property that could be transferred from one owner to another. This in turn meant that the mark did not simply transmit information about the origins of the good. It actually pointed to the good itself, independent of who actually made the product. As the court noted, "The use of the trade-mark does not imply that the medicine was manufactured by Jonas Blackman, but that it is the same article which he originally invented and manufactured." As a result, the younger Blackman was not allowed to manufacture the same good under that name, even if he indicated that he was in fact the actual manufacturer. This put the younger Blackman in an unenviable position: if he manufactured a product according to the formula given to him by his father, what could he honestly call the resulting product if not "Dr. J. Blackman's Genuine Healing Balsam"? Presumably he could make up some other name for it, but if he did so, he would not only lose the ability to draw on the reputation for the product that his father had developed. He would also be calling it by something other than what he believed was its true name.

There was another implication as well. During the 1870s patent medicine manufacturers typically acquired trademarks in the familiar style in which the name of the company, or some other indicator of the product's origin, modified the name of the good. As marks were increasingly assumed to point to goods themselves, as opposed to their origins, this naming formulation began to break down. Manufacturers began to realize that the names of goods themselves—names independent of any information about the origins of the product—carried economic value and were worth protecting. They also began to recognize that trademarks on these names could be used to prevent their competitors from manufacturing what they considered to be counterfeit goods. Manufacturers thus began to acquire trademarks on words or word-symbols that did not incorporate information about the company at all. In 1874, for example, a manufacturer named John Carnrick trademarked the name "Lactopeptine" for a lactated pepsin preparation.[83] Although the name suggested the ingredients of the product, it said nothing about who manufactured it. Anyone could have registered it for his own goods.

Here we see the origins of what were increasingly called "proprietary" remedies—neither commonly manufactured goods sold under names usable by all nor patent medicines in the traditional sense sold under the familiar formulation in which the name of the manufacturer modified the type of good in question. The term "proprietary remedy" and its variations had long been used as the equivalent of "patent medicine"—meaning monopolized and controlled, whether through patents or secrecy—and this sense of the phrase continued into the 1870s. However, the term was also increasingly used to refer to products that were given scientific-sounding names such as "Lactopeptine" and, frequently, controlled by trademarks. Manufacturers of so-called proprietary remedies occupied something of a middle ground between strictly ethical manufacturers and patent medicine manufacturers. Sometimes the term simply referred to familiar manufactured goods, such as fluid extracts, tinctures, or pills, that were sold under scientific-sounding names that contained no information about the origin of the good in question. Sometimes the term referred to preassembled remedies that were sold under the same types of names, and at other times the term referred to products that were made from new chemicals, mixtures, or other compositions of matter that went beyond the simple assembling of ingredients but were also given such names. Proprietary medicines were sometimes patented, and their ingredients were sometimes kept secret. At other times, their ingredients were well known. Whatever the case, they can generally be distinguished from both patent medicines and products made by strictly ethical firms by the fact that they were closely associated with a single manufacturer, they were sold under scientific-sounding names—frequently ending in "-ine," "-in," or "-ol,"—and that these names were frequently protected by trademarks.

Trademarks on these types of names were appealing to proprietary manufacturers in part because they appeared to have the ability to monopolize the manufacture and sale of these goods whether or not patent rights were involved. In the late 1870s, for example, a British resident of Fiji returned to England with a supply of a remedy called "tonga" that was used by the indigenous peoples of the islands to treat chronic pain.[84] The London pharmaceutical company Allen & Hanbury introduced the remedy to the English market, which prompted a number of physicians in Europe to publish their experiences using the drug. These articles caught the attention of physicians in the United States, and, sensing a potential market, Davis sent a representative of the firm to Fiji to secure a supply. Parke-Davis introduced tonga in the United States in late 1880; as one account noted, they sent it to hospitals across the country "for careful clinical test and report," and as a result "a moderate demand sprang up."[85] Meanwhile, however, Allan & Hanbury had secured a trademark on the word "Tonga" and Schieffelin & Company, which had acquired a license to sell the remedy, sued Parke-Davis for infringement. Davis fought the case as a matter of principle, arguing that "the only name of the article, being that only specification by which the article itself is known or described, is the common property of all and cannot be appropriated by any one individual to his own sole and exclusive use."[86] Allan & Hanbury withdrew from the case before it was resolved, and Parke-Davis went on to sell the remedy, proudly noting in its advertisements that as a result of its efforts "tonga is therefore free to science."[87] To both Davis and Stewart it seemed profoundly unethical to trademark the name of a substance because, as a result, "the article itself [is] monopolized forever in consequence, and the nomenclature of pharmacy ruined thereby."[88] Tonga, in their view, was a common name that belonged to all.

Davis and Stewart were not alone in their concern about the use of trademarks for monopolistic purposes. Over the course of the 1870s, therapeutic reformers in the medical community grew increasingly worried about the increasing use of trademarks on pharmaceuticals: by the late 1870s, according to one survey, about 10 percent of all prescriptions included proprietary articles.[89] Proprietaries troubled many physicians

because they were frequently made with unfamiliar ingredients or combinations of ingredients, their scientific-sounding names seemed to mask
their actual compositions, and they were often commercially introduced
before the medical community had come to a consensus about their utility. From the perspective of many critics proprietaries were really nothing
more than preassembled formulas, and as such their ingredients should
be public knowledge so that they could be properly investigated. From
this perspective, commercially introducing such products and giving
them a distinctive and scientific-sounding name before they had been
thoroughly investigated distorted medical science because it prioritized
the commercial interests of individual firms over methodical investigations of the scientific community. Furthermore, their popularity seemed
to depend as much on advertising as on anything else. As one frustrated
physician put it in 1880,

It [the use of proprietary medicines] is demoralizing to the profession because it is
ruinous to scientific nomenclature, and renders a classification of medicines utterly
impossible. What will the next generation of medical men know about Lactopeptine?
Maltine? Vitalized-Hypophosphites? Celerena? Bromidia? Iodia? Petroleum Syrup?
Soluble Phyenle? Mato-Cocoa? Hydroleine? Listerine? Caulocorea? Viburnum Compound? and a more innumerable host of mixtures? These are all of ephemeral existence, having no vitality other than what they derive from the advertising pages of
medical journals and the newspapers. They are for the most part the inventions of
tradesmen, and in no sense represent the growth and progress of medical science.[90]

By the late 1870s such concerns had become widespread in the medical
community, and therapeutic reformers began to turn against the use of
trademarks. "The evil of trade-marks on medicinal agents has grown to be
an abuse of gigantic proportions," noted one critic in 1881, "and is very
generally practiced by the manufacturing pharmacists of this country,
with a few honorable exceptions."[91]

Of course, not all physicians agreed. The growing use of proprietaries
points to the basic fact that many physicians considered them convenient and useful products. Moreover, legal observers had long pointed to
the role of trademarks in guaranteeing the authenticity of products by
linking them to the reputation of their manufacturers. Trademarks thus
protected the interests not just of the manufacturer but of the public
as well: as Francis Upton had put it in 1860, the "adequate security and
protection" of marks "is an imperative duty, as well as for the safety of
the interests of the public, as for the promotion of individual justice."[92]
During the late 1870s some physicians followed this line of argument and

suggested that prescribing remedies by the brand of the company was a way of ensuring the quality of goods that were dispensed by the pharmacist.[93] According to some physicians, proprietary remedies more generally served a similar function. Writing out formulas risked the fact that pharmacists might compound the formula with poor-quality ingredients—or even with different ingredients than the ones called for—while prescribing a proprietary remedy ensured that the pharmacist would dispense a product made from predictable ingredients. In this respect, trademarks on the names of goods protected the interests of manufacturers against unscrupulous competition and thereby protected the interests of the public. Of course, critics scoffed at such arguments and suggested that proprietary manufacturers could also alter their ingredients and that since they were sold under trademarked names, there would be no way to know. Was this not more dangerous than simply prescribing according to officinal names?

During the 1870s the orthodox medical community thus had mixed feelings about trademarks and the growing use of proprietary remedies. On the one hand, many critics believed that the use of trademarks acted as an unethical form of monopoly. Like patents, trademarks interfered with the practice of scientific medicine and were little more than another form of quackery. Other critics denounced proprietary remedies—whether or not they were sold under trademarked names—as quackish because they often had bizarre names that appeared to undermine scientific nomenclature, because they were made with unfamiliar and untested ingredients, and for other reasons. Still other physicians, however, considered proprietaries convenient and useful products and prescribed them accordingly. And some physicians even argued for a positive role for trademarks in medical practice, suggesting that they worked to ensure that the products physicians prescribed and those that were dispensed were the same. These and other contradictions swirled around the meaning of pharmaceutical trademarks, even as they increasingly acted as a vehicle for the accumulation of value.

Monopoly and the Dangerous Market

In 1874 a young boy named Jacob Bowen swallowed a small tin whistle and began to suffer terribly. His mother sent his older sister to the pharmacy to purchase some "opening powder"—a colloquial name for a laxative—which she hoped would help the boy pass the object. After she gave him the medicine, however, the boy became violently ill and died. An inquest was conducted, and it was determined that the pharmacist in question had misheard the boy's sister and given her opium powder instead. Both the druggist and his employer were then charged with criminal negligence, the first for not asking the girl what the drug was to be used for and the second, presumably, for hiring such a careless employee. They were both discharged after a judge ruled that they were not criminally liable, but two months later the boy's father sued the owner of the pharmacy for $10,000 in damages. The sources do not say whether he ever received the money or, if he did, whether it helped him cope with his loss.[94]

This sad story was far from unique during the 1870s. Accidental poisonings, suicides, even murder—these and other dangers seemed to swirl around the use of many drugs in the tumultuous years following the Civil War. Although local governments had occasionally passed pharmacy laws in the first half of the nineteenth century, these laws were poorly enforced and not effective. Beginning in the 1870s, however, a seemingly endless series of tragic stories involving the use of drugs prompted widespread efforts to regulate the practice of pharmacy. States and local governments passed a variety of laws intended to regulate the buying and selling of pharmaceuticals, including prescription requirements, labeling laws, antiadulteration laws, and educational and licensing requirements for the practice of pharmacy. Rhode Island, for example, passed the first comprehensive pharmacy law in 1870. It restricted the practice of retailing, compounding, or dispensing "medicines or poisons" to registered pharmacists or their employees; it also instituted educational requirements for pharmacists, established an examination system for acquiring a license, prohibited the adulteration of drugs or medicines, established a schedule of dangerous substances, and mandated that any bottle, box, or other container in which dangerous substances were sold be clearly labeled with the name of the product, the name and place of business of the seller, and the word "poison."[95]

In the following chapter I argue that these types of laws were intended in part to promote the interests of pharmacists. Here I want to suggest that they were grounded on the fundamental assumption that people are rational actors who pursue their own self-interest and that tragic stories of consumer suffering grow out of some type of error, deception, or other force that hampers the consumer's ability to act rationally. Historians have long noted that in the two decades following the Civil War, contractual relations were widely thought to embody the virtues of a free society.[96] Yet the logic of the market did not extend indefinitely—few people, for example, thought it legitimate to sell a dangerous poison to someone intent on killing himself. Deception on the part of unethical manufacturers; accidental distribution by poorly trained pharmacists; vengeful enemies or spurned lovers; consumer insanity, grief, or foolishness; an overpowering habit through which self-control was lost—these and other dangerous threats might contravene the ability of the individual to exercise the rational judgment and self-restraint necessary for personal success in an increasingly complex market society. Rather than restricting the freedom of the consumer, it seemed, pharmacy laws promoted the ability of individuals to participate in the market along rational lines by promoting both the free circulation of information and the authority of the pharmacist to guide consumer choices and, when necessary, to intervene between the forces of a dangerous market and the choices that a consumer might make as a result of error, deception, or other problem. The death of young Jacob Bowen was not just an accident, in other words, although it was that. The mistake was also an imposition, a type of distortion imposed on the market that prevented it from operating rationally.

Central to the effort to regulate the drug market was the assumption that names and things should correspond to one another. Therapeutic reformers were deeply concerned about what they perceived as a slippage between the two and believed that fixing the relationship between names and things in place was essential to protecting the ability of physicians to practice good medicine, of pharmacists to do their jobs, and of consumers to make wise choices for themselves. Well-known botanicals and other reputable goods, of course, should be prescribed and sold under names that were both familiar and scientific, including officinal names if listed in the *USP*. Other products were more troubling. The fact that patent medicines were made with secret ingredients was widely understood as a form of imposition, in part because secrecy made it impossible to know if the names of goods conformed to their underlying ingredients: how could a physician or pharmacist know what a medicine actually *was* if it was made with secret ingredients? Proprietary medicines were not much better. The proliferation of scientific-sounding names seemed deceptive, in part because they appeared to mask the ingredients used to make them, in part because they seemed to give individual names to amalgamations of discrete ingredients, and in part because they seemed the product of commerce as much as the product of science. What was Lactopeptine, really? Celerina? Bromidia? Shouldn't they simply be sold according to whatever formula was used to make them, thus revealing their true ingredients?

A related problem was the question of therapeutic equivalence. The reformers who passed the wave of pharmacy laws in the 1870s and early

1880s assumed that establishing equivalence between products that were sold under the same name was at the heart of their efforts. Secrecy was dangerous not only because it made it impossible to know what was in a product but also because it made it impossible to know if two products sold under the same name would have the same effect. Adulteration was dangerous for the same reason: if a pharmacists dispensed something called *opium*, it seemed essential that the product actually *be* opium. Even normal variation in the strength of goods was a problem: if a pharmacist dispensed a preassembled remedy, it seemed essential not only that the names of the ingredients be known but also that those ingredients have some sort of predictable strength so that one instance of the product could be considered equivalent to another. Although some states' pure drug laws prohibited adulteration without defining what the term meant, most linked the definition of adulteration to the standards and formula set out in the *USP*.[97] Drugs that were not included in the *USP* were more difficult to deal with, of course, because there were no agreed-upon standards by which to establish equivalence between different instances of the good in question and thus no way to establish whether or not goods sold under the same name were in fact the same thing. Still, the open circulation of information about products, the fixing in place of the relationship between names and things, and the establishment of equivalence between things that were sold under the same name were all essential to the project of therapeutic reform.

Yet there was also an important difficulty when it came to standardizing drugs that were monopolized in some way. The *USP* traditionally had excluded all remedies that were protected by patent or secrecy no matter how widely used; the revision committees followed this basic framework through the sixth revision (1882). This was based on the assumption that monopolized remedies were, by definition, both unscientific and unethical, but this assumption also reflected a very practical problem. If the *USP* decided on a particular standard, it tended to favor one manufacturer of that product over another by designating the product of that company as the standard according to which other products should be made. In the past, this had perhaps favored one manufacturer over another, but it was assumed that other manufacturers could simply change their methods and produce the product in question according to the new standard if they chose to do so. Including patented goods in the *USP*, however, would have meant that only the manufacturer who held the patent was able to deal in an officinal good. This would have forced physicians and pharmacists who wanted to act ethically to deal only in that particular manufacturer's product, and it would have meant that any manufacturer who produced a competing product was, by definition, selling a good that was not officinal. Linking the standards of the *USP* to the interests of a single firm in this way would have seemed both deeply unethical and profoundly unscientific during the antebellum period.

During the 1870s the revision committee found it increasingly difficult to avoid this problem. The issue first came up with Vaseline. Chesebrough's product was an important therapeutic advance because of its ability to soothe burns and prevent breakage in scabs. Chesebrough promoted Vaseline in both the medical and pharmaceutical presses and over the course of the decade his product quickly became a popular and widely used remedy.[98] A variety of other manufacturers also entered the developing market and introduced competing products under names such as Petroline, Saxoline, and Fluorine. These products were made through different manufacturing processes, many of which themselves were protected by patents, and as a result had different melting points and other characteristics.[99] Chesebrough does not appear to have made any effort to defend his interests through patent litigation, probably out of the belief that doing so would have been futile, and instead relied on heavy promotion and the quality of his product to develop and maintain his market. He was quite successful. Chesebrough's product was by far the most popular and widely used petroleum product in the country; as one account from 1881 put it, he "made vaseline *the* petroleum ointment."[100]

Despite the fact that it was patented, Vaseline rapidly came to occupy an important place in medical practice. During the late 1870s, the medical community thus debated whether—and how—to include some type of petroleum ointment in the *USP*. The question was surprisingly difficult. There were two related issues at hand: what temperature should the melting point be set at, and what should the product be called if it were to be included? Different consistencies of petroleum jelly were useful for different types of problems, and the revision committee did not want to limit practitioners to any one product. Equally important, if the officinal temperature corresponded to the melting point of any one manufacturer's product, then that product would, by definition, become officinal, and pharmacists and physicians would be obliged to use it instead of others. Moreover, if competitors of the original manufacturer changed their manufacturing processes so that their products matched the officinal temperature, they would then, in effect, be manufacturing the same product. In other words, it was assumed that if the temperature was set at eighty-five degrees, which was the most logical choice, since that was the temperature at which Vaseline melted, then *any* production of petroleum jelly that melted at that temperature would in fact be an instantiation of

Vaseline and would therefore violate Chesebrough's patent rights. As one pharmacist noted at the 1879 annual meeting of the New York State Pharmaceutical Association, "[Vaseline's] character is thoroughly known and understood; yet, by the peculiarities of patent law, we cannot direct it to be made."[101] Petroleum jelly, another pharmacist noted, is "just buried, almost out of reach, under a pile of Patent-Office rubbish."[102]

A closely related issue was the question of what name to use to refer to petroleum ointments in the *USP*. "Vaseline" was clearly unacceptable, since it would mean that Chesebrough's product was the only officinal petroleum ointment. In response, the revision committee decided to include petroleum ointment in a way that no one found particularly satisfying. For the sixth revision, which appeared in 1882, the committee created a new and vaguely Latin-sounding word—"petrolatum"—to refer to petroleum jelly. They also established a range of melting points for petrolatum, ranging from about 104 to 125. However—and this was important—the committee also set the melting point for petrolatum, when not otherwise specified, at 104.[103] This roughly corresponded to the product that resulted from a manufacturing process that had recently been described by Charles Toppan, and it was higher than Vaseline's melting point of eighty-five degrees. Toppan had recently published his process in the *American Journal of Pharmacy* in order to, as one observer put it, "make it public, and thus forestall any patent."[104] The revision committee appears to have been trying to promote Toppan's method, but this was not really a satisfactory way to address the issues at hand: it did not face the reality that the most widely used form of petroleum jelly was Vaseline and that the invented word "petrolatum" was not widely used, or even known, by most pharmacists, let alone physicians. No one prescribed or requested "petrolatum," and when confronted with a request for Vaseline, druggists would thus be placed in the unenviable position of having to either dispense a nonofficinal product or substitute a different good for the one requested. Neither was a good option for a druggist committed to an ethical practice.

The revision committee thus confronted a basic problem that would plague therapeutic reformers for years to come. Coining new officinal names for use in the *USP* in place of names that were associated with the products of specific manufacturers did not address the manner in which commercial names shaped the therapeutic market toward the interests of particular firms. The coining of new names seemed essential if products that were closely associated with specific manufacturers were to be included in the *USP*, yet given the seemingly artificial nature of these names, and their unfamiliarity to most physicians, it was not realistic to think that physicians or pharmacists would actually adopt their use. As a result, "petrolatum" could not really serve as the name of the thing itself: indeed, since petrolatum and Vaseline were defined as having different physical characteristics, it was not completely clear what the relationship between the two was. Yet despite such ambiguities, the effort to coin and promulgate the name "petrolatum" was tremendously significant. The revision committee asserted, perhaps for the first time, that a newly coined and nonproprietary name should be used in place of a variety of commercial names associated with the products of specific manufacturers. In doing so, the committee tried, albeit unsuccessfully, to both articulate and promote a *generic* name.

Therapeutic Reform and the Reinterpretation of Monopoly

This chapter examines the early stages of what would become a decisive shift in the ethical sensibilities of the orthodox medical community toward medical patenting. Although patenting played an important role in the broader growth and transformation of the economy during the 1880s, patents continued to be of only marginal significance in the domestic pharmaceutical industry. Some proprietary manufacturers began to patent their products around this time, but both "patent medicine" manufacturers and so-called ethical manufacturers continued to avoid the use of patents to protect their interests. Ethical manufacturers, of course, did so in part because they recognized that the charge of quackery could severely damage their reputation among the orthodox medical community. Yet this position also confronted them with a basic problem of how to innovate new products in a way that was commercially viable. Ethical manufacturers therefore began to consider the possibility of acquiring patents on their manufacturing processes.

Francis Stewart was a particularly important voice in this process. Stewart was remarkably prolific, and during the decade he became the leading expert on patent and trademark law in the medical and pharmaceutical communities. As such, he had a significant influence on the emergent debate about the ethical validity of both patents and trademarks. His thinking on these topics was shaped by the shifting basis of therapeutics, the impact of patent law on the behavior of manufacturing firms, and the introduction of a wave of powerful new synthetic drugs by German manufacturers that were typically protected by both patents and trademarks. German manufacturers had little concern for the ethical norms of the orthodox medical community, and their willingness to embrace both patents and trademarks posed a profound challenge to the framework of ethical thought that understood monopoly and quackery in overlapping terms.

These complex dynamics intersected with the broader project of therapeutic reform. During the 1880s reformers built on their earlier efforts and successfully passed numerous laws intended to rationalize the drug market and thereby advance the cause of scientific medicine and pharmacy. Yet as they did so, they confronted difficult questions related to the changing role of both patents and trademarks. Trademarks were increasingly useful to manufacturers as a means of developing the reputation of their goods, and as a result they began to acquire the ability to monopolize not just the names of things but also their manufacture. Trademarks also raised difficult issues related to therapeutic equivalence between products, the rights of pharmacists to compound goods, and other complex questions. At the same time, the price of pharmaceuticals emerged for the first time as a major source of concern. The low price of many drugs threatened the livelihood of pharmacists and endangered the ability of the industry to make a profit; ironically, other products were criticized as too expensive, and their manufacturers were denounced for using patents and trademarks to artificially inflate their price. The unethical use of both patents and trademarks was seen as one of many monopolistic practices that predatorial firms, and German firms in particular, used to distort the market and extract unfair profits from a suffering public. Thus, even as the prohibition on patenting itself began to soften, concerns about monopoly emerged in other domains.

Scientific Innovation, Patenting, and the Problem of Competition

During the 1880s a small number of ethical manufacturers began to innovate new products and introduce them to the therapeutic market before they had been formally adopted into the *USP*. Parke-Davis remained at the forefront of this trend, sponsoring botanical expeditions to distant lands and working to improve and develop new products in its laboratory.[1] In part as a result of these efforts, the fortunes of the firm grew dramatically over the decade: in 1880, just before Stewart launched his working bulletin system, the total sales of Parke-Davis stood at about $450,000. Ten years later sales had ballooned to almost $1.8 million.[2] This rapid growth was brought about in part by the success of the working bulletin plan, which allowed the company to generate, collect, and publicize scientific knowledge about new remedies in a way that assuaged much of the medical community's concerns about their commercial introduction; indeed, not a few physicians were grateful to the firm for its efforts. At the same time, the working bulletin system also generated demand for these new products as the company sent samples to physicians for testing, physicians published their findings, and the company, in turn, distributed these articles to physicians and pharmacists as a part of its promotional strategy. Although many physicians remained skeptical of the company's motives—and sometimes criticized the firm vociferously— the plan allowed Parke-Davis to build markets for its new products and to introduce them in a way that the medical community, for the most part, found acceptable. Stewart later noted the positive influence of the working bulletin system in this regard: "Before the introduction of the working bulletin system the publication of this information in medical journal advertisements was called the 'worst form of quackery.' When published in the form of 'working bulletins' it was credited as valuable research work and the demand for the new remedies grew rapidly to the advantage of all concerned."[3]

Stewart's development of the working bulletin system was tremendously important in the history of the American pharmaceutical industry. It combined scientific innovation and market promotion in a new way, allowing Parke-Davis to introduce new products commercially without significantly damaging its reputation within the orthodox medical community. Yet the commitment to a scientific framework in which knowledge is freely shared also meant that the company was unable to prevent its competitors from making use of the information that it circulated. This struck both George Davis and Francis Stewart as decidedly unfair because it meant that they had trouble profiting from the investment of time, labor, and financial resources it took to introduce new products. Following the introduction of cascara sagrada in 1878, for example, the company developed a cordial made from the plant and began to circulate it to physicians for clinical testing. In 1884 the company also issued a working bulletin on the plant that detailed its botanical characteristics, habitat, and other characteristics. Over the next several years the company worked assiduously to develop a market for the cordial by sending samples to physicians, distributing articles published in the medical press about it, and other means. Toward the end of the decade, however, a company was established in New York with the express purpose of manufacturing and selling an identical product. Davis was outraged. "I understand the parties in question propose to appropriate the literature which we have published in our working bulletin," he angrily wrote to Stewart. "Their purpose evidentily [*sic*] is to carry the impression to those of the public who have used our preparation and are satisfied with it that their preparation is identical." The hospital plan clearly had its downside. As Davis noted, "Herein lies a very apt illustration of the unpractical working of our ethical policy in the present condition of affairs. We introduced cascara sag. to the medical profession of America and Europe without protection of any kind, giving freely all the information we possess with reference to the drug." Now "a proprietary medicine co. proposes to step into the field at this hour and take advantage of all the literature which we have so patiently and expensively gathered together during the past twelve years and of all the money and effort we have expended toward creating a demand to reap the harvest which by right belongs to us and which would belong to us if we were properly protected."[4]

Parke-Davis was not alone in this problem. Although the company was the most innovative firm in the ethical wing of the industry during the 1880s, a handful of other manufacturers that hewed closely to the ethical norms of the medical community also began to develop new products and introduce them to market. These companies faced the difficult question of how to protect what they saw as their right to profit from their investments and at the same time remain within the boundaries of what they themselves considered ethical behavior. As a result, some of these manufacturers began to consider the possibility of using patents on manufacturing processes to protect their interests. In many cases, of course, this was not a viable strategy, if only because the innovation in question that deserved to be "properly protected," as Davis put it, was related to the discovery of a new plant. Yet even in cases where process patents might be applicable, ethical manufacturers treaded very lightly in this area. They only rarely secured patents in the 1880s; when they did, they were only on manufacturing processes and not on products themselves, and—as far as I have been able to determine—they did not actually enforce these patents to any meaningful extent. Parke-Davis, for example, acquired a patent on a manufacturing process for coating pills in 1880.[5] Whether the company actually used the method to make pills is not clear, but I have been unable to find any evidence that the firm ever tried to enforce the patent in court or even threatened to do so. Of course, this may have been because no one directly copied their method—in general, process patents probably offered only a relatively low degree of protectiveness because different manufacturing methods that produced the same result were relatively easy to develop. But it may also have been because the company did not want to tarnish its image within the medical community by developing a reputation for using a patented method. Despite the acceptance of salicylic acid the previous decade, physicians had little awareness of the differences between patents on manufacturing processes and patents on products themselves, and even if they were aware of it, the distinction would not have meant very much to them. From the perspective of the medical community, medical patenting was an unethical form of monopoly, and any manufacturer that engaged in it risked the label of quackery.

The low degree of protectiveness of process patents combined with the hostility of the medical community to their use sometimes pushed ethical manufacturers to use other techniques to protect their interests. The case of the Upjohn Pill and Granular Company is a good example of how this dynamic played out. In the early 1880s physician William Upjohn developed a method of making what he called a "friable" pill, which was a small amount of powdered ingredients enclosed by a gelatin skin. The easily crushable pills were easy to digest, and if patients refused to take pills—which many did, because pills were typically difficult to swallow and digest—they could be quickly converted to powder and taken in a drink. Upjohn's pills soon became quite popular. "The call [for pills] is beyond my ability to supply and is growing without any effort on my part," he wrote to his brother Henry in 1884. "Now what had I best do!"[6] Henry Upjohn was both an inventor and a physician, and he had acquired at least one patent on an earlier mechanical invention, so it is perhaps not surprising that he consulted with a lawyer on the prospects of obtaining a patent.[7] The lawyer thought that the pills themselves could not be patented because they lacked novelty, but he suggested that Upjohn "take the patent on the process, which will afford a splendid protection."[8] The brothers followed his advice, and William Upjohn was granted a patent on his manufacturing process in 1885.[9] The following year, the two brothers established the Upjohn Pill and Granule Company based on William's method, and their pills quickly gained a reputation as being among the best of the industry. Within five years, the company employed about fifty people and manufactured more than 60 million pills per year.[10]

Henry Upjohn died unexpectedly in 1887 after contracting typhoid fever, but the company flourished and maintained a commitment to high-quality products. The company also invested a tremendous amount of resources in promoting its "elegant" pills as "a rational idea" that would benefit both medical professionals and their patients.[11] In order to promote its pills, for example, the company maintained "an unusual number of men" to work on advertising, and employed at least ten full-time detail men and several contract workers to market its goods directly to pharmacists and physicians.[12] However, the company never used the patent on its manufacturing process to protect its interests. Instead, it kept its manufacturing methods secret. The company appears to have come to the conclusion that the patent would not be sufficiently protective to rely on. It may also have recognized that the medical community would react negatively if word about its patent got out. Yet it had to do something: the fortunes of the firm were built on its ability to manufacture a pill superior to those of its competitors, and once the company determined that relying on the patent was not a viable strategy, keeping its manufacturing process secret appeared to be the best alternative. Yet in doing so, the company was on dangerous ground. From the perspective of at least some critics, such secrecy was itself a form of quackery that undermined the progress of medical science. "Why should reputable physicians refuse to prescribe [Upjohn's pills]?" asked the *Buffalo Medical Press* in 1889. "Because the process by which they are made is kept *secret*. . . . It is a mean, selfish spirit, smacking of quackery, that will keep such a discovery secret for the purpose of making money; and the sooner such a spirit is stamped out of the profession, the better it will be for the profession itself and for humanity at large."[13]

Although manufacturers such as Parke-Davis and Upjohn only occasionally acquired patents in the 1880s, and did so only on manufacturing methods, manufacturers that were less concerned about their reputations among the orthodox medical community increasingly turned to patenting their goods. Patent medicine manufacturers occasionally obtained patents on what were essentially formulas, although they did so only in small numbers. More important, as I discuss below, German manufacturers introduced a series of important synthetic drugs during the decade, most of which were protected by both trademarks and product patents. Proprietary medicine manufacturers also began to patent their goods, even as they continued to rely on trademarks to protect their interests. This may have been in part a response to the increased sophistication of analytic chemistry and a corresponding reduction in the protectiveness of secrecy when it came to chemical products.[14] It may also have been a response to the fact that although trademarks increasingly acted as a vehicle for monopolizing the sale of goods, trademark law at the time was both complicated and in a state of flux; some proprietary manufacturers may have decided that patent law offered a more reliable means of protecting their interests. Whether or not this was the case, proprietary manufacturers increasingly began to patent their products during the 1880s.

The turn toward patenting took place in the wake of the 1874 Supreme Court case *American Wood-Paper Co. v. Fibre Disintegrating Co.* Following the decision, the idea that the purification of a substances was not, in itself, enough to justify a patent developed into an important legal doctrine. The most important decision here was in the 1884 Supreme Court case *Cochrane v. Badische Anilin & Soda Fabrik*, where the court ruled on the patentability of alizarine (an organic compound used as a red dye).[15] Alizarine had long been extracted from madder root, but in 1868 the German chemists Carl Graebe and Carl Liebermann synthesized the substance using chemical means and in 1871 secured US patents on both the process for making the dye and so-called artificial alizarine itself.[16] A series of lower court decisions in the late 1870s upheld the validity of the patent based on the idea that artificial alizarine was a new product because it was combined with anthrapurpurine, isopurpurine, and other chemicals that did not occur in pure alizarine and appeared to enhance the value of the dye. However, in 1884 the Supreme Court overturned the validity of the patent. The court ruled that there was no distinction between alizarine produced by "natural" and "artificial" means. The court noted that according to the descriptions in the patents, the article produced by Graebe and Liebermann's process was chemically identical to the alizarine produced using madder. The court also pointed out that the patent did not mention anthrapurpurine or isopurpurine and, based on expert testimony, ruled that the anthrapurpurine and isopurpurine gave the "artificial alizarine" its "practical success in the market." The court therefore found that the patent did not cover the inventive activity that resulted in the product's increased value or success in the market. In other words, the product patent covered chemically pure alizarine, which had long been used in manufacturing; moreover, its increased utility, which made it economically valuable, was the result of other chemicals mixed in with it, yet these were not covered by the patent. Considerations of economic utility therefore did not make it something new as opposed to the alizarine that was long in use.

Cochrane v. Badische Anilin thus upheld the doctrine that purification is not in itself enough to justify a patent; in the language of the time, simply producing something "artificially" does not confer patentability if the resulting product is "composed of the same constituents and possess the same properties."[17] At the same time, however, the decision also held out the possibility that if the alizarine produced by Graebe and Liebermann's method had been transformed in some essential way and this transformation had been the cause of its increased commercial value, then the substance itself might have been patentable. This possibility made sense according to the legal doctrine of the time. Following *American Wood-Paper Co. v. Fibre Disintegrating Co.*, courts increasingly ruled that inventions in which previously known substances were transformed into categorically new things through the creation or modification of "essential" characteristics could be patented. As one observer noted, "A composition of mater, in order to be patentable, must, like a manufacture, differ in its essential characteristics from any substance previously known."[18] Increased commercial value was sometimes understood as one determining factor in the question of whether or not an essential characteristic of a thing had been changed.[19] This grew out of the assumption that although effects themselves could not be patented, the embodiment of an effect or a means of doing something in practical form might be patented. Changes in color, minor changes in size, or other nonessential characteristics that did not alter a thing's use did not result in patentability because they were not related to the embodiment of an effect or means of doing something, nor did changes in the form of a thing itself. Sugar, for example, could not be patented simply because it was made into a powder, since both granular sugar and powdered sugar are used for essentially the same purposes.[20] Nor could simply making a familiar thing more cheaply be the basis for a patent on the thing itself, since commercial value was not *itself* an essential characteristic of a thing. However, if a new manufacturing process led to a change in a characteristic of a thing that in turn significantly changed its utility and led to increased commercial value, then the thing had been transformed into something truly new and was therefore itself patentable. In other words, the assumption that embodied effects or ways of doing things could be patented meant that in order to be patentable, a substance must have some sort of new essential characteristic that is linked to practical utility. Increased or new commercial value was one potential indicator that this type of transformation had taken place.

An important example of how these interrelated doctrines played out is the 1888 decision in *Ex Parte Latimer*. William Latimer filed for a patent on fiber made from the needles of a species of pine tree known as *Pinus australis*. The chief examiner in the case rejected the claim, and Latimer then appealed to the patent commissioner, Benton J. Hall. The following year Hall also rejected the application based on the fact that Latimer's manufacturing process had not changed the fiber of the pine needle in any significant way. Instead, the fiber had simply been separated from its surrounding material and, like a pebble freed from the sand of the beach by the ocean waves, could not be said to be something truly new. The commissioner reasoned that the issuing of such a patent would mean that patents could also be obtained "upon the trees of the forest and the plants of the earth, which of course would be unreasonable and impossible." He also noted that if the fiber from the pine needles had in fact been transformed by Latimer's manufacturing process, then it probably would have been patentable. As he put it,

If the applicant's process had another final step by which the fiber thus withdrawn or separated from the leaf or needle in its natural state were changed, either by curling it or giving it some new quality or function which it does not possess in its natural condition as fiber, the invention would probably cover a product, because the natural fiber, passing through the exigencies of such a process, would be treated and become something new or different from what it is in its natural state.

Hall admitted that he was anxious to grant a patent to Latimer because the "alleged invention" was "unquestionably very valuable," being stronger, more durable, and less expensive to produce than other similar fibers. In this case, however, the increased value of the product was understood as resulting from the invention of a new process that produced a fiber that was not categorically different from the fiber in its natural state. Yet if Latimer had gone a step further and "changed" the fiber by giving it "some new quality or function," then it would have been patentable.[21]

Courts thus began to expand the definition of novelty to include increased commercial value, but only if this value was the result of a change to some other essential characteristic of a thing that resulted in a new type of utility for that thing. The idea that patents could be granted only for truly novel things—that previously known or natural substances had to be transformed in some fundamental way to be patentable—played an important role in how proprietary manufacturers went about patenting their goods. During the 1880s the Patent Office appears to have assumed that for compositions of matter to be patentable, the ingredients must be intermingled or combined in some way that resulted in a truly new thing that was distinct from the simple assemblage of its ingredients. "A composition of matter, though generally regarded as a combination, is governed by rules peculiar to itself," noted one observer in 1890. "The invention is a substance possessing certain properties and formed by uniting certain other substances in a peculiar manner. Its identity depends upon the identity of its constituent elements, upon the identity of their co-operative law, and upon the identity of the properties exhibited in the composition as a whole."[22] Simple formulas, in other words, were now generally understood as being outside the domain of patentability because their ingredients retained their discrete character—they were not changed in any meaningful way by their assemblage—and their utility was assumed to be the same as the utility of their ingredients. Simply assembling ingredients, in other words, did not lead to the production of something truly new, and therefore few patents were issued for traditional patent medicines.[23]

As a result, in their patent applications proprietary manufacturers increasingly emphasized the ways in which the ingredients of their products blended together, transformed each other, and otherwise combined together into something with new characteristics. The products patented by John Carnrick are good examples. Born in 1837, Carnrick had taught himself chemistry and pharmacy as a young man. Carnrick developed a number of proprietary medicines in the 1870s, including a lactated pepsin called Lactopeptine and a malt preparation that he sold under the name Maltine. Although Carnrick does not appear to have patented Lactopeptine, he did patent Maltine in 1877—a relatively unusual move at that time, as proprietary manufacturers typically relied solely on trademarks during the 1870s to protect their goods.[24] Carnrick continued to invent new remedies over the next two decades, and in the late 1880s he secured patents on at least four additional products. One of these was for a "medical compound" made from caffeine, phosphoric acid, extract of celery, bromide of sodium, and either Antipyrine or Antifebrine.[25] In his patent, however, Carnrick emphasized the process through which these ingredients were "reduced to a powder," "melted or fused" together, and the resulting "magma" passed through a sieve, thereby resulting in a new substance.[26] Unlike patent medicine manufacturers in the past, Carnrick emphasized the changes in the physical characteristics of the ingredients that he used in his product. And, unlike Latimer in his failed effort to patent the fiber of *Pinus australis*, Carnrick successfully linked the changes in the physical characteristics of the ingredients to the utility of the resulting substance, thereby successfully claiming that his medical compound was a categorically different thing from the ingredients from which it was made. Both the transformation in the characteristics of the ingredients and the creation of new utility, in this case new therapeutic utility, were important factors in Carnrick's ability to claim that his medical compound was a novel substance and therefore patentable.

Not surprisingly, ethical manufacturers objected to these types of patents. Even as they very cautiously turned toward the use of process patents, ethical manufacturers remained deeply committed to a critique of patents on products themselves. Such patents contravened the norms of ethical manufacturing and prompted negative reactions among both pharmacists and manufacturers committed to what they considered ethical behavior. Indeed, both manufacturers and druggists sometimes questioned not only the ethics of such patents but also their legality. In 1887, for example, manufacturer Joshua Barnes of Brooklyn developed an improved method of manufacturing sugar-coated licorice. He secured a patent not only on the process but on sugar-coated licorice itself and warned competitors against making any products that might infringe his patent.[27] This provoked a strong rebuke from the pharmaceutical press. After consulting with patent lawyer George Lathrop, for example, the *Pharmaceutical Era* dismissed the patent as invalid and sarcastically noted that "we hope Mr. Barnes has a valid patent for we know of several *improvements* that we could then patent ourselves."[28]

The fact that the *Pharmaceutical Era* consulted with Lathrop points to an important issue. As Kara Swanson has noted, lawyers trained in patent law played an increasingly important role in mediating between inventors and the Patent Office during the late nineteenth century.[29] As the proprietary industry gradually turned to patenting, and as ethical firms flirted with the use of process patents, patent lawyers also played an important role in explaining the complexities of patent law to both manufacturers and the pharmaceutical press. These conversations, however, took place in the context of changing ethical sensibilities in the orthodox medical community. As a growing number of powerful new products were brought to market, physicians increasingly recognized that drugs could be effective even if they were monopolized in some way. Partially as a result, a small but growing number of physicians began to question the assumption that patenting was, by definition, both unscientific and unethical. In doing so, they called into question the ethical framework that had long understood patents and quackery in overlapping terms, thereby shifting the ground upon which ethical manufacturing firms operated.

Proprietary Medicines and the Conflict over the "Universally Known"

In 1879 the federal trademark law was struck down by the Supreme Court because trademarks had "no necessary relation to invention or discovery" and therefore could not be reasonably thought to fall under the patent clause of the Constitution, the presumed basis for the law.[30] The decision was by no means a complete surprise to legal observers, but it set off a brief period of panic among manufacturers who had come to rely on trademarks to protect their markets. Two years later, Congress passed the 1881 Trademark Act, which was based on the commerce clause of the Constitution and therefore able to pass constitutional muster. The law established a national registry and guaranteed access to federal courts for disputes involving marks that were registered. Enforcement of trademark rights did not depend upon registration, nor did the law supersede common law on the topic. Moreover, the law only regulated trade with foreign nationals, which was defined to include citizens of foreign countries and members of Indian tribes. Trade within states, or even across state lines, that did not involve foreign nationals was not protected, although manufacturers soon discovered a loophole that allowed them to use the law for domestic trade: as long as manufacturers shipped even a minuscule amount of their product overseas, they could claim to be involved in foreign trade and register their marks. Although seriously flawed from the beginning, the 1881 law thus set federal trademark law on solid statutory ground, establishing a permanent registry for marks and guaranteeing access to federal courts to those who used it.[31]

By the middle of the 1880s there were at least five thousand patent medicines on the market.[32] Many manufacturers continued to sell their products under names such as "Dr. J. Blackman's Healing Balsam" that combined both the name of the manufacturer and information about the product in question. These types of names could be, and frequently were, trademarked. However, patent medicine manufacturers also increasingly relied on marks that did not point to specific types of products. A good example is the Lydia E. Pinkham Medicine Company. In 1873 Lydia Pinkham began to sell a remedy for "female ailments" that she had recently invented. In 1876 Pinkham patented the label of her medicine and after several years of selling the product through pamphlets, began advertising in newspapers. Pinkham understood herself as a reformer dedicated to helping women overcome the problems of overwork, worry, and similar causes of women's poor health. She sold her remedy as a source of relief in a difficult and painful world and used her own story of personal suffering as a means of promoting her goods. In 1879 her company trademarked an image of Pinkham and made it the centerpiece of a relentless advertising campaign. The result, as the company's advertising agent put it, was that sales "boomed," and by the time of Pinkham's death in 1883 the company was grossing $300,000 per year. This success was built in part on the way the trademarked image of Pinkham communicated information about the company to the purchaser, assured her that what she buying was genuine, and promised that the products she bought would do for her what they had supposedly done for Pinkham herself. Advertisements, noted her agent, "can not help but have their effect on the general public." The image of Pinkham did not, however, distinguish specific products made by the company from one another. It could be used to sell any number of products made by the company, since it referred to the firm itself and, by inference, the general set of goods that the company manufactured.[33]

A small number of ethical manufacturers took a similar approach. Parke-Davis, E. R. Squibb & Sons, and other firms that hewed closely to the orthodox medical community's sense of propriety tended to avoid the use of trademarks, and some, such as Parke-Davis, explicitly rejected them as unethical. A handful of ethical firms, however, began to develop distinctive words and images that they used in their promotional efforts. The Upjohn Pill and Granule Company, for example, developed the logo of a thumb pushing down on a pill, indicating the company's pill's easily crushable nature. The company does not appear to have trademarked the logo, at least not at the federal level, but it stood at the heart of its efforts to promote its products as "rational" and "elegant" preparations and to "induce physicians to try our goods, knowing when once started they will continue their use."[34] The company's easily recognizable image served to distinguish its products from the supposedly inferior goods of their competitors. The logo itself operated at the level of the firm, in that it referred to the company in general and to all the different types of pills it made. It encapsulated and conveyed the reputation of the company, and, at least in the company's view, promoted both the fortunes of the firm and the practice of good medicine. "There is no question but we are producing *the best* goods that are made," Henry Upjohn noted in 1885. "We shall first and always keep in mind the fact that therapeutic value must be maintained no matter what."[35]

The turn toward this type of promotional strategy intersected with changes in trademark law in an important way. Even as marks were increasingly understood to be a form of property, courts continued to find that their primary use was to indicate the origins of goods. As the Supreme Court noted in 1879, "The object of the trade mark is to indicate . . . the origin or ownership of the article to which it is applied."[36] Courts thus continued to rule that while designating marks could be appropriated, marks that were descriptive in nature could not be. The Brown Chemical Company, for example, tried to prohibit the use of the term "Iron Tonic Bitters" based on the fact that it had trademarked the term "Brown's Iron Tonic Bitters." In at least two cases the company was unable to do so, including one case against Frederick Stearns & Company, in which a district court in Michigan ruled that it was "abundantly sustained by authorities" that "words which are merely descriptive of the character, quality, or composition of an article, or of the place where it is manufactured or produced, cannot be monopolized as a trademark." Bitter wine of iron, noted the court, is "a standard and recognized medical preparation," and therefore "no monopoly can be claimed of the words 'iron bitters,' which are indicative of the composition of the article."[37]

The long-standing formulation used by patent medicine manufacturers—in which names such as "Dr. J. Blackman's Healing Balsam" and "Brown's Iron Tonic Bitters" were designating in nature and therefore, ironically, could be used by anyone as long as they indicated the true origin of the product—thus began to fracture. On the one hand, some manufacturers turned to trademarks and logos that operated at the level of the firm and did not include information about the specific products manufactured by the company. On the other, names such as "Brown's Iron Tonic Bitters"—or at least parts of them—were sometimes found by courts to be descriptive in nature, thus suggesting that the portion of the name that referred to the *company* and the portion of the name that referred to the *thing* were distinct from one another. The portion of the name that referred to the company, in other words, had no relationship to similar products made by other manufacturers—anyone could manufacture iron tonic bitters and apply the name of his own company to its products. Certainly, numerous manufacturers continued to acquire trademarks on names that conformed to the older formulation, in which the name of the manufacturer modified the type of substance in question. Increasingly, however, names or images that pointed to the firm and names that pointed to types of products were distinct from one another. In a sense, the designating aspect of a product name and the descriptive aspect of it began to be separated. Increasingly, descriptive names of products no longer included the name of the original manufacturer.

This fragmentation also took place in relationship to proprietary goods. During the 1880s proprietaries were increasingly sold under arbitrary and sometimes bizarre names such as "Lactopeptine." These names were clearly arbitrary in the sense that they had no necessary relationship to the product in question, to its ingredients, or to its characteristics. They were obviously not descriptive in the same way that personal names, geographic names, or the common names of ordinary things were—after all, Carnrick could have easily given *another* name to "Lactopeptine" had he chosen to do so. Yet these types of names were not designating in the traditional sense either, since they did not necessarily point to any particular manufacturer. An important legal tension thus came to the fore: was it possible to trademark the names of things if those names were not descriptive in nature? By the end of the 1880s, a large body of court decisions had established that names that operated at the level of the product could in fact be trademarked as long as they were not descriptive in nature. Such names had to be arbitrary in the sense that they did not have any necessary relationship to the underlying good—after all, this would mean that they were descriptive terms that were "the common property of mankind, in which all have an equal share and character of interest."[38]

The question of what was an arbitrary name and what was a descriptive name was not always easy to resolve. Indeed, much of the drama around trademarks during the 1880s took place along these lines. For example, following the invalidation of Horsford's patent on acid phosphate in 1872, Rumford Chemical Works began to strategize about other ways of protecting its market in the substance. In 1877 the company registered its label for "Horsford's Acid Phosphate" with the Patent Office and trademarked the phrase.[39] The company then used its trademark to prevent other firms from using the term "acid phosphate" in their promotional efforts. In 1882, for example, the Supreme Court of Rhode Island enjoined the Hughesdale Manufacturing Company from selling any product called "acid phosphate" that was a substantial imitation of Rumford's good, and the following year it ruled that the company had violated the injunction by selling a product called "Hughes Acid Phosphate." Rumford pressed the point in its advertising, calling the attention of the pharmaceutical community to the decision and warning competitors from using the term "acid phosphate" in the names of their products or advertising.[40] In 1885, after consulting a patent lawyer about the possibility, the firm also secured a trademark on the term "acid phosphate" itself.[41] From the company's perspective, this probably did not seem odd: the term had become a shorthand way of referring to its product, which was quite popular at the time, and the firm had successfully prosecuted a suit that prevented a competitor from using it. The company probably felt as if it had every right to the name.

Yet Rumford soon ran into formidable opposition. In 1881 Parke-Davis had begun to manufacture and sell a preparation similar to Horsford's but made according to a different formula. The company initially called its product "Liquor Acidi Phosphorici" and advertised it widely as a substitute for Rumford's product. However, Parke-Davis soon came to believe that the phrase "acid phosphate" referred to all such preparations, and in 1887 the company changed the name of its preparation to "Liquid Acid

Phosphate" so that it would conform to what the company took to be its actual name. "We have hitherto labeled this preparation 'Liquor Acidi Phosphorici,'" the company noted in its explanation of the change. "The preparation has, however, come to be so universally known as 'Acid Phosphate' that we have thought it best to adopt that name on our labels."[42] In response, Rumford Chemical Works sued a druggist in Baltimore for selling the Parke-Davis preparation; Parke-Davis defended the suit, and the two companies ended up in court. A significant amount was at stake in the conflict. By this point Rumford earned more than $175,000 a year on the sale of acid phosphate, but these sales were heavily dependent on advertising, which was in turn dependent, at least in part, on its control of the term "acid phosphate." The loss of the term would clearly have significant implications for Rumford's ability to maintain its market.[43]

The case came to trial in 1888. In its decision, the Circuit Court of Maryland noted that deception was not an issue in the case, since Parke-Davis had explicitly identified itself on its labels as the manufacturer of its products. As a result, the "sole question" facing the court was whether or not the phrase "acid phosphate" was an "arbitrary name" or a "description" of the product. The court determined that the phrase "acid phosphate" was in fact commonly used in chemistry but also that the term was "inexact" in nature and could refer to a variety of chemical goods. However, the court also pointed out that other common phrases—such as "fresh bread"—are inexact as well, since such a term "does not indicate with precision whether the bread is made of wheat or rye, of bolted or unbolted flour, whether or not it contains salt, or with what character of yeast it is made." Therefore the issue at hand was not whether the words were "exhaustively descriptive" of the article in question but instead whether "they are commonly used by those who understand their meaning" and are "reasonably indicative and descriptive of the thing intended" to that group of people.[44] The court found that, according to this standard, the term was descriptive in nature and could not be appropriated. Rumford Chemical Works thus lost control of its trademark on the term, and Parke-Davis continued to manufacture and sell its product under what it considered the correct name of the substance. Rumford's sales declined almost immediately: within two years sales had dropped by more than $10,000 a year after more than a decade of steady growth.[45]

Rumford was unsuccessful in its efforts to control the term "acid phosphate," but other companies successfully managed to monopolize the sale of their products through the use of trademarks on arbitrary names. A notable example was John Uri Lloyd's monopolization of the term "Asepsin." Born in 1849, Lloyd was an eclectic manufacturing pharmacist who was highly regarded both for his dedication to the science of pharmacy and for his professional ethics. Over the course of his remarkable career Lloyd made a long series of important contributions to the science of pharmacy, and in 1887 he was elected president of the American Pharmaceutical Association. Yet Lloyd also struggled with the constraints on innovation facing ethical manufacturers. Sometime around 1884, for example, he had developed a method for isolating what he believed to be a pure form of the alkaloid of *Hydrastis canadensis*, commonly known as goldenseal, which he sold under the name "Lloyd's Hydrastis." Several other manufacturers quickly introduced similar products that they claimed were the same substance; Parke-Davis, for example, introduced a similar product under the commonly used name "hydrastine" around the same time. Lloyd responded by claiming that his product was superior to those of his competitors because it was nonirritating to the stomach and completely safe "internally, externally, or as an injection." These were strong claims that went beyond the accepted consensus about the potent nature of goldenseal and derivative products, and Lloyd's efforts attracted a storm of controversy. The *Bulletin of Pharmacy*, for example, denounced Lloyd's efforts as "bristling with unethical and quackish exaggeration and obtrusiveness" and unbecoming of "a house the chief exponent of which is regarded as a 'representative standard-bearer of the pure ethics of pharmacy.'"[46]

Lloyd thus faced the difficult problem of how to introduce new products to market in a commercially viable manner without violating the ethical norms of his trade. In response, he began to trademark his products in a way that clearly distinguished the name of the good from his own personal name. In 1887, for example, Lloyd produced a sodium form of methyl salicylic acid. Lloyd gave the substance the name "Asepsin," a term that an eclectic physician he knew coined because it sounded evocative of substance's therapeutic properties. Lloyd marketed the product under the name "Lloyd's Asepsin," but he trademarked the word "Asepsin" itself.[47] The strategy proved successful. Soon after Lloyd introduced Asepsin to the market, the William S. Merrill Chemical Company analyzed the substance, determined its chemical structure, and began manufacturing and selling its own version of the product under the same name. In 1889 Lloyd brought the company to court for trademark infringement. The court found that the term was arbitrary in nature and could therefore be rightfully adopted as a trademark; as one physician testified, "The word 'Asepsin' is utterly manufactured out of a suggestion, idea, that it may be antiseptic. . . . It is a name that has no meaning at all, only as it may be suggestive in regard to sepsis or antisepsis."[48]

The decision gave Lloyd an effective monopoly over the manufacture and sale of Asepsin. As a result of the decision no one could use the name "Asepsin" to refer to his own version of the chemical without Lloyd's permission, despite the fact that Lloyd did not have a patent on it and the manufacture of the substance was therefore, supposedly, open to all. Of course, this left an important question unanswered: was there some *other* name that could be used to refer to the substance? It was not at all clear that there was. Certainly, pharmacists and chemists might refer to the product as "the sodium salt of methyl salicylic acid" or some similar formulation, but the phrase was long and awkward and not at all suitable for either commercial or therapeutic purposes. More to the point, no one called it that. Once the substance was introduced, chemists, physicians, and druggists simply referred to it as "asepsin."[49] Indeed, the Merrill Chemical Company argued in its defense that the term had moved into public use and therefore it had every right to use it to refer to its version of the product. After all, what else could it be called? From a legal perspective, Merrill might have been able to coin some *other* name and give that to its preparation. Yet in doing so the company would have faced substantial risk: the difficulty in building a market for a good that was already known under a different name and, more significantly, the likely accusation of adulteration. Critics would certainly have argued that if the company was manufacturing "asepsin" then it should call the chemical by its proper name and not by some inaccurate name that concealed its true nature. It is conceivable that Merrill could have advertised its product under some other name and, simultaneously, tried to make clear that it was actually "the sodium salt of methyl salicylic acid," but even if the company had done so, it would have been both awkward and confusing to all involved, probably requiring some sort of explanation about why the firm was legally prohibited from referring to it by what everyone took to be its proper name. There was no real way out of this dilemma, and Merrill did the only thing it could reasonably do and remain within the framework of ethical manufacturing. It abandoned the market in the drug.

By the 1880s there was thus a range of perspectives among ethical manufacturers on trademarks that operated at the level of the product. For some manufacturers, most notably Parke-Davis, such marks were among the worst form of quackery. Others, such as Lloyd, disagreed and considered them a reasonable means of protecting their interests. Lloyd felt justified in monopolizing the term "Asepsin" in part because he considered the behavior of firms like Merrell to be a form of "piracy" that deprived him of the time and money that he had invested in the development of the drug.[50] Lloyd also felt justified in protecting the name of the drug because, as he put it to another manufacturer, "The chemical name is free to the world, there is nothing to prevent others using our discovery under the real name which we have given the world."[51] This argument, of course, was somewhat disingenuous in that "the sodium salt of methyl salicylic acid" was in no sense a functional name for either commercial or therapeutic purposes. Still, the argument may have helped protect Lloyd's reputation among his peers. As far as I have been able to determine, Lloyd was not criticized for his monopolistic practices, perhaps in part because Asepsin never became a particularly successful remedy.

Lloyd may have escaped censure for his introduction of Asepsin, but the growing number of proprietary remedies on the market prompted a tremendous amount of concern within both the medical and pharmaceutical communities. To many critics, proprietary medicines seemed quackish because they did not always make their formulas clear, their popularity often seemed to be a result of advertising more than their effectiveness, and manufacturers used trademarks on their names to prevent others from manufacturing them. All this seemed contrary to the goals of a benevolent medical science. Moreover, according to some critics, proprietaries appeared to undermine the very ability of physicians to rationally practice medicine because of the rapid pace of their introduction, the similarity and strangeness of their names, and the seeming impossibility of keeping up with the constant flood of new products.[52] As I discuss below, proprietaries were also frequently critiqued for artificially inflating the price of drugs. Trademarks, noted one critic, "must not be used to extort money from the sick and ailing, or to take advantage of infirm bodies and weakened minds. It is high time that the medical profession lodge their solemn protest against such proceedings and open the eyes of its poor deluded victims, as well as protect their sacred charge, suffering humanity, from the extortions of the patent medicine and trademark vulture."[53]

There was an extremely important and difficult question at the heart of the concern about proprietary drugs: to what extent could pharmaceutical products vary and still remain essentially the same thing? Once manufacturers began applying arbitrary new names to their products it became increasingly unclear what the relationship between similar products sold under different names actually was. If two products had very similar characteristics but one was sold as Asepsin and one was sold as something else, were they in fact the same thing? And if so, and if Asepsin was a trademarked term, then was not the other manufacturer guilty of misrepresenting his product by calling it something other than its proper name? From the perspective of many critics, the problem with trademarks on names such as "Asepsin" or "Tonga" was that they prevented competitors from manufacturing the same product, even if there were no patent rights involved. But the problem was actually worse than this: such names actually meant that goods with identical characteristics were either *not the same* product or, if the were called by two different names, that one of the names was incorrect. In other words, trademarked names that operated at the level of the product made it impossible for two substances to have identical characteristics and yet be called by different names. Equivalence between products that had different names was impossible.

This was a problem even when there was a nonproprietary name that was assumed to apply to the different goods in question. In many cases it was assumed that underneath the various proprietary names for similar goods lay a single substance that had its own nonproprietary name.[54] However there was still a serious problem here: if quinine preparations, for example, were sold under distinctive trade names, then physicians would need to refer to these names in their prescriptions in order to ensure that they had some knowledge of the substance that patients actually received—after all, quinine preparations could vary tremendously in terms of potency and other variables, and if a physician simply prescribed "quinine" a pharmacist might dispense any one among many such preparations. Yet this would invariably lead to a tremendous proliferation of names that would become unmanageable—how was a physician supposed to keep up with the seemingly endless number of new preparations? How was a physician supposed to know what the strengths of various preparations were, especially if manufacturers changed them at will? Indeed, how was a physician supposed to know that the product was *quinine* at all if it was sold under some other, scientific-sounding name? The problem was even worse for products that combined multiple ingredients. In these cases, trademarked names seemed to conceal the true ingredients of the product, and these types of goods sometimes seemed to be little more than patent medicines sold under impressive-sounding names. Indeed, according to many critics these types of products were really little more than combinations of other ingredients that could just as easily be compounded by a pharmacist—if, in fact, the ingredients could be determined. From this perspective, scientific-sounding names were nothing more than an advertising trick designed to make products sound like something they were not, and—even worse—there was no way to determine what they actually *were*. Proprietary medicines, noted one critic, "are for the most part the inventions of tradesmen, and in no sense represent the growth and progress of medical science."[55]

Lloyd appears to have escaped censure for his use of the name "Asep-sin," probably because he emphasized the fact that the product also had a scientific name that was not monopolized. Other ethical manufacturers that experimented with trademarks on the names of goods received more criticism. In 1886, for example, Frederick Stearns & Company trademarked a series of names based on the chemical term "alkamet-ric," some of which operated at the level of the product.[56] Critics assailed "Alkarits," "Alkametic Granules," and the other products the company sold under these names as examples of "protected medicines" and "per-nicious monopolies which permit the proprietor or manufacturer to charge an exorbitant price, far beyond a legitimate profit; and thereby obtain immense fortunes out of the ills of humanity."[57] Stearns quickly abandoned the names. Given his previous experience being charged with adulteration, this is not particularly surprising. Yet even his willingness to experiment with this technique and open himself to the possibility of once again arousing the anger of his peers indicates the extent to which ethical firms struggled to find some means to protect their investments in developing new products.

Francis Stewart and the Defense of Limited Monopoly

In his 1883 book *Medical Ethics and Etiquette*, Austin Flint provided a detailed and comprehensive commentary on the AMA's Code of Eth-ics. Flint was a highly regarded physician—he was the president of the New York Academy of Medicine, and the following year he served as the president of the AMA—and like many of his colleagues he considered the code to be a crucial part of maintaining the "purity and dignity" of the profession. Not surprisingly, he was also strongly opposed to the use of both trademarks and patents in medicine. Vending proprietary medi-cines under trademarked names is objectionable, he argued, because it interferes with the ability of physicians to adapt their treatment to the specifics of individual cases and because the ingredients of proprietaries are often unknown. Noting that "the grounds for the injunction not to patent remedies . . . and not to dispense secret nostrums, are not always appreciated by the public," Flint also denounced the use of patents in medicine by rhetorically asking what would have happened if the great medical discoveries of the past had been patented. Like many others in the orthodox medical community, Flint denounced both patents and trademarks as quackish forms of monopoly opposed to the benevolent practice of a truly scientific medicine. "Imagine Jenner to have applied for a patent giving exclusive property in vaccination, or keeping it a secret!" he wrote. "Here, as in all other instances, the restrictions of the code of ethics have reference to the welfare of the community, and not to the selfish interests of the medical profession."[58]

Flint's text was a part of a broad and heated debate taking place at the time about the Code of Ethics and its role in the organization of orthodox medicine. During the 1870s and 1880s medical reformers worked assiduously to pass a new wave of licensing laws in order to restrict the practice of medicine to those they deemed qualified; between 1873 and 1884 at least ten states passed new licensing requirements of some sort, while reformers in other states worked diligently to pass similar laws. The AMA was an active participant in this process, working closely with state medical societies to pass medical legislation and encouraging its members to fight against the specter of quackery. Debates about the Code of Ethics were an important part of this effort. State and county medical societies usually had their own codes that were modeled on the AMA's code, and physicians who violated these rules of conduct could be, and sometimes were, expelled. Whether this had direct legal consequences on the ability of physicians to practice medicine is unclear, but even if it did not, the result of being labeled a quack and expelled from one's medical society could be devastating.[59]

Not all physicians were happy about this process. Far from it. As John Harley Warner has argued, the 1880s also saw a broad revolt against medical orthodoxy among physicians who chafed at the stifling effect of the AMA's Code of Ethics. As Warner makes clear, the physicians who reacted against the code were also the ones who embraced the new physiological therapeutics and rejected the older therapeutic framework that grounded medical authority in the wisdom and experience of the individual practitioner.[60] The code's ban on consulting with heterodox physicians seemed particularly problematic to many critics because it appeared to prioritize factionalism—what David Hunt called "medical bigotry"—over science.[61] As Warner notes, the rebellion against the code was both a rejection of a punitive and unyielding code of behavior and an "intellectual and political maneuver that marked the contested emergence of a new order of *scientific medicine*—a 'scientific democracy'—in which trust in science was to be the best guarantee of technically and morally right conduct."[62]

For a small number of physicians a rejection of the prohibition on patenting was also a part of this process. Flint's essay makes clear that during the 1870s and 1880s orthodox physicians continued to see themselves as a distinct and noble class of professionals and that the rejection of monopoly continued to play an important role in this self-formulation.[63] In at least some cases the critique of monopoly translated into concrete actions against physicians who violated the ethical norms of their profession.[64] At the same time, however, physicians also increasingly prescribed medicines that were either patented or protected by trademarks: according to an 1885 survey of more than fifteen thousand prescriptions written in Chicago, about 2.5 percent of all prescriptions included a proprietary item, at least some of which were probably patented.[65] There was thus an important disjuncture between the daily practice of many physicians and the antimonopoly rhetoric of physicians such as Austin Flint, even if the number of proprietaries that were prescribed was still relatively small. Equally important, ethical manufacturers struggled to find a way to protect their interests in a highly competitive market. Manufacturers such as Frederick Stearns, E. R. Squibb, and George Davis increasingly chafed against an ethical framework that prevented them from introducing new products to science in a way that allowed them to commercially benefit from their efforts. Increasingly, the traditional framework for introducing new drugs seemed to hinder the advancement of medical science rather than protect it.

During the 1880s a small number of reformers thus began to rethink the traditional prohibition on monopoly. Francis Stewart was by far the most important figure in this trend. Following his development of the working bulletin system, Stewart quickly came to realize that a commitment to open science made it difficult for manufacturers to profit off of their investment in scientific research. In the early 1880s he began to argue that patenting, properly applied and understood, was both scientifically and ethically legitimate because it encouraged the process of scientific innovation. According to this view, as he put it in a pamphlet published by Parke-Davis in 1882, the purpose of patents

is to promote progress in science and the useful arts; and this end is secured by encouraging authors and inventors to write and discover, protecting capital invested in the product of their brains, creating a valuable industry, and making this industry a great knowledge-producing power engaged in original research and publishing the results for the benefit of humanity and the cause of truth in the world.[66]

Patents—properly understood and applied—guaranteed that scientific information could be freely shared among the medical community while simultaneously protecting the financial investment of the inventor. "The patent system secures the publication of full knowledge of every invention patented, and thus benefits science," Stewart noted two years later.

"It affords a just protection to inventors until the investment of capital in working and perfecting the invention becomes a remunerative one and the inventor rewarded for his labors."[67] Patents were the exact opposite of quackery. They promoted the cause of medical science rather than undermining it.

Stewart's argument hinged on the distinction between process and product patents. Stewart believed that patents on manufacturing processes encouraged scientific investigation among manufacturers, both by ensuring the publication of their methods and by protecting their financial investment in drug development. However, this was based on the assumption that competing firms would be free to investigate the new remedies in question and develop their own means of manufacturing them. Product patents, on the other hand, had no place in ethical manufacturing, since they would restrict the ability of other firms to investigate and manufacture these substances. Patents—by which Stewart meant process patents—were equated with openness and the circulation of scientific information; as he later noted, "A thing patented is a thing divulged."[68] This in turn was based on a distinction between open science and patents on the one hand and what Stewart sometimes called "commercialism" in drug manufacturing on the other. Stewart strongly critiqued the use of secret ingredients, adulteration, product patents, and other unethical practices that, as he put it, were the work of the "charlatan and quack."[69] He therefore argued that all new and useful medical inventions should be protected by process patents, that secret formulas should be abolished, and that strict laws should be passed in each state against adulterated goods.[70]

Stewart also strongly criticized the use of trademarks. Like many of his peers, Stewart was deeply disturbed by the flood of proprietary remedies on the market, and he was strongly opposed to efforts to monopolize the names of substances themselves. Beginning in the early 1880s Stewart therefore drew a distinction between trademarks that operate at the level of the company, or what he sometimes called the "brand," and trademarks that operate at the level of the good itself. Stewart believed that trademarks at the brand level were perfectly acceptable and furthered the interests of legitimate manufacturing. Trademarks on the names of things, however, were deeply unethical because they restricted the ability of other manufacturers to make the product in question and therefore distorted the practice of both pharmacy and medicine. "If the name of a thing can be legally held as a trade-mark on the thing itself," he suggested, "then can every invention be locked up to secrecy and an everlasting monopoly. Surely, such a system is, to the last extent, unscientific, and detrimental to the interests of the public."[71] At the annual meeting of the American Medical Association in 1881, Stewart thus introduced a resolution that would have modified the Code of Ethics to prevent physicians from dispensing products controlled by trademarks unless the mark was used to "designate a brand of manufacture" and the product was accompanied by "a technical, scientific name, under which any one can compete in manufacture of same."[72] The proposal generated a brief flurry of controversy and was tabled before being taken up, with some observers accusing him of quackery because the proposal also suggested allowing patents on manufacturing processes.[73]

Over the next decade Stewart produced a truly remarkable outpouring of articles and talks on these topics.[74] He became the single most important voice in both the medical and pharmaceutical communities arguing for the legitimacy of process patents in pharmaceutical manufacturing. He also became one of the leading critics of the use of trademarks for monopolistic purposes. In his view, both product patents and trademarks that operated at the level of the good were the equivalent of patent medicines made with secret ingredients because all three unfairly restricted the ability of competitors to manufacture the product in question, thus inhibiting the progress of medical science and undermining the health of the public. At the same time, however, both process patents and trademarks—properly used—guaranteed that reputable firms were able to profit from the investments they made in developing new products.[75] In both cases, according to Stewart, a limited monopoly actually promoted the cause of medical science rather than hindering it. Indeed, for Stewart the ability of manufacturers such as Parke-Davis to financially benefit from developing new products was an essential part of the promise of therapeutic innovation. Stewart's arguments in favor of medical patents and trademarks thus reconceptualized the relationship between commerce and science by reinterpreting the ethical meaning of monopoly. While certain forms of monopoly needed to be suppressed— secrecy, product patents, and trademarks on the names of things themselves—limited monopoly in the form of process patents and trademarks on brands advanced both the goals of medical science and the fortunes of private interests. From his perspective, there was no real distinction between advancing the ability of reputable manufacturers to commercially develop new goods and the promotion of medical science itself.

Yet there was an important irony here. For some, Stewart's arguments smacked of hypocrisy because they benefited the firm he worked for. In 1882, for example, the *St. Louis Clinical Record* declared the new remedies offered by the company "weeds" and attacked Stewart's arguments as a "gag" that was designed to fill the coffers of the firm.[76] Such comments point to a growing concern among many physicians about the impact of pharmaceutical manufacturers on the direction of medical science. The introduction of unfamiliar botanicals, the increasing use of manufactured items with strange-sounding names, and the growth of preassembled remedies manufactured by reputable firms all pointed to the growing influence of drug manufacturers on medical practice. Many physicians worried about the encroaching threat to their therapeutic authority, but few knew what to do about it. Refusing to prescribe the new remedies seemed pointless; patients would go elsewhere, and besides, they were both convenient and often seemed effective. Yet they also appeared to subordinate the interests of the physician to those of the firm. "There is a too successful effort being made by the leading drug houses to control, forestall, abridge, and render subservient to drug circles the whole medical profession," wrote one physician in 1884. "This is being effected by producing ready prepared compounds, not strictly patent medicines, yet answering the same purpose. . . . The trouble lies within the profession. They are owned by the druggists, and do not dare refuse [prescribing them] for fear of loss of influence."[77]

These criticisms struck close to home. Stewart was exceedingly cautious about protecting his reputation, and from his correspondence it is clear that he cared deeply about promoting a rigorous medical science. Yet he could not fully escape the contradictions of the times in which he lived. In 1883, for example, Davis asked Stewart to write a laudatory paper about one of Parke-Davis's new remedies and to publish it under his name "for advertising purposes." Stewart refused but tried to compromise by writing what he considered a scientific article about the product. The article was not sufficiently supportive to be used in the company's promotional efforts, and Davis asked him to revise it. In an angry letter in response, Stewart pointed out that as a physician he was bound by a set of medical ethics that he could not violate. "These reports," he argued, "must contain the truth, the whole truth, and nothing but the truth."[78] Davis fired him as a result of the incident, but he was soon rehired after the two reconciled. Shortly after the conflict, Stewart went to Philadelphia under the company's direction to serve as an "ambassador" to the city's medical community. While there he got to know many prominent physicians in the city, and in 1884 he was elected as a delegate to the Pennsylvania State Medical Society. Davis was quite pleased. "We read with interest . . . your successful personation in the double role of delegate to the state Medical Society and representative of our house," he wrote. "It is a pleasure to know that your professional colleagues will accept you in the former position without questioning your relationship to us."[79] In the same year, Davis gave him a substantial raise.[80] The simultaneous efforts to promote the interests of the firm and the cause of medical science were going well, and Stewart deserved to be rewarded for his work. Not surprisingly, Stewart did not discuss his good fortune with his peers.

The Challenge of the German Synthetic Drugs

By the 1880s the German chemical and pharmaceutical industries had become significantly more advanced than either their European or American counterparts.[81] Industrial firms worked closely with professors in the university system to develop advanced chemical techniques, leading to a host of important advances in chemistry and a reputation for cutting-edge scientific research. German chemical industrialists also became experts at using patent laws in other countries to advance their interests. The unification of Germany in 1871 had been followed by the establishment of a unified national patent law in 1877 based on a "first to file" system (rather than the "first to invent" system used in the Untied Sates). The 1877 law allowed patents on chemical processes but excluded chemical substances; beginning in 1888, a series of decisions by the Imperial Court expanded the definition of patentability to include some chemical substances, but between 1877 and 1890 medicinal substances could not be protected by German patents.[82] Indeed, the United States was highly unusual in its allowance of patents for medicines. With the exception of Great Britain, European countries generally prohibited acquiring patents on medicines, and some also prohibited acquiring patents on chemicals in general. This difference led to significant price differentials between German products sold in the United States and the same products sold in other countries, a dynamic that had significant implications for the shape of the drug market in the coming years.

During the 1880s the German chemical industry introduced a wave of powerful synthetic drugs to the American market. These included antipyretics such as Antipyrine (1884), Antifebrin (1886), and Phenacetin (1887); hypnotics such as Sulfonal (1888) and Trional (1889); antiseptics such as Aristol (1890); and sundry other synthetic drugs, such as Saccharin, which was first synthesized in 1878 and was being used medicinally within a decade. Most of these products were protected by patents on the chemicals themselves, although some were only protected by patents on the manufacturing processes. They were also typically introduced and sold under trademarked names that operated at the level of the product.

Some of these substances had rather straightforward chemical names, and in these cases the chemical name was sometimes used as a nonproprietary name and sometimes trademarked. Others had long scientific names that were too complex for commercial or therapeutic uses; in these cases German manufacturers typically trademarked a short and easy-to-remember name instead. Farbenfabriken vormals Friedrich Bayer und Companie (Bayer), for example, sold the chemical *diethyl-sulphon-dimethyl-methane* under the trademarked name Sulphonal and the chemical *diethyl-sulphon-methyl-ethyl-ethane* under the trademarked name Trional. Whether the chemical name and the commercial name were the same or distinct, however, both pharmacists and physicians almost always referred to the products by their commercial names. The result was that the trade names of these goods operated, as Francis Stewart put it, as a "perpetual monopoly."[83]

The antipyretics serve as an important example of how this worked. In 1884 Wilhelm Filehne of the University of Erlangen published the results of his clinical studies with a chemical that had previously been discovered by Ludwig Knorr, another chemist also at Erlangen. Filehne and Knorr began working with the German firm Hoechst, which named the substance Antipyrine. The chemical attracted an immense amount of interest in both Europe and the United States because of its ability to reduce fever and ease pain, and within a year Antipyrine was being used as a general treatment for fever and pain. In the United States, enthusiasm for the drug was such that Hoechst had trouble keeping up with demand. However, many physicians and pharmacists were deeply concerned about the way in which the company monopolized the sale of the chemical. Knorr had managed to secure a patent not only on the manufacturing process but on the chemical itself, which struck most observers as deeply troubling.[84] This was clearly out of the bounds of appropriate behavior for an ethical firm, and many critics compared the company's actions to those of patent medicine manufacturers.[85] It didn't help that the United States seemed to be alone in its willingness to grant patents on such important therapeutic discoveries. As E. R. Squibb noted, for example, while Hoechst was able to secure a patent on the drug in the United States, it was unable to do so in France. "In the political economy of France, and to the great honor of the nation," he observed, "it has long been held that the interests of suffering humanity are superior to the interests of inventors and therefore as a sanitary measure patents upon medicines are not granted and patented medicines from all sources are prohibited."[86]

Equally disturbingly, the firm had trademarked the name "Antipyrine." The problem with this was that there was no other name that could be used to refer to the substance. Filehne's initial publication had referred to the chemical only by this name, and among chemists there was significant confusion about its proper scientific name; Knorr had initially identified the substance in the German pharmaceutical literature as *methyloxychinzin*, but at other times it was referred to as *dimethyl-phenyl-oxypyrazol* or by some other formulation.[87] In any event, physicians and most druggists almost always referred to the product simply by its trade name. The chemical name—whatever that might be—was too complex for daily use, and physicians generally did not understand enough about organic chemistry for such terms to be particularly meaningful anyway. "The clinical name phenyl-dimethyl-pyrazolon," noted one physician with exasperation, "is calculated to produce luxation of the maxilla in one unversed in chemical lore."[88] Toward the end of the 1880s, there was a brief movement to use the name "methozin" as a nonproprietary name for the substance, but as the *American Druggist* explained, "As long as the substance is *sold* as Antipyrine, it will be generally prescribed for, or demanded by the public under this name, and it should be designated by the same name in the standard works of reference, and even in the Pharmacopeia."[89]

This dynamic took place even when there was a relatively well-known chemical name that was short enough that it could, at least theoretically, be used for commercial purposes. A good example is the antipyretic Antifebrin, which had been known since at least 1853 and referred to by chemists as either *acetanilide* (sometimes spelled without the final "e") or *phenyl-acetamid*. The chemical's antipyretic properties were first discovered in 1886, at which point the German firm Kalle and Company began selling it as a pharmaceutical. The company may have secured a patent on improvements in the manufacturing process—the historical record is somewhat unclear on this—but its real investment was in the name "Antifebrin," which it coined to be therapeutically suggestive and protected with a trademark. Following the discovery, manufacturers in the United States, including Parke-Davis, also began to sell the chemical to the therapeutic market but under the name "acetanilide." It certainly was no secret that the terms "Antifebrin" and "acetanilide" referred to the same chemical substance, but within the American medical community, physicians almost always prescribed it under Kalle and Company's name because the firm had been the first to bring the chemical to the therapeutic market. As a result the company was able to charge about double for its product what other companies were for theirs, despite the recognition that the two names referred to the same substance. At the same time, both Antifebrin and acetanilide were significantly cheaper than Antipyrine, a fact that was widely ascribed to Hoechst's patent: in 1888, according to one report, Antipyrine cost about $1.25 an ounce, whereas "Antifebrin" could be purchased for only 25 cents an ounce and "acetanilide" was available for just 10 cents an ounce.[90] This presented a dilemma for druggists. If given a prescription for Antifebrin, should they fill it with the drug manufactured by Kalle and Company, or could they fill it with a product manufactured by another party and sold less expensively under the name "acetanilide"? Pharmacists debated the issue, although they appear to have generally felt obligated to dispense the product made by Kalle and Company because they could not manufacture the chemical themselves.[91] The whole business struck domestic pharmaceutical firms as an unfair monopolization of a commonly made chemical substance. Parke-Davis, for example, had a history of manufacturing the chemical and argued that "the employment as well as registration of the name 'antifebrin' is in reality nothing more than the conversion of a scientific substance into a 'patent medicine.'"[92]

A third example illustrates how patents and trademarked names could be used to reinforce one another. In 1887 Bayer introduced another synthetic antipyretic. The chemical was initially referred to by two different names, *acetamidophenetol* and *acetphenetidin*. Shortly before it was introduced commercially, however, the name of the substance was shortened to *phenacetin*.[93] Bayer acquired a patent on the chemical itself and trademarked the shortened name.[94] The drug rapidly attracted a significant amount of attention because it was effective and had fewer side effects than either Antipyrine or Antifebrin. Initially, there was little concern about its patented or trademarked status; physicians often did not even realize it was patented, and many were pleased that they could pronounce and prescribe the product by its chemical name, *phenacetin*. "It is gratifying to note that the new drug is not compromised by a patent," wrote the *Medical Age* in 1888, "and that it comes before the profession under an honest chemical name, abbreviated and pronounceable."[95] As I shall discuss in more detail in the following chapter, however, Bayer promoted the chemical heavily under the shortened name—in effect transforming it into a trade name—and was quite aggressive in asserting its monopoly rights over the product. Pharmacists soon learned that Bayer controlled the patent on the drug itself and that any prescription—whether for "phenacetin" or "acetphenetidin"—was supposed to be filled by the product manufactured by that company. As the druggist community in United States soon realized, the patent on the drug allowed Bayer to set any price it wished for the product. However, because the company was not able to secure product patents in other countries, a significant price disparity soon developed between the domestic market and foreign markets. The fact that the price in the United States was substantially higher than in Germany, other European countries, and even Canada enraged druggists and led to a tremendous amount of legal conflict over the drug. Importers began to bring the drug into the country and sell it as either "Phenacetin" or sometimes "acetphenetidin," a practice that Bayer took to be a violation of its patent rights. By the 1890s Bayer had begun an aggressive effort to suppress the practice by suing both importers and retail pharmacists. Still, many pharmacists faced a difficult question: if confronted with a prescription for "Phenacetin," was it ethical to distribute a less expensive version of the drug under the name "acetphenetidin"? Did doing so violate the patent rights of the company, and even if so, was it justified by the fact that Bayer seemed to charge an unfairly high price for their product?

The use of coined names thus monopolized the use of these products, although in different ways for different goods depending on their patent status, the complexity of their chemical names, and a variety of other factors. In the case of drugs such as Antipyrine, either there was no distinction between the trade name and the chemical name, or the chemical name was too complex to serve commercial or therapeutic purposes. Antifebrin, on the other hand, had a relatively clear distinction between its commercial and scientific names because the substance was already known to the scientific community before its commercial introduction. Yet even there, the dispersal of the name "Antifebrin" through the medical and pharmaceutical communities led to its serving as the de facto descriptive name for the chemical: most physicians and druggists referred to the product as Antifebrin rather than by its chemical name, *acetanilide*. In the case of Phenacetin, the problem was even worse because of the immense popularity of the drug, the patent on the substance itself, and the willingness of Bayer to aggressively pursue what it considered to be its rights to monopolize the substance. The German drugs, noted one worried observer, "come to us covered all over with patents—patents covering the names, the process of manufacture, the ingredients. . . . In short, they are patent medicines in the very widest and strictest sense of the term."[96]

The German synthetic drugs presented therapeutic reformers with a difficult problem: they were clearly important drugs, yet they could not be condoned because of their monopolized status. This is illustrated by the debate that broke out about whether or not the drugs should be included in the seventh revision of the *USP*. It was an extremely difficult question. On the one hand, the German drugs were clearly effective and popular remedies. Some members argued that including the German drugs in the *Pharmacopeia* and establishing standard assay tests to determine their purity was the only way to ensure that these powerful substances were manufactured according to the standards that the manufacturers claimed for them. "These drugs are valuable drugs. They are used by the physicians, and they will continue to be used by the physicians," noted one physician involved in the deliberation. "It seems to me that by putting them in some form in the Pharmacopeia and putting in tests we shall, to a certain extent, have a control of them. That is to say, there will be some means of guiding the pharmacists and chemists as to their purity." Yet the majority argued against inclusion. Some suggested that the fact that they were patented meant that the German firms would simply ignore the standards set by the *USP*. Others argued that by including them the committee would be giving its seal of approval to drugs over which it had no real authority. "All we know about [these drugs] is what we see on this sales package," noted one. "We do not want to make ourselves responsible in this case for anything of that kind. We cannot do so, and I for one say that we should wash our hands of all responsibility."[97]

Stewart was one of the many critics of including the drugs. Stewart attended the convention as a delegate from the Delaware medical society, and he spoke out strongly against their inclusion during the proceedings. He also worked with George Davis to bring what influence he could bear on the process.[98] Stewart's premise was simple: the *USP* should only include remedies that everyone was free to manufacture, with patented goods being excluded until they reverted to the public domain and trademarked drugs being excluded altogether. Stewart in fact opposed not only their inclusion in the *USP* but even their use, noting that if the German companies had patented their drugs "and gone no further than this," he would not have a problem with their use, although he would have still opposed inclusion in the *USP* until they had reverted to the public domain. However, the "German syndicate" had also trademarked the names of their goods, thus making them "proprietary" as well as "patented" medicines. "What happens," he asked rhetorically, "when we endorse the patent and proprietary trades by using sulfonal and other preparations monopolized in this manner without protest?" The answer was clear: "We are making respectable a system of quackery."[99]

Stewart was only one voice in the debate about the inclusion of these products, but he was probably an influential one because of his growing reputation as an expert on the topic of patents and trademarks. In any event, the critics won the day, and the monopolized German synthetic drugs were excluded from the seventh revision of the *USP*, which was published in 1893. The *Pharmacopeia* also excluded trademarked names as a matter of policy.[100] Not everyone was happy about this: some critics dismissed what they saw as a "foolish rule" that excluded important remedies such as Antipyrine "and similar staples."[101] Yet Stewart and many other observers considered the outcome both reasonable and ethical. Thus, even as he worked to formulate a theory of intellectual property rights that collapsed the distinction between science and commerce, Stewart also resisted what he saw as the unethical behavior of the German firms. These were not contradictory positions but two different sides of the same general effort to "harmonize the scientific and professional ideal with the commercial ideal."[102] They also, perhaps not incidentally, promoted the interests of the firm he worked for.

Scientific Pharmacy and the Project of Reform

In late 1887 the parent of a nineteen-year-old boy wrote to a patent medicine manufacturer named Lucius Wood inquiring about one of his products. The letter speaks for itself:

I have a son who from a small child has been subject to epileptic fits—sometimes he will go for months without one and then he may have two or three at short intervals. He is nineteen—someway he heard of your "Neuralgic Remedy" and I believe he has taken six or seven bottles . . . don't know if this is helping him or not but propose to give it a trial—please write me what you expect from it, and if you know of its having effected any cure—also how cheap you can sell it to me, as I can't afford to pay the regular price for it—send me two bottles at once.[103]

We do not know if this letter was written by the father or mother of this boy—the name on the letter does not make it clear—but we do know that he or she wrote back several additional times to request more medicine. Wood sold other products as well, including a cough remedy and something he called "Liver Health." Based on his correspondence, some of these products appear to have helped, at least some of the time. Others did not. Yet he maintained a brisk business, and no wonder. Lucius Wood sold more than medicines made with secret ingredients to his customers. He also sold them hope.

Reformers in the pharmaceutical community were deeply concerned about these types of transactions. Albert Prescott, for example, was the dean of the School of Pharmacy at the University of Michigan, a well-known figure in pharmaceutical circles, and a frequent critic of patent medicines. "For the nostrum vendor," he wrote, "the symbol of the spider and the fly would be too tame a trademark, and his own service with chain and ball not too severe a retribution." Prescott's hostility to patent medicines was grounded in his critique of secrecy as a form of monopoly contrary to the benevolent aims of science. He made clear that public safety depended on all drugs being sold under accurate and clear descriptions, which could be either their "distinctive" name or a listing of their ingredients. The point was to stabilize the relationship between the name of a product and its underlying ingredients so that both pharmacists and customers could make rational decisions about what products should be used. "Let it be indispensable that every article is labeled with its distinctive name," he urged. "Let every article that anybody is to buy and use as a medicine have its constituents and their proportional quantities given, fairly and squarely." For Prescott, as for so many other therapeutic reformers at the time, the use of secrecy in pharmaceutical manufacturing was a dangerous form of quackery. The free circulation of accurate information was critical to the battle against a malevolent foe.[104]

Prescott was not alone in these beliefs. Building on their legislative achievements of the previous decade, therapeutic reformers during the 1880s passed a wide variety of laws at the state and local levels intended to control and rationalize the sale of pharmaceuticals; twenty-two states and territories, for example, passed comprehensive pharmacy bills during the 1880s. These laws were intended to protect the public from a predatorial and arbitrary market. They were also clearly intended to advance the interests of pharmacists. Laws that prohibited the selling of adulterated goods, restricted the trade in so-called poisonous drugs to pharmacists or required prescriptions to dispense them—these and other measures were intended to simultaneously protect the health of the public and to improve the practice and standing of pharmacy. These laws were thus grounded in part on a sympathetic and a paternalist understanding of human tragedy, one in which expert authority would protect consumers from the dangers of deception, error, monopolistic practices, and other impositions on the market. They were also grounded on the assumption that the proper response to the dangers of monopoly was the open circulation of information; this information would then be used by people to make proper choices for themselves—choices that would, not coincidentally, follow the recommendations of qualified experts. In other words, the promotion of professional authority and consumer autonomy were intertwined projects.[105]

Reformers thus worked to transform pharmacy along what they believed to be scientific lines as part of the broader effort to simultaneously reform the drug market and uplift their trade. The growing importance of scientific expertise to the practice of pharmacy was, in turn, intertwined with efforts by reformers to establish ethical norms that hinged on the rejection of unfair profit: the practice of both a scientific and ethical pharmacy, in other words, was assumed to lead to fair profit, while unfair means of earning a living were increasingly assumed to be not just unethical but also unscientific. For example, it was widely understood within the trade that in a highly competitive environment the pressure to adulterate products was high. Manufacturers regularly added inexpensive substances to their products in order to beat the prices of their competitors, and some claimed that they were unable to provide pure drugs at a price that retailers could bear.[106] Retailers, on the other hand, were regularly faced with a choice between purchasing drugs that they strongly suspected were adulterated at a low price and buying drugs that might be pure—or might not be—at a high one. By defining themselves as unwilling to adulterate because of both their ethical norms and their allegiance to science, pharmacists worked to define their profession as both reputable and specialized, thereby justifying their own developing authority over the drug market. On the other hand, manufacturers of adulterated products and pharmacists who gave in to the temptation to adulterate were ferociously criticized as betraying the ideals of pharmacy. As one critic put it, "The man who knowingly adulterates human food, drink, or drugs is an enemy to his race, and should be treated as such. He is simply a human wolf, fattening upon the bodies of his fellows, and deliberately poisoning them for the purpose of making his own vile existence a little more comfortable."[107]

Equally important, a growing concern about what critics called the "demoralization of prices" underlay the effort to both uplift the practice of pharmacy and reform the drug market.[108] By the 1880s declining prices for botanicals and other raw materials, increased production of manufactured goods, and heightened competition among retailers had led to a tremendous downward pressure on drug prices. Declining costs pushed retail prices to dangerously low levels, but at the same time, low prices meant that startup costs to enter the market were modest, leading to a growing number of practicing pharmacists relative to the population.[109] Large retailers also used proprietary and patent medicines as what would later come to be called "loss leaders," purchasing them in bulk and then selling them extremely cheaply, sometimes below cost, in order to attract customers to their stores. Some pharmacists did the same, both on preassembled goods and on botanicals, in an effort to remain competitive. This type of price cutting enraged reformers in the pharmaceutical community, since it further depressed already low prices for goods. Price cutting seemed contrary to the laws of fair competition; goods were sold under what was taken to be their natural price, reputable pharmacists were unable to match the price, and trade was thereby diverted from legitimate channels, creating "an artificial increase" in the profits of cutters.[110] Price cutting, like other unethical practices, distorted the natural functioning of the market and threatened the livelihood of reputable pharmacists. "The evil of 'cutting prices' is assuming enormous proportions," noted one observer in 1883, "and there are but few localities where it is not being felt."[111]

Given the fact that patent medicines and proprietary remedies might be adulterated, counterfeited, or made with inferior or even dangerous ingredients, many pharmacists considered it safer to compound these products themselves when possible. Doing so protected their customers from harm. It also benefited the pharmacists in question, since it protected them against accusations of distributing dangerous products and, simultaneously, frequently allowed them to earn a higher profit on the sale, since they could often compound the remedy more cheaply themselves. Yet it was not at all clear that it was ethically acceptable to substitute a compounded remedy that a pharmacist made in his own shop for the preassembled remedy sold by a manufacturer. Many pharmacists argued that if a customer asked for a product sold under a distinctive—and frequently trademarked—name, the ethics of their trade required them to dispense the product actually manufactured and sold under that name, even if they were able to personally compound the same remedy more cheaply. They also pointed out that compounding a trademarked remedy and selling it under that name would violate the trademark rights of the original manufacturer of the product. Others, however, resisted what they saw as an encroachment on their authority to practice their craft, arguing that they were completely justified in compounding goods and selling them under the name that the customer had asked for—even if that name was trademarked—since that was, in fact, the product's proper name. As one pharmacist noted,

Anyone has a right to manufacture Ayer's Pills, Jayne's Expectorant, Schenck's Pulmonic Syrup, Bromida, Helmbold's Buchu, Celerina, Lactopeptine, Hydroleine, or any other proprietary medicine, so-called, and to sell the same, each under its proper name as aforesaid, all claims upon the part of the alleged inventors of these compounds that these names are trade-marks as applied to the said articles to the contrary, notwithstanding; for by use these names have become the descriptive names of the articles, and therefore cannot be trade-marks—the use of the descriptive names as trade-marks being contrary to law.[112]

As proprietary remedies became increasingly popular, pharmacists were thus faced with a difficult choice: stock an exceedingly wide variety of proprietary goods and dispense them when they were called for, even if it went against their better judgment, or compound remedies themselves and potentially face legal sanction for violating a manufacturer's trademark rights. Not surprisingly, at times pharmaceutical associations passed resolutions declaring trademarked goods unethical altogether. In 1882, for example, the Pennsylvania Pharmaceutical Association declared that obtaining trademarks on the "common pharmaceutical names" was "one of the latest developments of quackery."[113]

These were very real risks. Sometime around 1876, for example, the Saint Louis manufacturing firm Battle & Company began selling a product made from chloral hydrate, extract of hyoscyamus, and bromide of potassium under the trademarked name Bromidia. The formula for the product was not particularly difficult to prepare, and when customers asked for "Bromidia" many pharmacists simply compounded it themselves. In the late 1880s, however, the company began legal action against druggists who did so for violating its trademark rights. In Saint Louis, the company actually arranged for the arrest and criminal prosecution of a handful of retail druggists under an 1876 law that had made the counterfeiting of registered trademarks a criminal offense. Some of the druggists pleaded guilty and paid fines; the rest were released after the judge threw out the case after determining that the company had not adequately demonstrated that it actually used the mark in a way that deserved protection.[114] The company continued to press its case and brought a number of other druggists to court. In 1891, the US Circuit Court in the Eastern District of Louisiana determined that druggists Finlay & Brunswig infringed on Battle & Company's trademark because the term "Bromidia" was in fact an actionable term, decreeing that "there can be no question in this case but that the complainants have a right to and a property in the word 'Bromidia' as a trade-mark, and that the defendants are infringing upon the same." Notably, the court ruled that the fact that Finlay & Brunswig had printed the words "Prepared by Finlay & Brunswig, Manufacturing Chemists" on their labels was not relevant. As the court noted, "The infringement of a trade-mark cannot be justified on the ground that it is accompanied by marks and advertisements showing that the goods so marked are manufactured by other parties."[115]

Promoters of scientific pharmacy thus juxtaposed their efforts to the monopolistic practices of people like Lucius Wood. Whereas Wood and other vendors of patent and proprietary medicines monopolized their goods and sold them at unnaturally low prices, scientific pharmacy depended on both the open circulation of information and the operation of a fair market. Fair prices ameliorated the corrosive effects of competition, allowing pharmacists to practice in both an ethical and scientific manner. Retail pharmacists thus worked ferociously to impose price controls and to punish price cutting through the development of regulatory schemes—one such effort was the so-called Rebate Plan, introduced in the early 1880s, whereby manufacturers would pay retailers a certain percentage on all items sold, provided that there had been no complaints filed against them in terms of price cutting. In another scheme, representatives from various industry groups fielded complaints of price cutting; once the complaint was authenticated, the manufacturer whose goods were being undersold would issue circulars to wholesale dealers asking them to refrain from supplying the goods in question to the offending merchant until he promised to restore the retail price.[116] Such efforts were about more than improving the ability of reputable druggists to earn a living, although they were certainly about that. They also promoted the practice of scientific and ethical pharmacy, thereby working to transform the drug market along rational lines. They were one part of the broader process of therapeutic reform.

Ironically, reformers were also deeply concerned about what they saw as the ability of unethical actors to artificially inflate drug prices. Adulteration, for example, was widely understood as a way to artificially inflate the value of goods beyond their natural price—by substituting or mixing in an inferior ingredient, unethical wholesalers, manufacturers, or even retail druggists could charge more for a drug than it was actually worth.[117] The use of secret ingredients evoked similar concerns, as did the monopolization of products through trademarked names and patents, both of which were believed to allow manufacturers to artificially channel commerce away from its proper ends, thereby allowing them to charge unfairly high prices. As one physician noted, after giving a product a "fanciful name" and gaining control over it, the manufacturer "is protected from legitimate competition, [and] holds it at a price double that which any reputable pharmacist would prepare it for."[118] The cost of "Antifebrin" as compared to "acetanilide," for example, was widely ascribed to the fact that Kalle and Company had both popularized and trademarked the name of its product, allowing the company to charge significantly more for it than the market price of the same chemical sold under its nonproprietary name. Pharmacists considered this an unfair manipulation of the market and deeply resented it—which was not surprising, given the fact that because of the competitive market in the nonproprietary form of the drug, they may well have been able to sell "acetanilide" at a higher profit than "Antifebrin," despite its lower price.

The passage of pharmacy laws was a complex and politically contested process. Retail pharmacists, wholesale druggists, and manufacturers often took competing positions on the passage of specific laws based on their own interests, as did physicians, moral reformers, and others involved in the bruising legislative battles that sometimes accompanied the effort to pass these laws. Manufacturers of both patent medicines and proprietary medicines, for example, were typically opposed to the passage of antiadulteration and labeling laws, in part because they frequently kept the ingredients of their products secret. Secrecy allowed unscrupulous manufacturers to make extravagant therapeutic claims for their products, but it was also an important technique for protecting markets during a period of intense competition—as far as I can tell from his correspondence, Lucius Wood did not intend to defraud his customers, but his use of secrecy to protect his recipes was undoubtedly an important part of his business strategy. A law requiring him to accurately label his ingredients may well have put him out of business. Wood ran a small operation, but in 1881 several of the largest patent medicine and proprietary manufacturers established the Proprietary Medicine Manufacturers and Dealers Association to advocate for their interests; the organization quickly developed lobbying relationships with legislators at both the state and federal levels and effectively blocked many efforts at reform, much to the consternation of therapeutic reformers.[119]

Ethical manufacturers had mixed feelings about these types of laws. For the most part, they had little reason to be concerned about labeling laws, since they did not rely on secrecy to protect their goods. Yet they were cautious about efforts to link the legal definition of adulteration to the standards in the *Pharmacopeia*. The problem was that once standards of the *USP* were given the force of law, drug companies would no longer be able to legally sell products under officinal names that varied in strength or other characteristic, even if they indicated that the product did not conform to the standards set forth in the *USP*. This gave the revision committee a tremendous amount of power, and if the committee made poor choices—either out of inexperience with a particular substance or for other reasons—manufacturers would be compelled to adjust to such changes, potentially causing great harm to their business, or face legal sanction. Following the passage of an antiadulteration law in Ohio in 1887, for example, John Uri Lloyd pointed to the dangers of the law for ethical manufacturing because it linked the definition of adulteration to the standards of the *USP*. Yet Lloyd recognized that there was little to do but go along with the changing times. The progress of science required it. "The privileges of the individual have been circumscribed," Lloyd wrote, "and in business matters we, as individuals, must bow to modern laws passed avowedly in the interest of the multitude."[120] Reflecting the new power granted to the Pharmacopeia through the passage of such laws, the 1890 revision convention voted to change the term "officinal" to "official" for all pharmacopoeial publications.[121] The *USP* no longer governed only those who voluntarily followed its dictates. It was now legally enforceable, at least in some parts of the country, and for those who lived in these areas conformity to scientific authority was no longer a choice. It was a requirement. The tide had begun to turn against manufacturers such as Lucius Wood.

CHAPTER FIVE

The Ambiguities of Abundance

"In their dreams of the ideal commonwealth, all reformers and statesmen have held that happiness involves not only freedom and intelligence, but abundance also," wrote Newell Dwight Hillis in his 1902 book *The Quest of Happiness.* "Now comes an age of abundance, when wealth is here, to build a highway of happiness for society, and to hasten all footsteps along this way that leads unto intelligence and integrity, to peace and prosperity."[1] Hillis was a minister from Brooklyn, a supporter of eugenics, and something of a philosopher. Like many other observers of his time, he believed deeply in the intertwined promises of economic prosperity and advanced science to improve the condition of humanity. "If science has lessened labor," he noted, "it has also lengthened and sweetened life." Among many other topics, Hillis wrote eloquently about the discovery of anesthesia, the reduction of infectious disease, and other benefits of modern medicine. He spoke of a bright future in which man would finally be free from the ravages of the body. As he put it, "These victories achieved in the past give promise that during the next century surgery and medicine may become exact sciences that shall discover the secret of length of days and the maintenance of life and happiness for all people."[2]

Historians have long been fascinated by the emergence of corporate capitalism in the decades between 1890 and World War I. And rightfully so: among an almost endless series of changes, the chaotic and sometimes violent transformation of the economy brought forth dramatic scientific and techno-logical advances, increased leisure time, a tremendous outpouring of new consumer goods, and numerous possibilities for a better and healthier life. These were the types of forces that animated Hillis's celebration of the coming order. Yet others viewed the rapidly changing world with ambiv-alence and concern. Dangers included the increasing concentration of economic and political power, social instability and conflict, a seemingly predatorial market, and the endless problem of corruption in politics and other domains. Reformers thus worked assiduously to improve the world around them, sometimes against the wishes of those they sought to help. In doing so, they drew deeply on the antimonopoly tradition, attacking privilege, graft, the concentration of power, and numerous other forces that seemed to unfairly distort both the market and the democratic pro-cess.[3] The corporate transformation of America brought both possibilities and dangers, hopes and fears, to those who lived through it and made it happen.[4]

Both patents and trademarks were an important part of this process. The legal ability to monopolize invention played a crucial role in the organization of entire industries, the financial stakes involved in pat-enting could be immense, and the rightful protection of the interests of inventors seemed to easily become an "odious monopoly" in which "fair competition is destroyed" and "the people are at the mercy of . . . the monopolist."[5] Yet despite such sentiments, manufacturers enthusias-tically embraced patenting, acquiring more than twenty thousand pat-ents a year during the 1890s.[6] The same can be said of trademarks. The ability to monopolize the names of goods was an important part of the ability of manufacturers to create distinctive identities for their products. Trademarks helped ensure repetition of purchasing by linking the repu-tation of the manufacturer to the product at hand, thus helping to make possible economies of scale and the commercial exploitation of new tech-nologies. Branding thus became increasingly important over the decade, as manufacturers increasingly relied on selling large numbers of goods at low cost and stimulating consumer desire for the acquisition of their products. While not yet as important as patents, trademarks were a cru-cial part of the emergence of corporate capitalism.[7]

The drug industry was one important part of the changing times. Manufacturers in all segments of the industry experienced rapid growth during the 1890s, in part based on the development of new products—many of which were commercially introduced before the medical com-munity had formed much of an opinion about their utility. Such efforts naturally intersected with the changing nature of patent and trademark law, including the legal construction of what were sometimes referred to as *generic* names. They also intersected with the broader social and cultural currents of the times. Pharmaceuticals were not neutral objects. They prompted what was sometimes a desperate hope among those who took them that bodily suffering could be alleviated. More broadly, they provoked both dreams of wealth among those who made them and a belief among therapeutic reformers that illness could be eradicated and man freed from the tyranny of disease. Yet pharmaceuticals also provoked tremendous anxieties about the dangerous nature of the drug market, the slippage between names and things, the apparent irrationality of consumers, and numerous other complex issues. Therapeutic reformers thus worked to rationalize the drug market and to promote an emergent epistemological framework in which knowledge about drugs generated through laboratory and clinical science would serve as the basis for medical practice. They also increasingly recognized that innovative products made by pharmaceutical manufacturers had an important role to play in the promotion of scientific medicine. They thus embraced the idea that corporate investment in the scientific process had the potential to solve difficult therapeutic problems, and they gradually came to accept the ethical legitimacy of monopolistic practices as a means to advance the public good. This was a complex and contested process, and even as therapeutic reformers embraced the possibilities of the market, new concerns about its influence on medical science emerged.

Patenting and Scientific Drug Development

Pharmaceutical manufacturing grew rapidly during the 1890s, one part of the broader economic and social transformations of the time.[8] According to the twelfth census, capital invested in manufacturing patent medicines more than doubled during the decade, with a corresponding increase in the total value of patent medicines on the market, despite what worried observers in the pharmaceutical community called "the demoralization of prices."[9] The rate of growth in the ethical and proprietary wings of the industry is more difficult to determine with any reliability because the classification systems used in the 1890 and 1900 censuses for these manufacturers are not comparable to one another.[10] However, it was clearly considerable. Parke-Davis, for example, more than doubled its total sales during the decade and by 1900 sold almost $5 million worth of goods, the bulk of which were sold in the United States.[11] Most companies did not do nearly as well as Parke-Davis, of course, but by all accounts growth in both sectors of the industry was rapid.

Ethical manufacturers continued to produce familiar and proven goods as a mainstay of their business, including fluid extracts, pills, and tinctures made from familiar ingredients. They also continued to produce chemical products such as sulfuric ether that were an accepted part of medical practice. However, by the 1890s there was little question that reputable manufacturers were able to introduce valuable new products to the medical profession without damaging their standing or their markets. By this time efforts by ethical manufacturers to innovate new products had become an accepted and important part of the broader effort to advance medical science. McKesson & Robbins, for example, had opened an analytic laboratory as early as 1880 to ensure the quality of its products.[12] By the 1890s the company had begun to introduce new products to both the orthodox and eclectic markets, such as a combination of bile and pancreatin that it introduced under the name *pancro-biline*.[13] "Since the day McKesson & Robbins established their laboratory," noted one observer in 1896, "there has never been a halt in the progress of the department. The sphere of the laboratory labors has been constantly widened to meet the requirements of fresh developments in therapeutics, and, as a result of those labors, new preparation after new preparation has been introduced to the notice of the medical profession."[14]

Other companies pursued a similar strategy and during the 1890s either established or improved laboratories to develop new products. Schieffelin & Company, for example, opened a "chemical laboratory" in New York City that was made up of five different departments, had multiple rooms for different types of scientific work, and housed extensive amounts of equipment, including a mammoth mixer that had the capacity to mix a thousand pounds of material at a time.[15] Manufacturers also increasingly distributed experimental products to physicians with invitations to test them on their patients, published the reports in the medical press, and used those reports in their promotional efforts. Francis Stewart, for example, ended his affiliation with Parke-Davis sometime in the early 1890s—probably after Davis was forced out of the company following a financial scandal—and in 1894 he helped establish the scientific department at Frederick Stearns & Company. Stewart stayed at the company for the next seven years, arranging for the company to publish a series of scientific monographs dedicated to "the free diffusion of knowledge" about new remedies.[16] In another example, in 1893 Smith, Kline & French established an "an analytic laboratory" under the direction of chemist

Lyman F. Kebler.[17] Within a decade the company was sending samples of its products to physicians for clinical testing and then distributing the results of tests to physicians interested in their products.[18] All this was necessary, as a journalist who described the new laboratory opened by Schieffelin & Company put it, to avoid "scientific inertia" and "commercial stagnation and ruin."[19] Investing significant resources in scientific and technological innovation, it increasingly seemed, was not optional for a firm that wanted to succeed in a difficult economic environment.

The history of therapeutics during this period is extremely complex, but broadly speaking there are two basic components that are relevant to this discussion. The first is that laboratory testing was increasingly understood as the proper basis for scientific knowledge about drug action. By this point the idea that drugs should be tested in laboratories to determine their physiological action was no longer controversial—indeed, it was widely considered the future of scientific drug development. At the same time, although sometimes compared unfavorably to laboratory techniques, clinical testing continued to be the primary basis for understanding the effects of drugs on actual patients. During the 1890s physicians thus began to systematize their experimental use of new products and to draw conclusions from these experiments that they believed applied across all similar cases. Of course, rigorous clinical testing protocols by today's standards did not yet exist: trials lacked comparison arms, investigations were not blinded, and treatment groups often included widely varying types of patients. Most clinical testing during this period resulted in little more than a case history or the presentation of information in very general terms. Still, a growing number of investigators began to draw conclusions about the utility of drugs based on clinical experiments that went beyond the constraints of the specific situation in which the test took place. Clinical experimentation, in other words, began to contribute to the production of a type of knowledge about drugs that was distinct from both the context in which the drug was used and the experience of the investigator. In both laboratory analysis and clinical testing, the goal was increasingly to establish facts about the effects of drugs on patients independent of the context of the individual case at hand.

During the 1890s ethical manufacturers embraced both laboratory and clinical testing in their efforts to develop and introduce new products. As a part of this process, they established working relationships with scientists and clinicians who worked outside of the industry. Over the course of the 1890s, a handful of universities established doctoral programs in pharmacology, chemistry, and other fields as part of their efforts to transform themselves into the types of research institutions that we are familiar with today. By the turn of the century significant networks had already been established between many of the leading firms and these programs, through which information, money, samples of materials, and sometimes jobs were cultivated and exchanged. At the same time, ethical firms also cultivated and maintained close relationships with independent schools of pharmacy and medicine, hospitals, asylums, and other institutions. These networks provided access to expert knowledge, samples of materials, professional connections, and, most important, the ability to have drugs clinically tested. Arranging clinical tests on their own was beyond the capabilities of even the largest and most sophisticated manufacturers during the 1890s, and as clinical testing became both more systematized and more important in the ability of firms to convince physicians of the utility of their goods, cultivating working relationships with clinicians who had access to patient populations became an important part of successfully developing new products.

Parke-Davis remained at the forefront of this movement, investigating new botanicals from around the globe, examining new botanicals and other substances in the company's laboratory, shipping samples to physicians for clinical testing, and collecting and publishing the reports in the company's working bulletins and descriptive catalogs of its "laboratory products."[20] In 1898, for example, researchers at Parke-Davis began studying a derivative of chloroform that the company named "Chloretone." Following animal trials, the company distributed the drug to physicians in Detroit and other areas, including several psychiatric hospitals, asking for their help in clinical trials.[21] Results were overwhelmingly positive. The drug appeared quite safe—one patient was anesthetized for three full days and recovered without any apparent problems—and it seemed promising as a treatment for a variety of conditions, including epilepsy and vomiting, and as a surgical anesthetic. By late 1899 the commercial promise of the drug appeared great, and the company had begun to market it in the medical press. Chloretone, noted one Parke-Davis memo, "is the most promising drug added to our List in several years."[22] The medical community responded positively to the commercial introduction of the drug. "It is too early . . . to prophesy what position chloretone will take in medicine," noted one physician in 1900, "but the results as a hypnotic and local anesthetic are very encouraging."[23]

Although Parke-Davis did not patent Chloretone, by this point the company had already begun to quietly embrace the use of patents on a small number of other products. Its willingness to do so reflected the basic fact that medical patenting no longer provoked the same types of concerns that it once had. Ethical manufacturers, for example, occasionally acquired patents on manufacturing processes during the 1890s without attracting criticism from the medical community. To take one example, in the late 1880s Parke-Davis had licensed a patented manufacturing process for dipping pills in liquids developed by a man named John Russell.[24] The device dramatically increased the output of the company—by the late 1890s, the firm manufactured almost 130 million pills a year using his method—and faced with this increased capacity from one of its chief rivals, Frederick Stearns & Company worked to improve its own manufacturing capabilities. Russell considered the machines that the Stearns company developed to be an infringement on his own invention and brought the company to court. In 1898, however, the Circuit Court of Appeals of the Sixth Circuit ruled that Russell's invention was nothing more than a "new application of an old device" and hence not patentable.[25] In many ways, this was a rather routine patent case. For my purposes, however, it is notable because the medical community—while aware of the case—does not seem to have particularly cared about the fact that two of the leading ethical manufacturers in the country were involved in a patent dispute. Reports of the case simply mentioned the outcome without commenting on the ethics of the companies involved.[26] Times were clearly changing.

I discuss the growing acceptance of patenting among the medical community in more detail below. For the moment, I simply want to point out that ethical firms such as Parke-Davis operated within a market that was increasingly shaped by firms that were not really constrained by the ethical norms of the medical community and that these companies tended to patent their products when they were able to. Chemical manufacturers primarily focused on other types of products continued to introduce a small number of patented goods to the therapeutic market: in 1898, for example, the Mallinckrodt Chemical Works patented a drug that it sold under the name Guaiamar.[27] More significantly, proprietary manufacturers continued to introduce a large number of products, a growing number of which were both patented and accepted by the medical community as therapeutically valuable. In the mid- to late 1890s, for example, John Carnrick patented a series of remedies made from animal organs and other organic materials that became fairly popular among orthodox physicians.[28] Carnrick died unexpectedly in 1903, and obituaries published in the medical press point to the fact that his products were widely considered to have benefited the advancement of medical science. As the *Medical World* noted, by "overcoming the faults of the old and bringing out new facts, which scientific research had brought to him, he developed Protonuclein, Peptenzyme, and Trophonine, which have received the world wide approbation of physicians."[29]

Perhaps most important, European chemical and pharmaceutical companies—and German companies in particular—increasingly dominated the pharmaceutical market. German firms secured numerous American patents on synthetic chemicals, intermediary substances, and manufacturing processes during the 1890s; by the end of the century more than 90 percent of American patents related to chemistry were held by foreigners, and foreign companies controlled about 40 percent of the total value of the domestic industry.[30] Foreign manufacturers also introduced a wave of powerful new drugs developed using advanced techniques in synthetic chemistry to the American drug market. Bayer serves as an important example: following the introduction of Phenacetin in 1887, Bayer introduced a series of powerful drugs during the 1890s, most of which were patented, including products sold under the names Piperazine (1892), Salophen (1893), Europhen (1893), Protargol (1898), Tanopin (1898), and Heroin (1898), which the company did not patent because it had been previously synthesized.[31] Phenacetin remained Bayer's most important product in the United States during this time: in 1896, Bayer sold fifteen drugs in the United States, but half of its sales came from Phenacetin.[32]

Given the growing importance of patented goods in the American drug market, it is perhaps not surprising that domestic manufacturers in the ethical wing of the industry began to cautiously—very cautiously— sell patented products. Parke-Davis was once again at the forefront of changing behavior among firms in the ethical wing of the industry. In the 1880s a Japanese chemist named Jokichi Takamine had traveled to the United States and studied patent law before returning to Japan and serving as the chief of the Patent Office in his home country. In 1890, for unknown reasons, Takamine returned to the United States and began working with the whiskey industry to develop a means of replacing the malt used in whiskey manufacturing with the ferments used in traditional Japanese brewing. In 1894 Takamine was granted a series of patents related to this work, including one patent that covered a diastatic enzyme that he named "Taka-Diastase."[33] Patenting in this general area was not particularly new—in 1873, for example, Louis Pasteur had obtained a patent on a form of yeast, and during the 1880s and early 1890s other inventors had obtained patents on various ferments, yeasts, and enzymes made from vegetable and animal material, some of which were used medicinally.[34] It is not completely clear how Parke-Davis became interested in the product, but beginning in 1895 the company marketed the enzyme in the medical press as a digestive aid. It quickly became a popular remedy for digestive problems.[35] Parke-Davis occasionally noted that it had an "exclusive license" to manufacture and distribute the substance, but it does not appear to have informed the medical community or other manufacturers that Takamine had actually patented the product.[36] This is not particularly surprising, given the long history of hostility to medical patenting among orthodox physicians. Still, the willingness of the firm to acknowledge its "exclusive license" to distribute Taka-Diastase points to the rapid softening in attitudes toward monopoly among the orthodox medical community.

The willingness of Parke-Davis to monopolize the sale of Taka-Diastase should be contrasted with its attitude toward another important product that it began manufacturing around the same time: diphtheria antitoxin. In 1883 the Swiss German pathologist Edwin Klebs had first identified the diphtheria bacterium; five years later the French physicians Émile Roux and Alexandre Yersin had shown that the toxin produced by the bacterium produces diphtheria in animals. In 1890 the German physiologist Emil Behring then used a modified form of the toxin to immunize guinea pigs against the disease. Within two years researchers had begun to use the antitoxin to treat cases of the disease in humans, and by late 1894 physicians in the United States were using the antitoxin to successfully treat people infected with the disease. By 1895 the New York Board of Health had begun to manufacture and distribute the antitoxin. Other boards of health followed over the next several years.[37]

A small number of manufacturing firms also began to commercially produce the antitoxin. In 1894 Parke-Davis opened a biological laboratory with the goal of developing the antitoxin and the following year recruited Elijah Mark Houghton, a pharmacologist at the University of Michigan, to lead the effort.[38] Within a year the company had begun to manufacture and commercially distribute the antitoxin. The second important firm to enter the market was H. K. Mulford & Company, a new drug manufacturer in Philadelphia. Harry K. Mulford had established the company in 1891 after studying at the Philadelphia College of Pharmacy, and the firm quickly developed a reputation for making a wide range of high-quality goods. In 1894 Mulford hired a scientist named Joseph McFarland to establish the company's biological laboratory and begin work on the antitoxin.[39] McFarland was a part-time lecturer on bacteriology and histology at the University of Pennsylvania; he had also recently been recruited to lead the effort to develop the antitoxin for the Philadelphia Board of

Health. Under his leadership, within a year H. K. Mulford & Company was manufacturing and commercially distributing the product.[40] In what historian Michael Willrich notes was a "display of public-private cooperation," the New York City Health Department bacteriologist William Park provided McFarland with the culture necessary to start production, while the company's product was initially tested on animals maintained by the University of Pennsylvania.[41] By the late 1890s, then, a small number of entities in both the private and public sectors were manufacturing and distributing the antitoxin in the United States. Cooperation between the two seemed to hold the promise of therapeutic reform.

The antitoxin also promised to generate tremendous profit. Behring recognized the commercial potential of the diphtheria antitoxin soon after his discovery and arranged for the German firm Hoechst to manufacture and commercially develop the product.[42] In late 1894 Behring also applied for a US patent on the antitoxin. This was undoubtedly with an eye toward the commercial introduction of the product; within a few months Hoechst had begun importing "Diphtheria Antitoxin (Behring)" into the United States, and Behring probably realized that other manufacturers were in the process of bringing the product to market.[43] In January 1895, however, patent commissioner James B. Littlewood rejected Behring's application for being "vague and indefinite," for attempting to patent a medical procedure, and for trying to patent a "general method" and "certain principles of general application which are not the result of the labor or invention of any one man, but of many, and which are not patentable."[44] Pointing to the fact that a number of different researchers had been involved in the effort to understand the relationship between the C. diphtheria bacterium, the toxin, and the disease itself, Littlewood noted that Behring was "at most only a joint inventor with several parties and cannot, therefore, legally claim to be the sole inventor as required by law."[45]

Behring was persistent. Over the course of the next three years he submitted four additional applications, each of which was also rejected by Littlewood.[46] Behring's lawyers successfully responded to most of Littlewood's early objections, but they were unable to convince him that Behring was the sole inventor of the antitoxin.[47] Behring and his lawyers acknowledged that he had not been the first to discover the general principles on which the development of the antitoxin was based, but they argued that he was in fact the first to reduce these principles to practice. Behring thus asserted that although others had discovered the "general principle of inoculating animals with special bacilli so as to render them immune against infection by said bacilli," no one before the invention of my process has ever gone beyond establishing general scientific principles. I was the first who succeeded in a practical process of making diphtheria-antitoxin on a large commercial scale and introducing it to the world as a remedy for diphtheria. I do not lay any claim to the underlying scientific principles as these were evolved by several discoverers, but I do claim the successful process by which the antitoxin could be produced on a large scale and placed into the hands of the medical practitioner as a remedy for diphtheria.[48]

Behring's lawyers also pointed out that he was widely credited with the discovery—he had received numerous prizes for it, after all—and that "everybody, with the exception of the Examiner in charge of this application in the United States Patent Office, acknowledges this remarkable invention to be the sole invention of Professor Behring."[49]

Littlewood was not impressed and continued to reject Behring's applications. Toward the end of March 1898, Behring appealed Littlewood's decision. Details of the process remain scarce, but on May 31, 1898, he was informed that his application for a patent had been approved. Unfortunately, records of the appeals hearing do not appear to have survived, so it is difficult to know on what grounds Littlewood was overruled.[50] However, numerous critics of the decision in the medical press mentioned that the ruling was reached based on the fact that the diphtheria antitoxin had dramatically reduced the rate of mortality from the disease. If this is true, it appears likely that the appeal was granted based on the utility of the antitoxin and the fact that Behring was the first, as he put it, to make the product available "as a remedy for diphtheria." In other words, it appears likely that the decision was based on a successful reduction-to-practice argument.

Whatever the legal rationale, Behring's patent provoked a storm of controversy within the medical community—as one editorial put it, Behring "now thinks he is in position, with loaded syringe, to demand of every defenseless babe its money or its life."[51] It also provoked an angry response from both Parke-Davis and H. K. Mulford & Company. Both companies had invested a significant amount of resources in developing their versions of the antitoxin and were not pleased with the thought of losing their markets in the product. Hoechst initially threatened to sue any manufacturer that violated Behring's patent, but Parke-Davis and H. K. Mulford both vowed to fight the patent and began to coordinate with state boards of health to bring a test case to determine the patent's validity. Both companies also retained legal counsel in preparation for the suit, and both publicly offered to pay the legal costs of pharmacists sued for distributing their products.[52] These efforts were successful. Both Parke-Davis and H. K.

Mulford continued to manufacture the antitoxin, and I have been unable to find any evidence that the dispute ever made it to court. Perhaps, after consulting its lawyers, Hoechst came to the conclusion that the patent would not stand up in court. Perhaps Behring decided that enforcing his monopoly rights over the antitoxin would do an unacceptable amount of damage to his reputation. Whatever the reason, Behring never enforced his patent rights in the United States. Still, controversy continued to swirl around the antitoxin, both because of the dangers that accompanied its use and because of the growing belief that government should not, as one medical journal put it, "get into a general manufacturing business, or enter into competition with private enterprise."[53]

Both the development of Taka-Diastase and the battle over the diphtheria antitoxin were important moments in the history of the American pharmaceutical industry. The fight over Behring's patent demonstrates that the critique of monopoly was still alive and well at the turn of the century. At the same time, however, the commercial development of Takamine's digestive enzyme suggests that, for domestic firms, the critique of monopoly was rapidly losing its bite. Parke-Davis's introduction of Taka-Diastase appears to have been the first time a major firm in the ethical segment of the domestic industry sold a patented medicine. This was neither a dramatic event nor one that caused controversy, perhaps because the patent was not well known in the medical community. Still, the willingness of Parke-Davis to sell a patented product points to the erosion of what had once been an iron-clad prohibition on dealing in monopolized goods, even as, simultaneously, the eruption of controversy over Behring's patent points to the continuing and deeply felt anxieties about the relationship between science and the commercial pursuit of profit. This was the shape of things to come. In the coming years, the critique of monopoly would be directed not at what were assumed to be patriotic and ethical domestic firms, even when they embraced patenting, but instead toward an apparently predatorial German industry that distorted the therapeutic market toward its own ends.

The Medical Acceptance of Monopoly

In 1894 a committee appointed by the American Medical Association to study the organization's bylaws made a number of suggestions about how to revise the Code of Ethics. Among other recommendations, the committee suggested changing the language prohibiting physicians from holding patents on medicines to declaring it derogatory to professional character for physicians "to hold patents for secret nostrums."[54] The proposal provoked a fierce debate. Some physicians supported the proposal, arguing that allowing patents on legitimate inventions would benefit the profession by encouraging new drug development. As one physician put it, "why should not the ingenuity of the profession be stimulated by the hope of reward in patenting new inventions, the same as in any other department or industry?"[55] Others, however, found the idea that the AMA might condone patents deeply objectionable. "Surely this would be progress," noted the venerable Nathan S. Davis, "but in what direction— that of science and honor, or that of mammon and dishonor?"[56]

Although the 1894 proposal was soon dropped, debate about the ethical legitimacy of patenting continued over the course of the next decade. Not surprisingly, one of the leading spokesmen for reform was Francis Stewart. Stewart was remarkably prolific, and he published a tremendous amount of material about both patents and trademarks during the 1890s; between 1895 and 1899, for example, he published an influential series of articles in the *Journal of the American Medical Association* in which he outlined his arguments and made the case for reform.[57] In Stewart's view, patent law—properly understood and applied—promised to rationalize therapeutics by ensuring the open circulation of scientific information, promoting commercial investment in new remedies along scientific and ethical lines, and abolishing unscientific and unethical manufacturing practices. Stewart argued for the ethical legitimacy of process patents, but he was also extremely critical of product patents, which he believed undermined the public good by retarding competition and monopolizing scientific progress to the benefit of single firms. He thus drew a clear line between the "wicked quackery" of product patents, secrecy, and other unethical forms of what he called "commercialism" and the legitimate use of process patents to stimulate commercial innovation by ethical firms.[58]

By the end of the century a significant number of reformers in the medical community had begun to make similar arguments, although they rarely distinguished between process and product patents in the same way that Stewart did. These reformers suggested that patents expire after a limited amount of time and are therefore not truly a form of monopoly, that patents might stimulate the "inventive genius" of chemists or physicians and thereby encourage pharmaceutical innovation, and that patented medicines should be distinguished from medicines made with secret ingredients. Certainly not all physicians during this period were persuaded by these arguments; the idea that patents might be used to monopolize medicinal compounds, and that this might be ethically legitimate, was still a controversial position. Most continued to believe that patents on medicines were ethically suspect because, as the president of Cooper Medical College put it, all medical knowledge is the "common property" of the profession, and "every fact which is discovered in medical science and art must by the finder be thrown into the common treasury."[59] Yet for a growing number of reformers the critique of patents seemed increasingly antiquated. "The term patent is grossly misapplied in the case of anything secret," noted the *Medical and Surgical Reporter* in 1896, directly echoing Stewart's arguments. "A thing patented is a thing divulged. The medical profession very properly objects to secret medicines as opposed to the ethics of their calling, but a medicine patented has its composition disclosed."[60]

None of this means that reformers in the medical community were sanguine about the drug market. Far from it. Therapeutic reformers spent a tremendous amount of time denouncing the growing trade in both patent and proprietary medicines, worrying about the dangers of secret ingredients, adulteration, duplicitous advertising directed at what was assumed to be a gullible population, and other problems. These types of concerns had long been central to the orthodox medical community's understanding of the relationship between medical science and the commercial pursuit of profit. As both the patent medicine and proprietary wings of the manufacturing industry expanded during the 1890s, physicians continued to attack the supposedly predatorial nature of nostrum manufacturers, articulating a critique of the market based on their own supposed altruism and the self-denying nature of their profession. "The principles which the members of the medical profession have voluntarily established for their guidance are the most unselfish known in all human affairs," noted the physician John Jay Taylor in a book written, ironically, to help physicians learn how to organize their finances. "This explains one reason of the intense hatred of all honorable physicians towards the patent medicine business—it is so essentially selfish in nature."[61]

The supposed distinction between orthodox medicine and quackery thus continued to be based on the juxtaposition of medical science to the pursuit of profit. However, as the prohibition on monopoly gradually eroded—and as drug manufacturers increasingly worked to introduce new products to market—the boundary between commerce and science became increasingly blurry. Increasingly, the practice of medicine seemed dependent on the commercial efforts of private firms, and debate about the proper relationship between science and commerce began to shift from the issue of monopoly per se toward other questions. For example, unable to keep up with the increasing number of new remedies on the market, physicians increasingly relied on educational material provided by manufacturers to guide their therapeutic choices. Manufacturers, in turn, significantly increased their sales forces: in 1885, for example, Parke-Davis maintained a sales force of twenty-three detail men. Ten years later, the number of detail men employed by the company had increased to over one hundred.[62] These salesmen used the working bulletins created by the firm to discuss new products with their clients; they also distributed numerous free samples and undoubtedly engaged in other activities that combined the goals of education and salesmanship. Numerous other companies pursued similar techniques, distributing an endless amount of material to harried physicians about their products. This material was appreciated more often than not—especially because it was free— but critics sometimes raised difficult questions about the relationship between the commercial interests of manufacturers and the practice of good medicine.[63] Reformers also worried about the growth of proprietary medicine advertising in medical journals and about the possibility that journals might shape their content to please their advertisers. Perhaps most alarmingly, many reformers worried about the growing willingness of their peers to prescribe trademarked goods that, while they *seemed* scientific, did not reveal their ingredients on their labels. "This is commercialism," noted one disturbed physician from Missouri. "Is that right? Is that ethical? Do you endorse it?"[64]

The changing ethical framework can also be seen in the controversy over Behring's patent on the diphtheria antitoxin. The medical press loudly denounced the patent as egregiously mercenary and violating the benevolent spirit of scientific medicine. Behring, noted one critic, was a "ghoul . . . robbing the tombs that his pocket may profit."[65] Much of this criticism pointed to the fact that it had taken Behring multiple applications to secure the patent, and editorials in the medical press often provided detailed descriptions of the dispute with Littlewood about whether or not Behring was the product's sole inventor. The idea that the antitoxin was not the invention of a single man but instead the result of the accumulated efforts of numerous scientists working for the public good was an important theme that ran through critiques of the patent, providing a key argument for the idea that Behring should never have received a patent on the antitoxin at all.[66] From this perspective, although Behring's patent might technically be legal, his "base spirit of commercialism" violated the intent of the patent law and unfairly exploited the labors of numerous scientists.[67] "Behring has strained the intent of the patent law in obtaining an instrument for appropriating to his own advantage the disinterested investigations of scientists who supplied the data and even the suggestions he adopted," noted one critic. "Science herself suffers from an abandonment of the liberal principles which are the glory of her votaries."[68]

Even as physicians railed against Behring's patent, however, they also subtly accepted the fundamental legitimacy of patenting itself. Critics argued that the patent would impose a great injustice not only on the American public but also on the domestic manufacturers who had invested time and money in the development of their own brands of the antitoxin. As one critic noted,

Simple justice to the firms which have so invested funds and brains, simple justice to the physicians who have preferred the home product because it was a better product, simple justice to the numerous patients who are able to purchase the remedy at a fair, just price, because business competition has gradually lowered the price, ought to arouse a public sentiment sufficiently powerful so that it shall oppose and prevent a foreign firm from reaping the pecuniary benefit which will come to them from securing an unjust monopoly of a very necessary and useful remedy.[69]

The key term here is *unjust* monopoly: Behring's patent seemed highly unethical because it violated the spirit of the patent law, because it would allow him to extract unfair profits from a suffering population, and because it would deny legitimate profits to domestic manufacturers who were perfectly capable of making it themselves. The reaction to Behring's patent, while vituperative, did not presuppose that all such patenting was equally unethical. Indeed, numerous patented drugs were introduced around the same time without attracting such criticism. Nor did it assume that diphtheria antitoxin itself was quackish in nature. Unlike in earlier times, when patenting and quackery were overlapping categories, the patent on the antitoxin was considered unethical in part precisely *because* of its effectiveness and its presumed ability to produce profit for those who manufactured it.

In 1901 Behring was awarded the first Nobel Prize for Medicine for his work on serum therapy. The award was controversial because of his patent on the antitoxin. The *Journal of the American Medical Association*, for example, noted:

The awards of the [Nobel] prizes for the first time have given universal satisfaction, with the possible exception of that to Professor Behring for the greatest recent discovery in medicine. There are many who think that Professor Behring's position in securing a patent on his invention in order to reap the benefit of it for himself should preclude him from award for humanitarian advances made.[70]

Yet even as the *Journal* raised the question of the patent, we see an important transition: the grudging willingness of the medical community to accede to the narrative of the lone inventor and embrace a model of scientific innovation based on supposedly sudden discoveries and heroic science. Half a century earlier the medical community had rejected the idea that William Morton deserved credit for the discovery of sulfuric ether's anesthetic powers based on the idea that therapeutic progress is a gradual process that grows out of the shared efforts of a community of investigators. Physicians had made a similar argument about the development of the diphtheria antitoxin when Behring had first announced his patent. As this editorial indicates, however, that critique faded from view rather quickly, even among those who in the coming years remained critical of the patent. The development of the diphtheria antitoxin was instead articulated through the lens of the mythical lone inventor, a heroic narrative that, perhaps not inconsequentially, dovetailed with the requirements of patent law. Still, questions about Behring's behavior lingered. Many continued to believe that he had acted unethically, despite the fact that it was increasingly understood as his own invention. The growing acceptance of patenting within the medical community and the growing acceptance of the myth of the lone inventor that was intertwined with this shift in ethical sensibilities were thus accompanied by the ongoing concern that medical science might be corrupted by the commercial drive for profit. Yet it was no longer completely clear how to distinguish between the two. The distinction between science and quackery was not as strong as it had once been.

The changing ethical framework can also be seen in the controversy over including patented goods in the eighth revision of the *Pharmacopeia*. The revision convention met in 1900, with the revision finally becoming official in 1905 after five years of protracted debate about the issue. Pharmacists—who by this point dominated the revision process—generally saw the *USP* as a compendium of standards to guide manufacturing and dispensing practice, and although some pharmacists believed that it was time to rethink the prohibition on including patented goods, most thought that including them would unfairly benefit certain manufacturers and shape their own dispensing practices.[71] Pharmacists were also deeply concerned about the use of patents by German manufacturers to maintain what they considered unethically high prices for their products. From this perspective, including patented drugs in the *USP* seemed to condone the unfair pricing schemes of foreign patent holders.

By this point, however, most physicians who paid attention to such things fully believed that important drugs should be included in the *USP* irrespective of their patent status. In 1899, for example, the American Medical Association passed a resolution calling on its delegates to the 1900 convention to support the inclusion of useful synthetic drugs "without regard to patents." This was not a particularly controversial position; as one observer noted, "Not a shadow of a protest arose against the action."[72] The convention ignored the AMA's position and passed a prohibition on including products that were "controlled by unlimited proprietary or patent rights." During the revision process, however, supporters of including patented goods argued that because patents expire and then revert to the public domain, goods protected by patents alone are not protected by "unlimited" proprietary rights and could legitimately be included. The Subcommittee on Therapeutics—which was the physicians' main stronghold on the revision committee—thus recommended including a variety of products currently protected by patents, including Phenacetin and Trional.[73] Opponents scoffed, and a bitter and complex dispute broke out about the issue that occupied a significant amount of time over the course of the next five years. The Subcommittee on Therapeutics eventually won the battle, and a handful of drugs that were controlled by patents were included, including at least three (Phenacetin, Sulphonal, and Trional) that remained under patent protection for a short time after the eighth revision became official in 1905.

Of course, the inclusion of these products raised the difficult question of what to call them. It seemed obvious that these and other products protected by trademarks could not be included under their commercial names.[74] In some cases scientific names could be used instead, if they were short enough. Phenacetin, for example, was included under the name "acetphenetidinum," a Latinized version of the scientific name *acetphenetidin*. However, the scientific names of most chemical products were simply too complex to serve therapeutic purposes. The revision committee therefore coined new names to use for these products, just as an earlier committee had done for Vaseline. Sulphonal, for example, was included under the name "sulphonmethanum," while Trional was included under name "sulphonethylmethanum."[75] These names were designed in part to offset the dangers of commercialism associated with the use of coined trade names. The problem, of course, was that physicians could not really be expected to use them. They were unfamiliar to the vast majority of physicians, they were awkward to use in prescriptions—or even to pronounce—and there was no real reason for physicians to bother. Why write a prescription for "sulphonethylmethanum" when one could simply write a prescription for "Trional"?[76] Still, members of the revision committee hoped that the medical community at large would adopt the new names. "Though somewhat unwieldy," noted two of the physicians involved in the revision process, "they certainly should be given the preference over coined and arbitrary titles, especially since these are usually protected by registration as trademark or by copyright."[77]

Both the controversy over Behring's patent on diphtheria antitoxin and the victory of the Subcommittee of Therapeutics pointed to the basic fact that by the late 1890s the orthodox medical community had largely accommodated itself to pharmaceutical patenting—but also that significant concerns remained about the relationship between commerce and science. The long-standing equation between quackery and monopoly in orthodox medical thought was thus fractured. At the same time, however, anxieties about patenting continued in new forms as physicians continued to understand medical science as threatened by the dangers of commercialism. The result was something of a compromise: while it would still be considered unethical for a physician to *hold* a patent on a medical compound or remedy, whatever prohibition there had once been on the *use* of patented goods would be left aside.[78] Thus, when the Code of Ethics was finally revised in 1903, the new language declared it "derogatory to professional character" for "physicians to hold patents for any surgical instruments or medicines," while the prohibition on dispensing or promoting quack remedies was revised to include only medicines with secret ingredients.[79] The compromise made sense to most observers at the time. From the perspective of most physicians—even elite physicians intent on reforming the market—it was obvious that many patented remedies had become indispensable to good medical practice. Yet the prohibition on physicians actually *holding* patents continued to be understood as a defense against the temptations to profit that might corrupt the practice of medical science. "We are not permitted to . . . patent a medical invention or discovery, however meritous in itself it may be," noted one physician in 1903 about the new code. "This negation is not for the benefit of defending ourselves against each other, but to protect the community from the chance of our yielding to those ordinary temptations that surround all classes of men, and to ensure to it the full measure of every stream of beneficence of whose source we may perchance obtain the key."[80]

Trademarks and the Invention of the Generic

"We are a bitters-and-pill-taking people," declared the surgeon John Shaw Billings in 1892 in a sarcastic overview of the role of patents in promoting innovation in medical science. Billings had been asked to provide

an overview of the benefits of the patent system to medicine for a celebration of the centennial of the patent law held at the Smithsonian. His
talk made it clear that he thought there were none. For example, Billings
sarcastically suggested that the patent system promoted the fortunes of
the drug industry by allowing the use of trademarks in advertising. Noting that most patent and proprietary medicine manufacturers were "too
shrewd" to obtain patents for their goods, since doing so would force
them to reveal their ingredients, Billings suggested that they instead
relied on trademarks to advance their interests. "From a commercial and
industrial point of view the great importance of patent and proprietary
medicines is advertising," he noted, before rhetorically hoping that "the
modern professional expert in advertising" would explain exactly how he
managed to convince gullible people to buy "Jones's liver pills" through
the use of the numerous advertisements he saw for the product on the
sides of barns. "I suppose there must be such people," he joked, "for I have
a high estimate of the business shrewdness of the men who pay for these
abominations."[81]

Billings's belief that advertising by the patent medicine industry was
used to dupe unwary customers into purchasing worthless products was
not unusual at the time. Nor was his recognition of the importance of
trademarks to the flourishing industry. By the end of 1900 the Patent
Office had issued more than six thousand trademarks for medical compounds under the 1881 law.[82] Some of these trademarks went to foreign
manufacturers, and a small number may have gone to domestic manufacturers in the ethical wing of the industry. However, the large majority
of them went to manufacturers of either old-style patent medicines or to
manufacturers of the growing number of proprietary remedies. In August
of 1894, for example, 21 trademarks were registered for medicinal products
(out of 105 total marks registered during the month). Some of these were
in the traditional naming formula of patent medicines, such as "Scallin's
Headache Killer," while others were arbitrary in nature, if sometimes suggestive of the medicine's origins or properties, such as "Prunola" (a trademarked name for a laxative) and "Unguentine," an antiseptic ointment.
Some of the marks also combined words and images, and some operated
at the level of the firm and covered multiple products; one trademark, for
example, combined the word "Egyptian" with a picture of a supposedly
Egyptian woman and was used for a variety of medicinal products made
by the Emin Pasha Drug Company of Chicago.[83]

The increasing use of trademarked names continued to arouse concern among therapeutic reformers in the medical community. Much
of this concern echoed long-standing criticisms of the patent medicine

industry as predatorial and dangerous to the health of the public. Billings was certainly not alone in his views, and as the volume of products produced by the industry grew, these criticisms grew increasingly shrill. At the same time, however, the distinction between quackery and legitimate medicine also began to blur in complex ways. By the 1890s manufacturers of old-style patent medicines had begun to move away from the traditional naming formulation in which the company name was used to modify the name of the product; they had also begun to incorporate synthetic chemicals and other increasingly powerful ingredients into their goods. Many products were still sold under names such as "Dr. Williams' Pink Pills for Pale People," but a growing number were sold under scientific-sounding names that operated at the level of the product itself. A variety of manufacturers, for example, combined acetanilide with sodium bicarbonate, citrated caffeine, and other ingredients and sold the resulting products under names such as "Phenolid," "Exodyne," "Antisol," "Pyretin," "Phenetol," and "Kaputine."[84] Whether or not these types of products were properly characterized as "patent medicines" was very much open to debate. Certainly, the fact that their names said little or nothing about the actual composition of these products suggested that they were little more than patent medicines made with secret ingredients. On the other hand, it was also obvious that such products were sometimes quite effective, they often appeared to be made according to advanced scientific techniques, and they were frequently advertised in medical journals. As a result, the distinction between so-called patent medicines and proprietary medicines began to break down. The phrase "patent medicine" remained a term of opprobrium in the medical community, but increasingly it was not always clear which products qualified for the designation. At the same time, the term "proprietary medicine" was increasingly applied to any product that was protected by a trademarked name, whether or not the ingredients of the product were secret and independent of the reputation of the firm in question. As a result, the distinction between proprietary medicines and strictly ethical medicines began to break down as well. The numerous synthetic drugs manufactured and sold by the German pharmaceutical industry, for example, were sometimes characterized as proprietary medicines, despite the fact that their chemical structures were known and the fact that they were sold by highly reputable companies. Increasingly, what had once been seen as distinctly different types of products were understood in overlapping terms.

Reformers in the medical community expressed a significant amount of concern about the growing trade in patent and proprietary remedies.

Physicians frequently prescribed proprietary products because it was easy to do so, they were frequently exposed to the advertising and promotional efforts of firms, and they apparently believed that the products helped their patients. By the middle of the 1890s proprietary goods thus made up a significant number of all prescriptions written; according to one survey, in 1895 just over 11 percent of all articles prescribed by physicians in Illinois were proprietaries.[85] This troubled many reformers in the medical community. Trademarked names often sounded alike, and accidental substitution of one product for another could result in terrible consequences for the patient. Trademarked names seemed to conceal the ingredients in the preparation; they sometimes suggested the problem the product was supposed to cure—a type of therapeutic advertising physicians found objectionable—and, of course, the willingness of many manufacturers to advertise directly to the public was deeply offensive to reformers. Yet there was little that they could do about the use of proprietaries, given that they were easy to prescribe and, at least in some cases, therapeutically beneficial to patients. When the detail man came knocking, it was undoubtedly difficult for the physician to turn him away based on what was probably at most a vague sense of concern about the changing nature of the times.

Pharmacists also continued to be concerned about the increasing use of trademarked goods. By 1900 retail druggists sometimes estimated that more than half of the value of their business came from the sale of proprietary medicines.[86] Most pharmacists continued to believe that the best practice was for physicians to prescribe official drugs whenever possible and to write out the complete formulas of their prescriptions. Doing so ensured accuracy, allowed the pharmacist to dispense the highest-quality drugs of the type called for, and frequently meant a lower price for the customer, since products sold under official names tended to be cheaper than products sold under trademarked names. Even if there was no official name available for the item in question, pharmacists still frequently argued that that scientific names should be used in prescriptions rather than commercial names, both for scientific rigor and so that they would not be obligated to dispense the product of one manufacturer over another. Debate also continued to swirl around the question of whether or not pharmacists were obligated to dispense a particular manufacturer's product when they could easily—and, quite possibly, more safely and cheaply—compound the same remedy themselves from ingredients they already carried. Why dispense the product made by Battle & Company when a pharmacist could just as easily compound Bromidia himself? These types of questions were complex and provoked significant controversy among retail druggists, but in general reformers in the pharmaceutical community believed that physicians should simply prescribe according to official names.[87]

Yet this position ignored several important problems. Complex chemical names could not really serve therapeutic purposes, and as chemical structures became more complicated the problem only grew worse. The use of scientific names for trade purposes—when such names were short enough that this was even possible—was also problematic, since manufacturers could monopolize the sale of the product through the use of trademarks, a practice that struck most pharmacists as deeply unethical.[88] Even when a product had a scientific name that was both distinct from its commercial name and not trademarked, and even when this scientific name was short and easy enough to spell that physicians might actually use it in their prescriptions, physicians *still* had a tendency to prescribe according to trademarked names because of the influence of advertising and other promotional strategies by manufacturers. The result was that the trademarked names of popular products had a tendency to act as the common name for those goods—why prescribe *acetphenetidin* when "Phenacetin" was easier to spell and more familiar to all concerned? Indeed, the trade names of popular products tended to act as the name for all products of a similar type, in effect giving manufacturers who controlled these names a de facto monopoly over all goods that were roughly equivalent to their own product.

Vaseline serves as a good example. Chesebrough first registered the name of his product in 1892, although he may well have asserted his rights to the name under common law doctrine previous to this. By this point physicians had already begun to use the term as a general name to refer to any of the many different petroleum jelly products on the market.[89] "Petrolatum" was not a particularly difficult name to use, and proponents of the *USP* sometimes tried to promote it, but it could not compete in popularity with the Chesebrough name. Nor could the names of other petroleum jelly products on the market. Not surprisingly, Chesebrough expected his product to be dispensed whenever someone used the term "Vaseline" to request a petroleum jelly product from a pharmacist, even if the customer did not necessarily mean the product manufactured by his company. At the same time, the company actively worked to discourage the prescribing of "petrolatum," advertising that competing brands were of lower quality and informing physicians that by prescribing under the official name, and thereby permitting pharmacists to dispense supposedly inferior products, "you are much more likely to injure than to benefit your patient and may do him serious harm."[90] Critics responded by arguing that the company promoted the term in order to monopolize the sale of all petroleum jelly products. As a committee of the American Pharmaceutical Association noted in 1897, "The object of the manufacturers has been, not to employ the word 'Vaseline' for the purpose of distinguishing their brand of petroleum jelly from other brands of the same article, but to obtain a monopoly in petroleum jelly itself by educating the people to call it 'Vaseline.'"[91]

Tensions around these issues intersected with trademark law in two important ways. During the late 1880s and the 1890s courts continued to understand trademarks primarily as indicators of the origins of goods and thus to define infringement in terms of deception and the unfair passing off of one product for another. Courts also continued to rule that the descriptive names of things could not be trademarked.[92] However, trademarks were also increasingly understood as a form of property—as something that embodied the reputation of a company, that financial resources could be invested in, and that could accumulate and hold value—and courts began to decide that names deserved legal protection even if they could not be adopted as trademarks for technical reasons. As a result, a series of court decisions held that although descriptive names could not, in general, be trademarked, if such names acquired a "secondary" significance indicating the origin of manufacture, then they deserved protection under the broader law of unfair competition.[93] A legal distinction between what would come to be called "technical trademarks" and what would come to be called "brand names" thus began to emerge. On the one hand, arbitrary names could be trademarked by manufacturers and thereby appropriated for their sole use; these types of marks were protected under the provisions of trademark law, and they were juxtaposed to descriptive names that could not be trademarked because they operated as the common name of things or for other reasons. On the other hand, names that were closely associated with a particular company—and therefore embodied the reputation of that company and accumulated value as such—could still be protected under the broader law of unfair competition, even if they could not be adopted as technical trademarks. As one court noted, competitors were not permitted to "put up" their goods in "a peculiar form previously adopted by another so as to give to his goods the general appearance of the originators, and this notwithstanding the originator may not have a technical trademark in his packages."[94]

At the same time, the question of whether or not trademarks could be used to monopolize the sale of goods was clarified. By this time it was well established in case law that the descriptive name of a thing could not be trademarked but coined or arbitrary names could be. But what happened when a coined name *acted* as the descriptive name of a good, whether as a result of frequent use or because there was no other name available? As demonstrated by the Asepsin case, during the 1880s and early 1890s courts sometimes ruled that trademarks on arbitrary names could not be infringed, even if there was no other name available to use to refer to the product in question and even if the product was not under patent. The result was that competitors were unable to manufacture the good in question, even if there were no patent rights involved. Sometimes, however, courts disagreed and ruled that trademarked names could not be used to monopolize the manufacture of a product if the underlying product was not patented.[95] In other words, by the middle of the decade there was an important legal question at hand about how to handle the fact that arbitrary names—which could be trademarked—had a tendency to act as descriptive names, which could not be.

The issue was finally settled by the Supreme Court in 1896. In the landmark case *Singer Manufacturing Co. v. June Manufacturing Co.*, the court ruled that the trademarked word "Singer" had passed into common use and acted as a "generic designation" of the type of sewing machine manufactured by the Singer company rather than a name "indicating exclusively the source or origin of their manufacture." The Singer company's patent on its machine having expired, other manufacturers had every right to manufacture the same type of sewing machine and, the court ruled, had the right to do so under the term that acted as the common name of that type of machine. As the court noted, following the expiration of the patent "and the falling of the patented device into the domain of things public," there "must also necessarily pass to the public the generic designation of the thing." To rule otherwise, the court reasoned, would be to hold that although "the public had acquired the device covered by the patent" the owner of the patent "had retained the designated name which was essentially necessary to vest the public with the full enjoyment of that which had become theirs by the disappearance of the monopoly."[96]

The *Singer* decision unequivocally established the doctrine that if a trademarked name moves into common use and thus becomes the descriptive name of the product in question, it reverts to the public domain following the expiration of the patent on the underlying good. Although the implications of the decision would take years to fully play out, the *Singer* decision was the beginning of the end of the ability of manufacturers to legally monopolize the production and sale of goods through the appropriation of descriptive names. Following the decision, trademarked names that became descriptive in nature reverted to the common domain following the expiration of the patent on the underlying product. Competitors to the original manufacturer thus gained the ability to produce and sell goods under what was increasingly referred to as the product's *generic* name. In a series of decisions from 1898 and 1899, for example, federal courts ruled that the manufacturer of a medicinal product named "Castoria" did not retain exclusive control over the name of its product because its patent had expired and its name had entered the common vocabulary.[97]

Not surprisingly, Francis Stewart was one of the first observers in the medical and pharmaceutical communities to appreciate the significance of the *Singer* decision. Stewart almost immediately recognized that as a result of the decision, manufacturers had lost the ability to indefinitely monopolize the descriptive names of goods. As he put it, shortly after the decision,

When a baby is born into the world, a name is given it. Does the name of the baby belong to the baby or to the one who gave the baby its name? Every new thing born into the world must have a name; and that name belongs to the thing, not to the one who named it. While the patent is in force the use of the name is restricted to the patentee along with the invention, but when the patent expires both should, and I hold that they do, become common property.[98]

From Stewart's perspective, the name of a product belonged to the thing itself, and this name reverts to the public domain at the expiration of the patent. The *Singer* decision thus reinforced what Stewart had already come to believe: that the names of things and trade names applied to those things should be kept distinct from one another. For Stewart, then, trademarks could legitimately be taken out on "brand names" but not on names that operated at the level of the product itself. According to Stewart, trademarks that were taken out on brand names served legitimate and even important commercial purposes by protecting the interests of manufacturers, while leaving "the product and the name of the product open to science."[99] Importantly, however, Stewart also believed that this meant that all new goods must be given names by their manufacturers that are distinct from the names used for commercial purposes. As he put it, "Manifestly, the products themselves must have names under which all may deal in them, before there can be any such thing as brands."[100]

Price Control and the Problem of Monopoly

In 1896 an anonymous writer in the pages of the *Paint, Oil, and Drug Review* surveyed the rapidly changing economic landscape. He did not like what he saw. He pointed to an "economic craze" in which "the rapid concentration of capital and business energy in the hands of a few" was accompanied by an "epidemic of economy" in which "the pruning knife" was used to reduce expenses through the reduction of wages, the closing of plants, and other means. Declining prices were at the heart of the troubles. Noting that that prices had fallen "from 45 to 50 per cent" over the past two decades across multiple industries, the author suggested that "every drop in prices swallows up millions of profits" and that "consolidations, pools, curtailment of production and expenses [and] failures" all followed naturally as businesses tried desperately to reduce costs. The problem, however, was that cutting expenses also "diminishes the consuming power of the country, destroys demand, so that more pruning follows, and this in turn lessens consumption, and so on *ad finem*, each succeeding state being worse than the former." From this perspective economic consolidation was both a cause and an effect of declining prices, and like numerous other critics at the time, the writer believed that the solution to the troubles of the day lay in a properly functioning market in which prices were set at their natural levels, economic consolidation was curtailed, and consumer spending naturally followed. As he noted, "The solution of the problem must be found in a reversal of the influences which have sunk prices into the depths; the setting into activity of forces that will rescue values to normal base where the general tendency will be for a rising rather than a falling market; and the revival of a fair and natural price-level for all products; each procedure cannot fail to induce a quickened demand and healthy commercial conditions."[101]

The passage of the 1890 Sherman Antitrust Act was the most important effort of the time to structure the market along what were taken to be rational and fair lines. Growing out of profound concerns about both economic concentration and declining prices, the law was an effort to restrain monopoly toward the goal of a just market in which prices would be stabilized at their supposedly natural level and all participants could compete with one another on an equal footing. The law thus made illegal "every contract, combination in the form of trust or otherwise, or conspiracy, in restraint of trade or commerce" in interstate trade. Almost immediately, however, debate erupted about what the phrase "restraint of trade" actually meant. Courts initially interpreted the Sherman Act narrowly, ruling that the law only applied to restraints of trade enforced by combinations of businesses; the actions of individual firms, no matter how large, were considered to be outside the scope of the law. Courts also initially interpreted the law according to the common law understanding of the phrase, in which reasonable restraints of trade were permissible. In 1897, however, the Supreme Court declared both reasonable and unreasonable restraints of trade illegal, thereby constructing the 1890 law as superseding common law doctrine. Over the next fourteen years the court interpreted the Sherman Act from this literalist position, promoting a vision of the economy in which, as Martin Sklar has put it, "the liberty of the citizen did not include the right to make a contract directly and substantially restraining or regulating interstate commerce, a power that Congress reserved exclusively to itself and denied alike to the states and to private persons." [102]

The relationship between patent law and antitrust law has long been complicated and fraught with difficulty. In the first two decades following the passage of the Sherman Act, however, courts generally assumed that business practices related to patenting could not be restrained through the application of 1890 law. Patents, after all, were granted under constitutional authority, and even after the Supreme Court turned to a literalist interpretation of the law, restraints of trade based on patent rights were assumed to fall outside its domain. Thus, for example, in the 1902 case *Bement v. National Harrow Co.*, the court ruled that a variety of different restrictions placed on the harrow industry through an extensive patent pool—a pool that covered more than 90 percent of all manufacturing and sales in the industry—did not violate the Sherman Act, since the basic point of patent law was to establish monopolies. Patents, the court ruled, allowed manufacturers to engage in a wide variety of monopolistic practices without running afoul of the 1890 law.[103]

Despite its importance in other sectors of the economy, antitrust enforcement under the 1890 law had little impact on the pharmaceutical industry in the first decade after the law's passage. Domestic manufacturers do not appear to have established cartel arrangements to any significant extent, or at least not in ways that attracted concern. European pharmaceutical companies, on the other hand, frequently entered into cartel agreements with one another in order to maintain prices on their products, in part because of the restrictions on obtaining patents on pharmaceuticals and chemicals in various European countries. Bayer and other German companies were particularly adept at this strategy and formed numerous cartels with one another that covered European

THE AMBIGUITIES OF ABUNDANCE

markets. In the United States, however, foreign companies were typically able to secure product patents on their goods and as a result only occasionally entered into cartel arrangements with one another that covered American markets.[104] When they did, they tended to be founded on patent-based practices and as a result were—if they were even noticed—probably assumed to be outside of the scope of the Sherman Act.[105]

During the 1890s antitrust law also initially had little to say about efforts to fix prices. Declining prices were a tremendous source of economic instability in the 1890s, and many manufacturers sought to curtail expenses through the "pruning knife" of reducing wages, closing plants, and other means. Manufacturers also sometimes established minimum resale prices for their goods and forced both wholesalers and retailers to conform to these price schedules through contractual means—a practice known as resale price maintenance (RPM). Courts had few objections to their doing so.[106] Part of the reason for this was the initially narrow interpretation of the 1890 law, which restricted its application to the behavior of combinations, as opposed to single firms. Part of it was the fact that in many industries RPM was tied to patenting and that licensing practices related to patents were generally understood to be exempt from antitrust action as a result of the legitimacy of patent monopolies.[107] More important, at least for my purposes, the legal tolerance of RPM in the decade or so following the passage of the Sherman Act grew out of a general assumption that the owners of a product who had a lawful monopoly over that product—whether gained through a patent, the use of a trade secret, or some other means—had what one later commenter called an "exclusive dominion over the goods which he *owned*" and that along with this dominion came the right to "vend or not to vend, according as he chose, and, as an extension of that right, to specify the conditions under which he would sell—one of those conditions being the maintenance by the dealer of the resale price suggested by the manufacturer."[108]

It is difficult to know the extent to which drug manufacturers engaged in RPM during the 1890s. Manufacturers probably had mixed feelings about the practice. On the one hand, they worried about declining prices just as other manufacturers did, and for companies with well-established brands, instituting RPM probably helped protect the reputations of their products. On the other hand, manufacturers also understood that low prices allowed them to move into new markets, and in the fiercely competitive drug business it may not have seemed reasonable to try to prevent price cutting through contractual means. Price cutting could take place at both the wholesale and retail levels, and manufacturers may have been hesitant to establish minimum prices out of fear of losing business to their competitors. Still, at least some manufacturers did clearly engage in the practice, sometimes working with the National Wholesale Druggists Association (NWDA) and other organizations to enforce minimum resale prices for their goods. As I discuss in the following chapter, for example, the Dr. Miles Medical Company used a contract system to set minimum resale prices for its popular products. The system allowed the company to enforce minimum prices for its goods, but it also provoked the anger of both wholesalers and retailers who objected to the practice, and the company was involved in numerous lawsuits about the issue in the first decade of the new century, including a landmark 1911 Supreme Court case that declared the practice illegal.

Whatever the extent of resale price maintenance among manufacturers, retail druggists actively pursued cooperative schemes to stabilize prices during the 1890s. Over the course of the decade the economic position of retail druggists grew increasingly precarious, as prices remained dangerously low and so-called price cutters continued to engage in what many druggists considered unfair practices. By the turn of the century the belief that price cutting was both unethical and contrary to the natural operation of the market had become a central assumption among retail druggists concerned about the state of their trade. The goal of price cutters, as one critic put it in 1899, was "to attract trade from its natural sources" through the use of "guerrilla practice, a procedure which, at least in pharmacy, should not be allowed to exist another year."[109] Price cutting, like other supposedly unethical practices such as adulteration and the use of secret ingredients, distorted the natural price of goods and endangered both the health of the public and the livelihood of the reputable druggist. Unfair competition from other types of retailers, particularly large department stores, had a similar effect. Department stores, for example, often used patent medicines and proprietary drugs to attract customers to their stores and regularly sold them at prices that retail druggists were unable to match. Pointing to the "commercial degeneracy of the American drug store," one critic thus denounced both "the establishment of the department store with its drug and prescription department" and "the cutting of prices on so-called patent medicines" as intertwined evils.[110]

Retail druggists responded to their problems in a variety of ways, including by working to pass laws that regulated the practice of pharmacy, mandated labeling requirements, and otherwise uplifted the practice of their trade. They also directly attacked the problem of price cutting by organizing themselves into cooperative associations and establishing a wide variety of price control programs. The most important of these was the "tripartite plan," organized by the National Association of

Retail Druggists (NARD) around the turn of the century. The NARD was organized in 1898 in response to the deteriorating conditions facing the retail drug trade. Its tripartite plan required manufacturers of proprietary goods to establish minimum retail prices for their products and to sell their goods only through participating wholesalers. Wholesalers were required to sell only to approved retailers, and retailers were required to not sell below the established price, to give preferential treatment to goods protected by the plan, and to report violations to the NARD. The association kept a blacklist of retailers known as price cutters and prohibited approved wholesalers from dealing with them, thus cutting off their access to manufacturers who had agreed to participate in the plan.[111] It was a powerful strategy, and the NARD quickly became one of the most influential organizations in the drug trade.

Both retailers and wholesalers accused of price cutting sometimes tried to defend themselves through legal means. Following the general doctrine that manufacturers have the right to fix the resale price of their goods, however, during the 1890s courts supported the rights of the NARD and other organizations to establish and enforce resale price maintenance schemes. In one widely watched case, for example, a wholesaler named John D. Park & Sons sued the NWDA over its so-called Detroit Plan, in which a standard contract was used to establish minimum resale prices. In 1896 the company asserted its right to sell proprietary goods at any price it chose, and after a number of manufacturers boycotted the company, it filed suit against the NWDA, claiming that the organization's standard contract was an illegal restraint of trade under the Sherman Act. The court disagreed and in 1898 ruled that the NWDA had the right to enter contractual relations with manufacturers that set minimum resale prices for their goods and that the association and its members had the right to refuse to trade with both wholesalers and retailers that did not agree to enter into contracts with them. The case was widely taken among druggists to indicate the legality of cooperative resale price maintenance plans; according to the *Midland Druggist*, which closely observed the case, "jobbers can legally refuse to sell to cutters."[112]

Of course druggists were not only concerned about prices that they considered to be unnaturally low. They were also deeply concerned about prices they considered unnaturally high and, in particular, the use of patent monopolies by foreign firms to maintain what they considered artificially high prices. Bayer and other foreign companies frequently charged significantly higher prices in the United States for their patented products than they did in other countries. In Canada, for example, prices on drugs were significantly lower than in the United States, a fact that critics widely ascribed to differences in the patent laws of the two countries. In 1897, Antipyrine cost $1.40 an ounce in the United States but $1.10 in Canada, Trional cost $1.50 an ounce in the United States but $1.00 in Canada, and Sulphonal cost $1.35 in the United States but just 30 cents in Canada.[113] Pharmacists, like physicians, were increasingly tolerant of patenting during the 1890s, seeing it as a legitimate means for manufacturers to recoup their investment in the development of new products.[114] Yet the ability of foreign manufacturers to charge higher prices in the United States than they could charge in other countries, presumably because of their patents, enraged them. This anger was directed against all foreign manufacturers that maintained these types of price disparities, but most of its focus was, not surprisingly, on German firms. American druggists believed that under the 1891 German patent law medicines could not be patented in Germany, that discrepancies between American prices and prices in other countries for the same product could be explained through the ability to secure American patents on their drugs, and that German companies should set their prices in the United States to match their prices in their home country.[115] Patent law, like price cutting, seemed to unfairly distort the market and artificially change the price of goods. It struck druggists as deeply unfair and desperately in need of correction.

The most important example of this dynamic was Bayer's antipyretic Phenacetin, which was commercially introduced in 1888 and patented the following year.[116] Bayer initially sold Phenacetin in the United States for a dollar an ounce, which was roughly the same price as in Germany. In late 1888, however, Bayer was denied a German patent on the drug, other German firms began to manufacture it, and the price in Germany collapsed. By 1895 the price in Germany had declined to about twenty-five cents an ounce. Bayer reduced its prices in Canada and Europe to roughly the same price, presumably because of the lack of patent protection in those countries, but did not adjust the price of the drug in the United States and insisted that druggists continue to pay a dollar an ounce. Many pharmacists balked at the price, and wholesalers began to purchase the drug in Canada or England and then sell it to jobbers in the United States at a price lower than what Bayer mandated. By the end of the decade a robust trade had developed in Phenacetin that was purchased in bulk from manufacturers in other countries and brought into the United States contrary to the wishes of the company.

In the view of many pharmacists this was a perfectly reasonable response to the company's unfair pricing scheme. Many pharmacists assumed that it was legal to resell imported versions of the drug that had legally been manufactured in countries where there was no controlling patent. Druggists probably also believed that even if doing so was technically illegal, it was ethically legitimate because Bayer was acting in a predatorial manner. Reputable pharmacists had long believed that druggists were ethically obligated to dispense the preparation called for in a prescription; however, they also believed that druggists had the right to compound prescriptions themselves and to sell these prescriptions under their true names, even if those names were trademarked. A similar logic probably applied to selling a product made by a different manufacturer than the one that held the patent on the product or selling what might be considered a smuggled version of the patent holder's product: such behavior was probably seen as an extension of the traditional right to dispense the proper drug to a customer without concern for the monopoly rights of the manufacturer. In other words, if a prescription called for Phenacetin, many pharmacists felt comfortable selling their customers a product made by another manufacturer, or even the product made by Bayer that had been brought in from another country against the company's wishes, because they were providing the product that was called for at the best available price.

From Bayer's perspective, of course, this practice violated its patent rights and amounted to little more than illegal smuggling. In response, the company launched a massive effort to try to suppress the practice. The company sued numerous importers, and when that did not end the trade turned to bringing suits against retail druggists.[117] Druggists frequently settled with the company for token fines and an agreement to honor its patent rights. In cases that went to court, however, the pharmaceutical community often rallied to the support of the accused druggists, raising money for their defense and publicizing their situation. According to an internal legal document prepared for the company, by the time the patent on the drug expired in 1906, the company had prosecuted between seven hundred and eight hundred cases, some of which were ongoing, and had settled with thousands of druggists out of court. The company also maintained active files on over seven thousand retail druggists and importers.[118] The scope of the operation was unprecedented, and it provoked a tremendous amount of anger among retail pharmacists. The ability of Bayer and other foreign firms to enforce these types of pricing schemes struck many pharmacists as deeply unethical, and many druggists found Bayer's willingness to sue retailers shocking.

One strategy to respond to Bayer's behavior was to attempt to invalidate the patent. In late 1896, Bayer sued a druggist named Conrad Maurer. The Pennsylvania Pharmaceutical Association paid the cost of hiring the patent attorney Hector T. Fenton, who argued that Bayer's patent was invalid because it was essentially similar to a substance described in 1879 by the chemist Edward Hallock.[119] Bayer hired the well-known patent lawyer Anthony Gref, who in turn retained Charles F. Chandler of the Columbia Institute, one of the country's leading experts on the coal-tar industry, to serve as an expert witness. Gref's key legal strategy in the case was to show that Hallock had accidently manufactured the chemical as part of his efforts to produce something else and that even if his product did contain Phenacetin, it was not therapeutically useful if it was "buried" in what was otherwise a "poisonous" substance.[120] The question of commercial utility was central to this argument: citing the *Wood Paper Patent* case, Gref pointed to a distinction between "the discoveries of a merely scientific chemist, and of a practical manufacturer who invents the means of producing an abundance, suitable for economical and commercial purposes." Quoting the words of Vice Chancellor Stewart, Gref argued that "what the law looks to . . . is the inventor and discoverer who finds out and introduces a manufacture which supplies the market for useful and economical purposes with an article which was previously little more than an ornament of a museum."[121] The judge in the case agreed and in 1901 ruled that the patent was valid. In 1902, the Phenacetin patent was upheld in the US Circuit Court for the Eastern District of Pennsylvania, and soon after, Maurer lost his appeal to the Supreme Court.[122]

Following the Supreme Court decision it became clear that efforts to overturn the Phenacetin patent were not viable. In response, under the leadership of the NARD retail druggists began to advocate for changes in patent law that would prohibit patenting medical substances and limit the rights of foreign patent holders to what they were able to obtain in their home countries.[123] In 1903, Joseph W. Errant, the general attorney of the NARD, drafted a bill that included these provisions, and in the same year Representative James R. Mann from Chicago introduced it to Congress.[124] The NARD organized a comprehensive lobbying effort in support of the bill, and druggists flooded Congress with impassioned pleas for its passage. "The present [patent] law is the most outrageous, discriminating measure ever forced on the American public," wrote the president of the Erie County Pharmaceutical Association. "Our laws now foster forming monopolies and enrich foreign capital at the expense of our own sick. The 'Mann Bill' is aimed to correct this inhuman and ridiculous condition."[125]

The House passed a version of the Mann Bill in late 1904, but after much debate it was tabled in the Senate and never reached a floor vote.[126] Opposition to the bill came from a variety of sources, including the domestic chemical industry, foreign drug manufacturers, and the Patent Office itself.[127] More generally, the bill never attracted the support of the pharmaceutical community as a whole. Manufacturers split on the bill, with some opposing its passage—possibly with an eye toward their own future efforts to patent medicinal products—and others supporting it, possibly as a result of intimidation by the NARD.[128] Many local and state pharmaceutical associations strongly supported the bill, but national organizations outside of the NARD were more cautious and refrained from endorsing it.[129] The American Pharmaceutical Association, for example, generally supported efforts to reform patent law along the lines advocated by the NARD, and in 1903 Stewart testified to Congress as a representative of the organization about the need for significant patent reform.[130] However, members of the organization disagreed on the specifics of the Mann Bill itself, with some taking the side of retail druggists and others arguing that both inventors and manufacturers needed to be justly rewarded for their efforts.[131] As a result, the organization did not formally endorse the bill or organize on its behalf.

The failure of the NARD to secure the passage of the Mann Act probably explains the organization's tepid response to trademark reform. Retail druggists were deeply concerned about the role of proprietary drugs in driving the price of goods downward, and trademarks were sometimes linked to the problem of patenting because they were thought to allow manufacturers to continue to monopolize the name of a good even after the patent on that good had expired, thereby acting as "perpetual protection" on the drug.[132] Organizations of retail druggists sometimes lobbied Congress to change trademark laws.[133] Despite such efforts, however, the NARD and other organizations of druggists never devoted significant resources to reforming trademark law. In part, this was probably because their attention was occupied with passing a law to ban the use of product patents. It may have also been because, given the 1896 *Singer* decision, knowledgeable druggists realized that trademarks no longer had the potential to effectively provide a permanent monopoly over the name of the good in question. After initially lobbying Congress to abolish the ability of trademarks to act as indefinite monopolies, for example, the NARD was pointedly informed by the commissioner of patents that as a result of the *Singer* decision they no longer had the ability to do so.[134] Perhaps ironically, the emergent category of the generic thus worked to legitimize trademarks, rendering them free from the taint of monopoly.

Rational Therapeutics and the Dangers of Commercialism

At the turn of the century, therapeutic reformers faced a flood of patent medicines, nostrums, poorly manufactured and adulterated products, deceitful advertising, and other problems—what one physician
bemoaned as a "swamp of speculation . . . filled with all sorts of absur-
dities and frauds."[135] Reformers increasingly approached this chaotic
market from the perspective of what, following Harry Marks, might be
called *rational therapeutics*: a strong belief that knowledge about drugs
derived from laboratory and clinical science should serve as the basis for
medical practice.[136] Closely associated with this belief were a variety of
other assumptions about how the drug market should be organized: an
assumption that drugs should be manufactured according to formally
established standards and that, as a result, drugs sold under the same
name should be relatively equivalent to one another; a belief that drugs
should be sold in an open market based on the norms of disclosure and
honest advertising; and the assumption that dangerous and irrational
forms of consumption should be suppressed. Reputable physicians were
thus supposed to prescribe drugs that had been thoroughly investigated
and had their therapeutic properties scientifically established. Whenever
possible, reputable pharmacists, in turn, were supposed to compound
and dispense drugs in accordance with the standards of the *Pharmacopeia*
or other authoritative texts. And consumers, increasingly, were only supposed to purchase and use drugs under the guidance of expert authority.

This process was not antagonistic to the pursuit of profit. Far from it.
Powerful drugs promised to heal the world of terrible ills, and therapeutic
reformers worked to rationalize the market so that these products could
do their work. Even among reformers who were not directly connected to
the manufacturing industry, there was a growing assumption that ethical
firms should be able to invest resources in scientific drug development
and earn a legitimate profit off of that investment in order to produce
new remedies. As a result, ethical manufacturers often worked in concert
with therapeutic reformers in the medical community and with officials
in government agencies to promote their goods and thereby improve
the health of the public. Ethical manufacturers certainly did not always
share goals with reformers in the medical or pharmacy communities, in
state or local governments, or in other institutional or professional ven-
ues. Indeed, they sometimes worked to advance their interests in ways
that other reformers found troubling. Still, along with other therapeutic
reformers they shared the same basic goals of rationalizing the market,

ensuring that useful products reached the people who needed them, and suppressing the trade in what they all considered dangerous nostrums. A rational therapeutic marketed would benefit everyone concerned— except, of course, those who manipulated the market and corrupted the practice of medicine toward their own selfish ends.

An important example of this dynamic is the passage and immediate impact of what historians refer to as the 1902 Biologics Control Act. In 1901 contaminated lots of diphtheria antitoxin and smallpox vaccine led to outbreaks of tetanus in Saint Louis and Camden, New Jersey. At least thirteen children died as a result of the outbreaks. In response, reformers quickly passed a comprehensive law regulating the manufacture of vaccines, antitoxins, and related products. Among other provisions, the law required that companies obtain a license from the Hygienic Laboratory of the US Public Health and Marine-Hospital Service (USPHMHS) in order to manufacture so-called biologics, that these types of products be manufactured according to standards established by the Hygienic Laboratory, that manufacturers be subject to unannounced inspections, and that these goods be clearly labeled with the accurate name of the product, the license number under which they were manufactured, and an expiration date after which they should no longer be considered safe. The law was surprisingly strong for the time, reflecting an awareness of the dangers of these products and the difficulty in making and selling them safely. Yet despite its teeth, in one fundamental respect the law was not that different from other efforts to rationalize the therapeutic market. Biologics now had to be made according to promulgated standards and to be clearly labeled and sold under their true names; like many other efforts to rationalize the buying and selling of pharmaceuticals, the 1902 law thus sought to stabilize the relationship between names and things. The licensing requirement, inspection provision, and other parts of the law that gave it an unusual degree of regulatory strength were extensions of this basic goal. They seemed eminently reasonable given the potentially devastating effects of selling products that were poorly manufactured or otherwise did not conform to the developing norms of rational therapeutics.[137]

The 1902 law significantly restructured the market in biologics. Following its passage, the USPHMHS began to issue licenses to manufacturers and enforce standards for the production of vaccines, antitoxins, and related products. Numerous small manufacturers were either put out of business or withdrew from the biologics market as a result of the law. Those that remained complied as best they could with its requirements. By 1904, thirteen companies had been licensed under the act, including both Parke-Davis and H. K. Mulford, which secured the first and second licenses, respectively.[138] Over the course of the next decade the number of licensed manufacturers grew steadily. The safety of biologics also improved significantly. Both Parke-Davis and H. K. Mulford, for example, temporarily lost their licenses in 1908-1909 for manufacturing contaminated vaccines. The companies quickly recalled the products, destroyed their remaining stores, and improved their manufacturing facilities in order "to make a clean start," as a 1910 report issued by the USPHMHS put it. By early 1909 the two companies had regained their licenses, having fully complied with the 1902 law.[139]

The consolidation of the biologics market under licensed manufacturers was accompanied by a growing diversity in the volume and type of available goods. Companies such as Parke-Davis and H. K. Mulford invested significantly in developing new vaccines, antitoxins, and similar products, a small number of which they patented. Foreign manufacturers did as well. As I discuss more fully in the next chapter, Parke-Davis turned toward the use of product patents around this time, most notably by patenting Adrenalin, and the company acquired at least two patents on vaccines following the passage of the 1902 law.[140] Foreign manufacturers also patented a small number of biologics in the decade before World War I.[141] Not surprisingly, patenting activity in this area was relatively modest. Profits, however, were undoubtedly substantial: by the outbreak of World War I, forty-five manufacturers were licensed to manufacture biologics, about half of which were domestic companies.[142] Together, they produced a tremendous number of goods intended to counteract the ravages of disease. Parke-Davis alone had more than thirty vaccines, toxins, antitoxins, serums, and "modified bacterial derivatives" on the market, including anthraxoids—"little pills or pellets containing a preparation of killed anthrax germs"—antidiphtheric, antimeningitic, and antitetanic serums; and vaccines for typhus, gonorrhea, and acne.[143] For complex reasons, biologics were typically sold under descriptive names that included the names of diseases, such as "diphtheria antitoxin," and were almost never sold under arbitrary names that could be trademarked. As in other segments of the drug market, however, financial investment in scientific research led to the development of new products that were then marketed through advertising and other promotional efforts that linked the reputation of the company to the presumed value of the product.[144] The major difference between the developing market in biologics and the broader drug market was simply that what reformers took to be quackish behavior had been effectively suppressed and ethical manufacturers were subject to relatively strict regulation governing the production of their goods.

At the same time, the idea that government agencies had a legitimate role to play in the manufacturing of biologics began to erode. This was in part the result of a concerted effort by manufacturers to delegitimize the role of state and municipal laboratories in the production and marketing of these products. Parke-Davis, H. K. Mulford, and presumably other firms active in the biologics market strongly opposed the manufacture of vaccines and similar products by government bodies, sometimes working behind the scenes to undermine such efforts.[145] The orthodox medical community also frequently opposed the public manufacture of biologics, attacking it as monopolistic, corrupt, and quite possibly dangerous to the health of the public. The efforts of the New York City Health Department to manufacture the diphtheria antitoxin, for example, were denounced as "municipal socialism" almost immediately after the organization began distributing the product.[146] Of course, therapeutic reform was an internally variegated and at times inconsistent movement, and numerous reformers continued to promote the public manufacture of biologics as a reasonable response to the dangers of infectious disease. Yet over the course of the decade, reformers increasingly insisted that the government had no place competing with reputable manufactures in the pharmaceutical market. As one physician put it shortly after the passage of the 1902 law, "When we trust large pharmaceutical houses to prepare for us aconite, digitalis, strophanthus, ergot and every other imaginable kind of valuable but dangerous medicine, there is surely no reason why we should go to amateurs under the control of political machines when we want such preparations as vaccine virus and antitoxins."[147]

Therapeutic reformers thus worked to shape the market along what they considered rational and scientific lines by suppressing the sale of what they took to be quackish products, standardizing the production of reputable ones, and delegitimizing the efforts of state and municipal authorities to compete on the market. They also confronted a public that was at times skeptical and even hostile to their efforts. Despite increased safety, significant risks continued to be associated with biologics following the passage of the 1902 law, and local political factors, a long history of suspicion toward vaccination, and other complex dynamics fueled popular criticism of and resistance to the use of these products—particularly to compulsory vaccination laws that were sometimes enforced through violent means.[148] Frequently articulated in the language of the antimonopoly tradition, these critiques were part of the broader concern about the apparent ability of powerful actors to distort the natural and fair operation of the market. Opponents of mandatory vaccination laws, for example, saw themselves as fighting against the monopolistic practices of a "vaccine trust," a supposedly shadowy network of vaccine makers and physicians who contracted with municipal governments to administer vaccinations, thereby artificially inflating demand for these products and unfairly diverting public money into private coffers.[149] Critics railed against this "dangerous, authoritarian medical monopoly" as a threat to individual freedom, simultaneously rejecting the corporate pursuit of profit, medical and governmental authority, and a scientific framework that allowed physicians and other experts to prioritize the health of the public over the concerns of individuals and families.[150] These concerns echoed the critiques made by Thomsonians in the early nineteenth century about orthodox physicians and licensing laws; in both cases populist sentiments confronted a form of medical authority that drew on the power of the state to enforce its claims to benevolence and expertise. The difference was that now orthodox physicians promoted a form of profit-driven medicine that linked the interests of commercial manufacturers to the promulgation and enforcement of rational therapeutics.

From the perspective of most therapeutic reformers, populist resistance to vaccination was little more than an obstacle to be overcome, an irrational form of behavior that needed to be counteracted so that the progress of medical science could benefit everyone. Most therapeutic reformers believed that promoting the use of these products, through coercion if necessary, was a reasonable response to the dangers of infectious disease. They understood their efforts as promoting science, not commercialism. Yet the linking of medical science and private profit on which this project increasingly depended also provoked suspicion within the medical community itself. Reputable manufacturers were of course increasingly understood as a vital part of the effort to provide needed drugs to the public—by this point no one questioned their ability to innovate new products, to introduce them to the market, and to advance medical science by doing so. Yet the distinction between ethical behavior and dangerous forms of commercialism was not always clear. The growing number of detail men, the swelling amount of promotional material sent to physicians, advertising in medical journals masquerading as science—these and other promotional efforts seemed to bend the benevolent goals of medical science toward overtly commercial ends. Outright quackery—such as products made with secret ingredients—was still easily recognized, at least most of the time, and reformers worked to suppress it as best they could. But there was a new danger as well. Even ethical firms that made useful products sometimes seemed to threaten the integrity of medical science through their increasingly relentless advertising. At the same time, dangerous or inadequately tested products were often marketed under scientific names, advertised in medical journals, and promoted in ways that appeared to conform to the norms of times. It was not always easy to distinguish between useful new products and nostrums, between educational material intended to help physicians make wise choices and what was taken to be improper advertising methods. One physician thus decried the "ever increasing mass of so-called 'literature' relating to patented pharmaceutical products," dismissed detail men as "agents ignorant of anything save the words that are put into their mouths by their employers," and critiqued as "bribery" the numerous "'presents' in the shape of paperweights, calendars, penholders, etc." that drug companies sent to physicians. The promotional strategies of even the most ethical firms seemed to corrupt the practice of medicine. As this physician put it, "Let us by all means strive to put a stop to this prostitution of our profession."[151]

As a result, even as the orthodox medical community embraced the importance of commercial drug development to the advancement of medical science—and at times supported forcing people to use the industry's products against their will—they also faced a new and surprisingly difficult problem: how to protect medical science from the corrosive effects of the market. The turn toward rational therapeutics and the acceptance of both patenting and the commercial motive with which it was intertwined required a new way of relating to the seemingly endless number of new drugs on the market: neither an ultraethical stance that rejected the use of monopolized goods out of hand nor an overly credulous one that embraced every new product that came along. The reconciliation of medical science and private interest, of medical ethics and monopoly, was thus articulated through both a skeptical attitude toward new products and the growing irrelevancy of patenting to the question of therapeutic utility. In 1904 a physician named William J. Robinson thus argued that while nostrums made with secret ingredients should be discounted without trial, new products introduced by reputable firms should be tried—and yet one should still be wary of the tendency of manufacturers to exaggerate claims about their products. "The commercial instinct, even at its best," he noted, "is apt to exaggerate the good qualities of a product and minimize the bad ones." For Robinson, as for a growing number of other physicians, the blending of science and commerce required a critical stance, one that was skeptical yet not overly dismissive of new goods. As he noted, "This, then, is the right attitude toward proprietary remedies; unbiased, yet cautious; unprejudiced, yet skeptical."[152] At the same time, however, the willingness to critically yet respectfully engage the claims of others was less frequently extended to those who disputed or otherwise challenged medical authority. Forms of behavior that therapeutic reformers took to be irrational, whether in terms of using the wrong products or refusing to use the proper ones, simply needed to be suppressed.

The Embrace of Intellectual Property

By the first decade of the twentieth century the pharmaceutical industry had grown into a major sector of the new industrial economy. Therapeutic reformers looked upon the "ever-changing, but never-ceasing flood" of its products, as one observer put it, with mixed feelings.[1] Powerful and effective medicines were among the many bounties of the new industrial order, yet the drug market also continued to seem irrational and deeply flawed, filled with what reformers believed were dangerous and fraudulent products. Something clearly needed to be done. In the first decade of the twentieth century, therapeutic reformers managed to establish a variety of mechanisms for shaping the market in drugs along what were increasingly considered rational lines, including the Council on Pharmacy and Chemistry (established in 1905) and the 1906 Pure Food and Drug Act. The Bureau of Chemistry, which enforced the 1906 law, and the Council on Pharmacy and Chemistry wielded different types of power; the former carried the weight of the federal government and was able to legally force manufacturers to adjust the claims they made about their products, while the latter primarily enforced its will through its ability to influence the reputation of a drug. Yet both played a significant role in shaping the market along what reformers considered more rational lines.

This process was intertwined with both the shifting nature of therapeutics and the ongoing transformation in orthodox medical thought about patents and trademarks.

Laboratory and clinical science were increasingly used by a variety of actors, including both the Bureau of Chemistry and the Council on Pharmacy and Chemistry, to develop, evaluate, and pass judgment on pharmaceuticals. As this took place, the older equivalence between monopoly rights and quackery was fractured, and the scientific validity of a drug was rendered independent of its status as a patented or trademarked product. This was a profoundly important realignment of the relationship between science, ethics, and commerce that would reshape the nature of scientific drug development in the coming years. By the outbreak of World War I, ethical manufacturers had begun to recognize the possibilities of the new framework and embrace the use of product patents to defend their scientific innovations. They also grew increasingly comfortable taking out trademarks that operated at the level of the product itself. The modern pharmaceutical industry had been born.

Pharmaceutical Patenting and the Dominance of German Synthetic Drugs

During the first decade of the new century patenting among the ethical wing of the domestic pharmaceutical industry remained rare. Patents had long been assumed to be contrary to the norms of ethical manufacturing, and firms such as Parke-Davis, E. R. Squibb & Sons, and H. K. Mulford & Company continued to base their business strategies on marketing to the orthodox medical community. Following the successful introduction of Taka-Diastase, however, Parke-Davis appears to have recognized that it could deal in patented goods without damaging its reputation. In 1899, for example, the company introduced a line of metallic germicides, at least two of which were patented.[2] Around the same time, as I discuss in detail below, the company introduced a patented supernal preparation developed by Jokichi Takamine that it sold under the trademarked name Adrenalin. Unlike the company's germicidal preparations and its patented vaccines, Adrenalin was a major therapeutic breakthrough and rapidly became one of the most important drugs of the time. In its willingness to commercially introduce these products —and to defend its patent rights to Adrenalin in court—Parke-Davis once again worked to redefine the relationship between science and commercial markets for companies that self-consciously conformed to the ethical norms of the orthodox medical community.

This change in orientation grew out of the basic fact that the therapeutic market was increasingly shaped by patented goods. Manufacturers continued to produce a tremendous number of so-called patent medi-cines, of course, although by this point virtually no patents were granted for "mere physicians' prescriptions," as the commissioner of patents put it.[3] However, both domestic chemical manufacturers and specialty companies dedicated to manufacturing proprietary drugs continued to introduce new products to the market, some of which they patented and some of which gained reputations as useful remedies.[4] More important, German manufacturers and, to a lesser extent, manufacturers in other European countries introduced a large number of new products in the late nineteenth and early twentieth centuries, virtually all of which were both patented and trademarked. By 1912 about five thousand synthetic chemicals had been found to have therapeutic properties by the German chemical industry alone, and while only a fraction of these reached the market, those that did represented a large number of new products.[5]

Many of these drugs were rapidly incorporated into widespread medical use. Beginning in 1906 the American Medical Association's Council on Pharmacy and Chemistry issued regular reports on new reme-dies that it considered to have been adequately tested from a scientific perspective and useful enough to be adopted by physicians. Beginning in 1907, the council issued an annual list of all the newly accepted prod-ucts under the title *New and Nonofficial Remedies* (*NNR*). The 1911 edition, which covered new remedies introduced through 1910, listed about 375 products. Close to half of these were made by foreign manufacturers, all but sixteen of them made by a handful of German companies. Perhaps more strikingly, just over 20 percent of all the drugs listed were protected by patents at the time, and another 5 percent had been protected in the past by patents that had since expired. Bayer, for example, had twenty-five drugs listed that were currently protected by patents.[6] Of course, the products listed in *NNR* represented only a small portion of the total number of drugs on the American market, but the text still illustrates the fact that, by this point, patented remedies were an inescapable part of medical practice. "However much we may prefer to use unpatented drugs," noted one observer at the time, "the fact remains that practically all the modern, synthetic drugs are so patented, and among this number are many remedies in constant, daily use, which could not be dispensed with without the serious crippling of our therapeutic resources."[7]

The growing reliance of physicians on patented drugs meant that complex technical issues in patent law increasingly shaped the thera-peutic market, linking the scientific and business strategies of firms that monopolized their products to the daily practices of both pharmacists and physicians. A good example of this dynamic is the popularization of acetylsalicylic acid, introduced by Bayer in 1900 under the trademarked name Aspirin. Chemists had been trying to derive acetylsalicylic acid since the early 1850s, and in 1869 the German chemist Karl Johann Kraut had produced a relatively pure form of the chemical. Almost thirty years later, in 1897 a young chemist working at Bayer named Felix Hoffman synthesized a pure form of the drug. Bayer was unable to secure a patent for the chemical in Germany, but the company did secure patents in both Great Britain and in the United States. It also secured a US trademark on the name "Aspirin" in 1899.[8] The drug rapidly became an extremely popular treatment for rheumatic disorders, fever, and then for pain more generally. As it had done with Phenacetin, Bayer charged a significantly higher price for Aspirin in the United States than it did for the chemical in other countries, presumably because its patent allowed the company to enforce a vending monopoly. Many druggists believed that the price differential was unethical and purchased versions of the drug manufactured in other countries. Bayer responded by threatening importers and druggists with legal action, just as it worked to suppress the trade in smuggled Phenacetin before the patent on that drug expired in 1906.[9] The stakes in the drama were high: by 1910 Aspirin accounted for about a quarter of Bayer's total sales in the United States.[10]

Bayer was quite aggressive in its efforts to suppress Aspirin smuggling, and although the company preferred to settle out of court, it was not afraid to sue druggists that it felt had violated its patent rights. There were numerous complex legal issues involved in these cases, but for my purposes the most important ones related to the question of purification. By this point it had been well established that purification itself was not enough to justify patentability.[11] However, over the last two decades courts had also sometimes ruled that new uses for familiar things could be patented.[12] The issue was especially pertinent in the chemical and dye industries, where utility was sometimes the only way to effectively distinguish between products. In 1900, for example, the Second Circuit Court of Appeals upheld an infringement claim after finding that the two dyes in question were equivalent based on the fact that they both had increased utility over previous products.[13] Novelty, in other words, was sometimes understood to be a function of increased or transformed utility; as the Supreme Court noted in 1908, in the creation of a new manufacture "a new and different article must emerge, 'having a distinctive name, character or use.'"[14] As a result, if purification led to significant new utility, then the resulting product might be considered novel enough to be patentable.

This doctrine played an important role in how the Aspirin situation was resolved. Sometime around 1901 a Chicago wholesaler named Edward Kuehmsted began purchasing Aspirin in bulk, either from a London importer or from a supplier in Canada, and selling it to peddlers in the Chicago area, who then resold it to retailers at a price significantly below what Bayer charged. Bayer then sued Kuehmsted for violating its patent rights. The company was initially confident that it would have little trouble winning the suit, but in 1905 a British court ruled that the British patent on the drug was invalid. The judge in the case found that in Hoffman's process, purification followed the chemical synthesis of acetylsalicylic acid itself and that as a result it did not lead to a new substance that was distinct from Kraut's earlier product.[15] The ruling meant that Bayer lost control of the British market, and following the decision the company grew concerned about the potential for a similar verdict in its case against Kuehmsted.[16] Bayer responded by delaying the case as long as possible, both to forestall a negative verdict and to avoid abandoning the case altogether and thereby giving free reign to smugglers. The tactic did not really help. Some companies misunderstood the situation and began to manufacture or import the drug under the assumption that the American patent on the drug was no longer valid. Others ignored the patent or threatened to have it voided, while some such as Parke-Davis simply waited for the resolution of the American case. Meanwhile, retailers continued to ignore the patent, and smuggling continued.[17]

To the surprise of most observers, in 1910 the patent on Aspirin was upheld by the Seventh Circuit Court of Appeals.[18] The court ruled that although acetylsalicylic acid had in fact been previously produced in an impure state, the question at hand was not the relationship of purification to the chemical substance, as the British court had decided, but instead the question of whether or not the purified substance was sufficiently different from the unpurified substance to render the two products distinguishable from each other. The court determined that it was:

Kraut and his contemporaries, on the other hand, had produced only, at best, a chemical compound in an impure state. And it makes no difference, so far as patentability is concerned, that the medicine thus produced is lifted out of a mass that contained, chemically, the compound; for, though the difference between Hoffmann and Kraut be one of purification only—strictly marking the line, however, where the one is therapeutically available and the others were therapeutically unavailable—patentability would follow. In the one case the mass is made to yield something to the useful arts; in the other case what is yielded is chiefly interesting as a fact in chemical learning.

According to the court, the key difference was that acetylsalicylic acid, in its purified form, was something "therapeutically different" from antecedent substances; this difference meant that Hoffman's product was new and worthy of patent protection.[19] As a later judge noted, "A pure compound may, under certain conditions, be patentable over the same compound in an impure form."[20]

Following the decision, Bayer settled the numerous suits against druggists that they had accused of smuggling for nominal fees. The company also began to work closely with state and municipal authorities to prosecute recalcitrant druggists who continued to smuggle the product, with at least some druggists spending time in jail as a result.[21] The combination appears to have convinced organized pharmacy to abandon its support for smuggling. Following the decision, leaders in druggist organizations began to encourage their colleagues to follow the law and to refrain from importing Aspirin and other products against the wishes of their manufacturers. This was a complicated process, of course, but leaders in organized pharmacy appear to have concluded that resisting the price differentials between the US and foreign markets through smuggling was no longer a justifiable position. Although smuggling remained an important dynamic in shaping the market in German drugs—and in the drug market more generally—the collapse of support among pharmacists for smuggling Aspirin marks an important turning point. No longer would retail druggists throw their weight behind the use of smuggled products. Following the initial ruling upholding the Aspirin patent, for example, the National Association of Retail Druggists noted that the organization had for years opposed product patents but it did not "endorse, defend or countenance in any manner royalty theft, smuggling and kindred offenses."[22]

The medical community reacted to the decision in a similar way. Observers in the medical press had sometimes criticized the patent on Aspirin in the decade following its introduction, generally in ways that focused on the role of the patent in Bayer's ability to charge what seemed to be an unfairly high price for the drug.[23] Notably, however, few critics attacked the patent on the drug as unethical in and of itself. This should not be surprising: even as therapeutic reformers continued to rail against both adulterated drugs and products made with secret ingredients, by this point even the most conservative physicians had come to accept the basic legitimacy of patented goods—after all, some of the most important drugs in use at the time were patented, and synthetic drugs made by German manufacturers were widely recognized as staples of good medical care. Indeed, a growing number of physicians believed that pharmaceutical patenting was not just ethically tolerable but in fact a positive good because of its supposed role in promoting corporate investment in the drug development process.[24] At the same time, however, the way in which the German industry apparently used patents to artificially inflate the price of its goods continued to seem troubling to many observers, even if it was increasingly assumed that physicians and pharmacists should respect their patent rights. From this perspective, the Aspirin decision was regrettable, but it did not point to the illegitimacy of medical patenting per se. "It is our duty to respect the decrees of our courts," noted an editorial in the *Journal of the American Medical Association* about the decision, "even though in this case the decision is unfortunate and worse an injustice."[25]

With the coming of World War I, the assumption that German patents should be respected came under increasing strain as a result of the sudden shortage of imported drugs and a resulting increase in price. Much of this concern centered on the new antisyphilitic drug Salvarsan. Beginning around 1906 Paul Ehrlich and his team of researchers in Berlin began investigating arsenic compounds hoping to discover a chemical treatment for sleeping sickness. In 1909 a researcher in his laboratory named Sahachiro Hata discovered that one of the compounds that the group was studying—sometimes called "606" because it was the sixth substance in the sixth group of compounds that the team tested—was effective against the bacterium that causes syphilis. This was a tremendously important discovery because it was the first time a specific chemical compound could be used to target a specific microbial agent.[26] It was also important because of the way in which the discovery took place: Ehrlich had developed a method of systematically testing a large number of chemical substances, and as a result much of the actual labor in the laboratory was conducted by his subordinates. The discovery thus pointed toward the future of drug research, a future in which powerful pharmaceuticals are developed through a research process in which laboratory science resembles factory production, routine tasks are carried out by workers not in control of their own labor, and both credit and profit primarily accrue to the head of the organization.[27]

Following the discovery, Ehrlich acquired a US patent on the compound and assigned it to Hoechst AG, which marketed the drug under the trademarked name Salvarsan.[28] Salvarsan—and the closely related Neosalvarsan, which was introduced by Ehrlich not long after—quickly became the standard treatment for syphilis. The patent on the drug attracted a modest amount of criticism from the medical community, in part because of the idea that Ehrlich did not deserve a patent because the drug had been developed through a routine procedure that required little inventive activity. Once war broke out in 1914, concerns about the patent began to increase. The naval blockade instituted by the Allied powers against Germany cut off supplies of the drug, and as prices jumped sharply upward, critics began to argue more strongly against Hoechst's monopoly on the drug. Increasingly, it seemed deeply unfair that such an important drug was unavailable to those who needed it because of the patent rights of a foreign manufacturer. By 1915, medical associations were beginning to petition the federal government to address the situation, pointing to the "shortage of salvarsan and other drugs manufactured in Germany under patent rights, greatly to the detriment of the sick and suffering in the United States."[29] Some solution would need to be found.

The 1905 Trademark Act and the Genus/Species Distinction

"When a man gets a trademark he wants to put it all over everything everywhere," noted a editorial in *Printer's Ink* in April 1903. "If you could figure out a scheme for putting the trade mark of Syrup of Jigs on the moon . . . your everlasting fortune would be made." Such sentiments were not uncommon in the first decade of the new century. Although trademarks had not yet assumed the central place in marketing that they hold today, the creation of distinctive brand identities for consumer goods was already an important part of the promotional strategies of many manufacturers. Trademarks were increasingly important to the ability of firms to establish economies of scale and scope because they helped ensure repetition of purchasing by linking of the reputation of the company to an easily identifiable name and image. As the editorial in *Printer's Ink* noted, "This overwhelming fondness for the name and the trade mark has to be taken into consideration in advertising. . . . The name, or the brand, or the trademark must stand in the minds of the public as representative of the quality and merits of the goods."[30]

It thus came as something of a shock when, six months after the editorial quoted above was printed, the Supreme Court invalidated the practice of registering marks under the 1881 law through nominal international trade.[31] The decision threw manufacturers into chaos because it effectively nullified the utility of federal registration for domestic commerce; as one manufacturer put it, the decision "virtually destroys the [1881] law so far as we are concerned."[32] In response, in 1905 Congress passed a revised trademark law that extended registrability to goods that were shipped across state lines.[33] Registration under the new law ensured access to federal courts and a presumption of ownership that could be rebutted in case of disputes. It did not, however, establish either a legal right to the mark or the validity of the mark itself, and as a result manufacturers continued to face a significant amount of risk as they invested resources into popularizing their brands. Still, registration under the 1905 law and the use of federal courts to resolve disputes was generally considered preferable to relying on state courts because of the possibility of conflicting rulings, differences in state law, and other issues.

In many ways, the primary significance of the 1905 law was simply that it placed federal trademark registration on solid statutory ground. However, the law also significantly contributed to the construction of the *generic* in the way that we understand it today. The law prohibited the appropriation of descriptive words as trademarks, whether personal names, geographical names, or words that are descriptive of the character or quality of goods. It also excluded both flags and words that that are "immoral or scandalous" from registration. For all other marks, the law established that trademark infringement could only occur in relation to goods that were in the same "class" of product. A trademark on watches, for example, could be registered by a manufacturer even if it was very similar to a mark that had already been registered by another manufacturer who produced furniture, since watches and furniture were different classes of goods. The law also declared that registration might be declined if the mark so nearly resembled an already registered mark in the same class of goods as to likely cause confusion in the mind of the public. In other words, the law divided the universe of goods into different classes of products and limited infringement to marks that applied to products within the same class.[34]

This classification system was soon linked to the common law doctrine that descriptive names cannot be monopolized if there are no patent rights involved. In the years since the *Singer* decision, it had become a central assumption of trademark jurisprudence that, as the Supreme Court put it in 1901, "when the right to manufacture became public, the right to use the only word descriptive of the article manufactured became public also."[35] Following the passage of the 1905 law, a series of court decisions linked this doctrine to the idea that the universe of goods is divided into multiple classes and that trademark infringement occurs between similar marks that are affixed to goods in the same class but made by different manufacturers. Courts increasingly used the language of *genus* and *species* to refer to this classification system, and descriptive names that operated at the level of product class were increasingly referred to as "generic" in nature because they operated at the level of the "genus" of goods rather than a particular "species" of a product made by a specific manufacturer. According to these decisions, descriptive names that operated at the level of the "class" or "genus" of goods could only be monopolized as long as the product was under patent; once the patent expired and the product reverted to the public domain, these *generic* names became available to all. Names that operated at the level of the *species*—or manufacturer brand—could, however, be trademarked independent of whether patent rights were involved because they were not descriptive in nature.[36]

The categorization of goods into classes was also intertwined with the doctrine of what was sometimes called "secondary" meaning. Designating marks that operated at the level of the brand—or the "species" of good—could be monopolized even if the product was not under patent. These types of marks were sometimes referred to as "technical" trademarks in the years immediately following the passage of the 1905 law, although the term did not become popular in legal discourse until after World War I.[37] Since trademarks could not be acquired for descriptive words, if these marks began to operate at the level of product class or genus, then they reverted to the common domain if the product in question was not protected by patent. Technical trademarks, in other words, could not be descriptive in nature, a fact that conformed to the long history of trademark jurisprudence. At the same time, however, words that were descriptive in nature *could* be protected under the broader law of unfair competition if they had acquired a "secondary" meaning and therefore operated at the level of the species. Personal names, for example, might deserve court protection if used in brand names, even if they could not be adopted as technical trademarks. As the Supreme Court noted in 1911, "It is true that the manufacturer of particular goods is entitled to protection of the reputation they have acquired against unfair dealing, whether there be a technical trade-mark or not, [and] the essence of such a wrong consists in the sale of the goods of one manufacturer or vendor for those of another."[38]

The 1905 law thus interacted with both common law and the behavior of manufacturers in complex ways. Some manufacturers, for example, used the 1905 law to bar competitors from registering marks that were similar to their own.[39] Others found that their ability to appropriate names in familiar ways was curtailed. To take just one example, as I have noted in preceding chapters, patent medicine manufacturers had long referred to their products by highly designating names that combined the personal name of the manufacturer and the name of the object in some way, and competitors had, in the past, been able to freely adopt these names as long as they indicated the true origin of their goods under the assumption that trademarks were essentially transparent in nature. This doctrine had gradually eroded as trademarks began to acquire the characteristics of property, but it fully collapsed under the assumption that technical trademarks, which operated at the level of the species, were distinct from generic names, which operated at the level of product class. In 1911, for example, the Supreme Court ruled that Thomas Beecham had the right to monopolize the name "Beecham's Pills" for his product, even if a competitor used the same formula for his own product and even if the competitor indicated the true origins of the product on his label. The court ruled that the name was not "generic" in nature and was instead "the highest degree individual and means the producer as much as the product"—in other words, the name operated at the level of product species or brand, rather than the level of product class. As a result, "one calling his product by the same name is guilty of unfair trade even if he states that he, and not Beecham, makes them."[40]

The division of the world of goods into classes, and into species within those classes, also intersected with the branding strategies of manufacturers. During the first decade of the twentieth century ethical manufacturers in the domestic pharmaceutical industry continued to develop their brand identity in order to distinguish themselves from their competitors. Developing and maintaining a reputation for high ethical standards, for manufacturing quality goods, and for promoting the cause of medical science were all important parts of this strategy, and as companies invested resources in their promotional efforts, they linked these qualities to their company names, abbreviations of their names, and images that they used to represent them. In 1906, for example, one advertising page in the *Pharmaceutical Review* included advertisements from multiple firms instructing pharmacists to purchase according to brand. "Specify Merck's Codeine Sulphate," noted one of the advertisements. "Stearn's Antitoxin pays you well," noted another. Mallinckrodt Chemical Works, meanwhile, advertised morphine under its "M.C.W." logo, an abbreviated and easy-to-remember form of the company name. "'M.C.W.' Morphine. Standard in Quality, handsome in appearance, and always as low in price as other makes. We invite specification for our brand."[41]

Such efforts combined descriptive names that operated at the level of the genus (such as "codeine sulphate" and "morphine") and designating names that pointed to particular species—or brands—of goods ("Merck's" and "M.C.W."). Ethical manufacturers also, however, increasingly sold products under short and memorable arbitrary names. There were significant legal and economic forces pushing them to do so: short and easy-to-remember names could be effectively popularized within the medical community, and the ability to acquire trademarks on these names meant that they could be used to promote the interests of the firm by linking them to the reputation of the company as a whole through advertising and other promotional strategies. In the past, these types of names had been assumed to refer to all instantiations of the product in question, and manufacturers had sometimes been able to use them to monopolize the sale of the underlying goods even when no patent rights were involved. Increasingly, however, these types of names were understood as brand names that operated at the level of the species (i.e., as names that referred to specific lines of a product made by specific manufacturers) and as a result were increasingly understood as distinct from their scientific names, which were frequently too long and complex for commercial or therapeutic use. As a result, the ability of these types of names to monopolize the sale of the underlying product receded. Competitors were increasingly able to manufacture what was assumed to be the same substance under different names that operated at the level of the product.

The commercial naming of *hexamethylenamine-tetramine* is a good example. The chemical had first been synthesized in 1860 and was made official in the eighth revision (1905) of the *USP* under the name "hexamethylenamina" (with "hexamethlyenamine" listed as a synonym).[42] Some manufacturers sold it under its official name, but others sold the chemical under a variety of names that operated at the level of the product, including Urotropin, Uritone, Uristamine, Formin, Hexamine, Cystamine, Cystogen, and Aminoform. The official name was, of course, quite "cumbersome," as one observer put it—and the scientific name was even worse—and these types of "coined names" allowed manufacturers to popularize their versions of the chemical through advertising and other promotional strategies that associated the reputations of their companies with their own particular product names.[43] One result was that different manufacturers were able to charge different prices for what many observers assumed were essentially the same product. In 1906, for example, the chemical cost $1.75 a pound when sold under its official name. Sold under a trademarked name, however, it cost substantially more: Parke-Davis charged $12 per pound for Uritone, while Schering & Glatz charged $7.50 per pound for Urotropin.[44]

Branding thus began to take place at two levels: the level of the company, in which brand names could be applied to multiple products made by the same manufacturer, and the level of the product itself as arbitrary names that operated at the level of product species began to adopt the characteristics that we now associate with the brand names of goods—

these names pointed to a specific line of products made by a specific manufacturer, typically without incorporating any actual information about the manufacturer in the name itself, and as a result they both drew on and contributed to the reputation of the company with which they were associated. They were also juxtaposed to names that could be used to refer to all instantiations of the product in question independently of its origin, whether complex scientific names or some type of other common name (such as names used in the *USP*). The product manufactured by Parke-Davis, for example, could properly be called by one of a variety of names that operated at the level of genus, including the scientific name *hexamethylenamine-tetramine* or the *USP* name "hexamethlyenamina" (or its variation). It could also be referred to by the name that operated at the level of the product species and referred to its own specific brand of the drug, "Uritone." Other manufacturers, however, could not use "Uritone" to refer to their own products, although they could of course use the names that operated at the level of product genus.

The use of arbitrary coined names that operated at the level of product species allowed firms to invest resources into the development of brand names that were linked to specific products. However, there was also a significant danger here for manufacturers. By the first decade of the new century the implications of the *Singer* decision were beginning to be clear to many observers in the pharmaceutical industry. As a result, competing manufacturers began to consciously adopt the trademarked names of popular products as those products went off patent, sometimes even under threat of legal action by the original manufacturer. They did so even when other, nontrademarked names for the product were available, such as an official name used in the *Pharmacopeia* or a suitably short scientific name used by the scientific community. In order to distinguish their own versions of these products from those made by other manufacturers, they also sometimes attached the names of their companies to these newly public names in some way. This dual strategy recognized the fact that the formerly monopolized names of popular drugs continued to carry an immense amount of value because of their familiarity and the fact that physicians frequently, and sometimes exclusively, prescribed according to these names. At the same time, it also recognized the fact that their own brand identity could serve as a site of investment and the accumulation of value in their efforts to develop markets.

The most important example of this process was Bayer's loss of control over the name "Phenacetin." The patent on acetphenetidin expired in early 1906, much to the relief of druggists across the country.[45] Other manufacturers and distributors quickly entered the market, including Lehn & Fink (founded in 1874) and Monsanto Chemical Works (founded in 1901). Lehn & Fink began selling the product under the name "phenacetin" almost immediately after it went off patent, although the company also clearly identified itself as the manufacturer of its product by placing its own name immediately adjacent to the term in their advertisements. They did the same thing with three other products that had been introduced by Bayer and had recently gone off patent—Sulphonal, Trional, and Aristol—in each case selling the product as "Phenacetin Lehn & Fink," "Sulphonal Lehn & Fink," and so on.[46] Bayer soon brought a lawsuit against the firm, claiming that it retained a common law right to these names despite the fact that the patents on the products had expired.[47] Despite the legal threat, other companies followed Lehn & Fink's lead. In 1909, for example, Monsanto announced that it would begin selling "phenacetin." Bayer also threatened Monsanto with legal action, noting that "the product has names enough of its own without adopting the one which indicates our manufacture."[48] It is not clear whether Bayer ever brought a case against Monsanto, but in 1913 the company finally dropped its suit against Lehn & Fink, supposedly for procedural reasons. More likely, the company probably recognized that it had lost control of the names in question and that it would not prevail in court.[49]

The battle over the name "phenacetin" took place in the context of the drug's rapidly declining price. Following Bayer's loss of the patent, the price of phenacetin collapsed, falling from a dollar an ounce to just thirty-three cents an ounce by the middle of 1907. Bayer and Monsanto then became involved in a brutal price war, and by 1913 Monsanto was selling the drug under the name "phenacetin" for less than eight cents an ounce in bulk.[50] Bayer continued to sell the product under the same name for thirty-three cents an ounce, with its ability to charge a higher price than its competitors resulting from the widespread assumption among pharmacists that when confronted with a prescription for Phenacetin, they were obligated to dispense the Bayer product.[51] At the same time, however, Bayer *also* sold the drug under the name "acetphenetidin" at prices designed to match Monsanto's price, a fact that critics found both confusing and infuriating. Some pointed with anger to the fact that the same drug could be sold under two different names but at widely different prices—Bayer sold "acetphenetidin" for just thirteen cents an ounce in 1912—while others wondered if the two were in fact the same drug.[52]

By the outbreak of World War I the names of pharmaceuticals were divided into those that operated at the level of product *genus* and those that operated at the level of product *species*. Manufacturers sometimes used hybrid combinations of these two types of names in their promotional strategies (such as "M.C.W. Morphine" and "Phenacetin Lehn & Fink"). Increasingly, however, they coined arbitrary names for their products that operated at the level of product species and thus were designating in an important sense without specifically including information about the manufacturer in question. As long as these names were sufficiently arbitrary in nature, they could be adopted as technical trademarks. As the example of P/phenacetin indicates, however, these types of names could also begin to act as the generic name of the product in question and as a result revert to the common domain after the expiration of the underlying patent. Investing significant resources in these names thus carried a substantial risk to manufacturers. Indeed, Bayer appears to have learned this lesson as a result of its debacle with the Phenacetin trademark and, facing the looming expiration of the patent on acetylsalicylic acid, began to work feverishly to associate the name "Aspirin" with the company in an effort to prevent it from moving into common usage. Bayer was the first company to clearly recognize the threat of genericide, but it would not be the last.

Antitrust Law and the Origins of Fair Trade

Despite the fact that patents are, by their very nature, a form of monopoly, in the two decades following the passage of the 1890 Sherman Act, patent law was assumed to trump antitrust law. This doctrine was overturned by the Supreme Court in the groundbreaking 1912 case *Standard Sanitary Manufacturing Co. v. United States*, in which the court ruled that a patent pool that involved fifty manufacturers of bathtubs was an illegal restraint on trade because it was used to fix prices and thereby restrict competition beyond what was reasonable. Between 1890 and 1912, however, patent pools and other economic arrangements based on patenting were considered exempt from antitrust law. As a result, monopolistic practices both in the domestic chemical industry and among foreign drug manufacturers that were based on patent pools, cross-licensing schemes, and other patent-based techniques were not subject to antitrust enforcement, even if they led to significant market consolidation. In the first decade of the century, for example, the Dow Chemical Company gained control over virtually the entire output of the bromine industry, in part through the use of licenses on a patented technology that increased its ability to manufacture bromine products efficiently and in part by establishing cartel arrangements with German manufacturers to keep them out of the domestic market.[53] Foreign drug manufacturers also sometimes used cartel arrangements to escape the cost of patent litigation or costly price wars, although the extent to which this took place is difficult to determine. In 1904, for example, Bayer acquired a US patent for a process of preparing diethyl barbituric acid. The following year, the German firm E. Merck acquired a patent that covered both a different process of making the chemical and the substance itself. Rather than fight the issue out in court, the two companies established a cartel on the product and split profits on the drug, which was sold under the trademarked name Veronal. After some legal wrangling, Schering was also included in the cartel, which controlled the manufacture of Veronal through the outbreak of World War I.[54]

Although patent pools and similar arrangements were exempt from antitrust enforcement before 1912, price fixing at the retail level was not. Even as druggists continued to press for a law that would abolish product patents, the National Association of Retail Druggists (NARD), the National Wholesale Druggists Association (NWDA), and other druggist groups also continued to establish and enforce plans intended to stabilize prices.[55] However, during the first decade of the century the courts became increasingly skeptical of these efforts and began to interpret them as violating the 1890 law. At the same time, federal officials began to target druggist organizations that engaged in such schemes for antitrust enforcement. In 1906, for example, a federal court in Philadelphia ruled that the tripartite plan was an illegal trust, noting that fixing the minimum retail price of drugs and then restricting their sale to retailers who conform to this "arbitrary standard" was a "clear restraint of interstate commerce . . . and is in violation of the [1890] act."[56] In the same year, William Henry Moody, the attorney general, filed suit against the NARD, the Proprietary Association of America, and the NWDA in the Circuit Court of the District of Indiana, asserting that the tripartite plan violated the Sherman Act.[57] Under pressure from the federal government, and facing a shifting legal environment, in 1907 the NARD abolished the tripartite plan.

During the first decade of the century, the presumed priority of patent law over antitrust law provided some ability for manufacturers to legally enforce price maintenance schemes. The Dr. Miles Medical Company was the most important manufacturer in this regard. The company was tremendously successful—by 1906 it spent about $50,000 a year on advertising—and by 1911 the company had allegedly entered into resale price maintenance contracts with more than twenty-five thousand retail druggists across the country.[58] The company also instituted a series of court actions against both wholesalers and retailers for violating the terms of its contracts.[59] In the process, it drew on the 1902 *National Harrow* decision to argue that it had a right to set minimum resale prices for its goods. According to this argument, the use of secret ingredients conferred a legal monopoly just as patents did and, with it, a corresponding right to establish minimum resale prices. Lower courts generally agreed with the logic; as one court noted in 1906, "The right of a patentee, owner of a copyright, or owner of a secret process is merely the right of exclusion or debarment. The holder of such a property right . . . is a czar in his own domain. He may sell or not, as he chooses. He may fix such prices as he pleases. He may sell at one price to one person, and another to another person. He is not required to give reasons or deal fairly with purchasers."[60]

In 1911 the Supreme Court rejected this argument. In the landmark case *Dr. Miles Medical Co. v. John D. Park & Sons*, the Court held that resale price maintenance was illegal for goods that are not patented. John D. Park & Sons was a wholesaler notorious for cutting prices. The Dr. Miles Medical Company accused the firm of purchasing its products from other dealers and selling them below the minimum resale price established by its contract agreement. The principal question at hand was the validity of the restrictive agreement because, as the court noted, "that these agreements restrain trade is obvious." In defense of its contract system, the Dr. Miles Medical Company asserted the same right to control its goods granted to patent holders under the *National Harrow Company* decision. The strategy had worked with lower courts, but here it was ineffective. The court ruled that the monopoly granted by patents had a statutory basis that had been conferred through their presumed benefit to the public; the use of secret ingredients, however, conferred no such benefit, and therefore the two could not be equated. Finding that "the complainant having sold its product at prices satisfactory to itself, the public is entitled to whatever advantage may be derived from competition in the subsequent traffic," the court ruled that "agreements or combinations between dealers, having for their sole purpose the destruction of competition and the fixing of prices, are injurious to the public interest and void."[61]

Dr. Miles Medical Co. v. John D. Park & Sons clearly established that resale price maintenance was illegal. It was also the first of a series of decisions that established what has come to be known as the "rule of reason." Shortly after the ruling, the court issued two additional landmark decisions—*Standard Oil Company of New Jersey v. United States* and *United States v. American Tobacco Company*—that affirmed the power of the federal government to dissolve large trusts that acted as monopolies in their industries. In these three cases, the court recognized that, "taken literally," the language of the Sherman Antitrust Act prohibiting "restraint of trade" could refer to a large number of practices, including contracts, that do not in fact injure the public. The court also concluded that the law had originally been passed under a common law interpretation of the phrase and that this interpretation held that only unreasonable restraints that "unduly" interfered with trade and resulted in public harm should be prohibited. As the court noted in the *Standard Oil* decision, "The Anti-Trust Act contemplated and required a standard of interpretation, and it was intended that the standard of reason which had been applied at the common law should be applied in determining whether particular acts were within its prohibitions."[62]

The Supreme Court also rejected the primacy of patent law over antitrust law. In 1912—ten years after the *National Harrow* decision—the court ruled that a patent pool in the ironware manufacturing industry was an illegal restraint of trade because it was used to fix prices and limit competition.[63] The following year, in *Bauer & Cie v. O'Donnell*, the court ruled that patents did not grant manufacturers the right to fix minimum resale prices on their goods. The case involved the German drug manufacturer Bauer & Cie and a patented product it sold under the name "Sanatogen."[64] In 1907 the company had begun attaching labels to its products that stated that Sanatogen could not be sold for less than one dollar and that doing so would infringe its patent rights. A druggist named James O'Donnell refused to comply, and in 1911 the company sued him. The question facing the Supreme Court was thus whether or not a patentee may limit the price at which future retail sales of the patented article can be made. The court ruled that although the intent of the patent law was to secure to the inventor an exclusive right to manufacture, use, or sell a product and that this right could be extended to his agents, "there is no grant of a privilege to keep up prices and prevent competition by notices restricting the price at which the article may be resold." The court thus found that the patentee's right to control the price of a good stopped once it passed beyond the domain of his agents. "The right to vend conferred by the patent law has been exercised," noted the court, "and the added restriction [of fixing resale price] is beyond the protection and purposes of the act."[65]

The Supreme Court's articulation of the rule of reason in 1911 and the decisions immediately following related to patent law were tremendously important in the political economy of the nation. The court both reasserted a common law interpretation of federal antitrust law and ended the primacy of patent law over antitrust law. In doing so, the court both conformed to and helped create a managerial vision of the relationship between the federal government and the market. The court negotiated a path between the possibilities of complete and unfettered competition on the one hand and statist domination of the market on the other, helping to institute a regime of managed competition in which both private parties and jurists would exercise significant influence on the shape of the market. At the same time, by establishing that patent pools can be unreasonable restraints of trade and ending the ability of manufacturers to enforce minimum resale prices for their goods based on patent law, the court strongly curtailed the ability of manufacturers to use patent law to shape the market. The Court thus asserted the primacy of a common law interpretation of reasoned competition over the authority of federal patent law. Indeed, in the coming years, *Bauer & Cie v. O'Donnell* would play a central role in the developing doctrine that although a patent grants the patentee the right to exclude others, it does not convey any per se right to sell or use his own invention; those rights are granted under common law, not the federal patent law.[66]

The coming of the new order was widely noted in the pharmaceutical press.[67] Still, retail druggists continued to see price fixing as the best possibility for stabilizing what they saw as a chaotic and deteriorating market. Following the *Dr. Miles Medical Company* decision, retail druggists worked feverishly with manufacturers to legalize resale price maintenance. In 1912 they helped organize the American Fair Trade League, and by 1914 the organization had managed to introduce a bill to Congress that, had it been made law, would have legalized the practice. Written by Louis Brandeis, in consultation with both the American Fair Trade League and the NARD, the so-called Stevens Bill was designed around the idea that the reputation of manufacturers was embodied in their trademarks and that they did not lose their right to their reputation simply because a product was transferred to another party. The bill was thus written to allow a manufacturer to "prescribe the sole, uniform price" for trademarked goods that were resold, provided that—among other conditions—the manufacturer affixed a notice on the product indicating its price. The goal was to prevent the deterioration of the market by stabilizing prices and, in doing so, to battle the influence of price-cutters, department stores, and other predatorial actors that distorted the market through monopolistic and unfair practices. Louis Brandeis, in his testimony in support of the bill, thus argued that the Stevens Bill grew out of the "widespread consideration of the trust problem" and that if enacted it would "further supplement the Sherman antitrust law." Resale price maintenance would help ensure the proper functioning of the market and prevent both the consolidation of economic power and the market distortions that came with it. As Brandeis put it, "Monopoly is the natural outcome of cutthroat competition."[68]

Brandeis and other reformers were not able to legalize what they called "fair trade" in the years before World War I. They were, however, successful in passing legislation establishing the Federal Trade Commission (FTC) to battle the problem of monopoly. Section 5 of the 1914 Federal Trade Commission Act declared "unfair" methods of competition to be illegal and empowered the newly formed commission to order the "discontinuance" of such methods. The FTC was also given the power to investigate business conditions and the extent of anticompetitive practices within different sectors of the economy and to disseminate its findings to government authorities and the public. Shortly after the FTC was established, Congress also passed the Clayton Antitrust Act. It defined a series of activities as anticompetitive in nature, including discriminatory pricing practices, and empowered the FTC to restrain their use. Taken together, the two laws established the FTC as an organization intended to counteract the dangers of monopoly by working to ensure a fair market. As the Federal Trade Commission Act noted, "The most certain way to stop monopoly at the threshold is to prevent unfair competition."[69] In the coming years, the FTC would play a central role structuring the market in therapeutic goods, both through its efforts to curtail what it considered unfair forms of competition and, with the entry of the country in the Great War, the seizure of German pharmaceutical patents and trademarks. In both cases, the FTC dramatically shaped the market in pharmaceuticals and the therapeutic possibilities available to a medical profession increasingly dependent on the products of massive corporate organizations.

Reputation and Profit: The American Medical Association and the Battle over Legitimacy

In 1905 Lewis McMurtry, the president of the American Medical Association, declared that the profession of medicine had entered a "new era." No longer wedded to the ways of the past, medicine was now a scientific endeavor in which "laboratory research and clinical investigation" had taken the place of "tradition and authoritative opinion." McMurtry pointed to the recent establishment of the AMA's Council on Pharmacy and Chemistry as an important step forward in this regard. Established earlier that year, the council was made up of a group of pharmacists and chemists who would examine proprietary remedies on the market and determine which were "honestly made and ethically advertised" and had therapeutic value that "deserve the approval of the medical profession." The council would then publish the results in the pages of the association's *Journal* and thereby suppress the use of nostrums by encouraging physicians to use only scientific remedies. Noting that "the use of proprietary medicines in the treatment of diseases has become one of the most confusing and demoralizing questions of the day," McMurtry claimed that distinguishing between "legitimate pharmaceutical preparations" and "fraudulent nostrums" was extremely difficult and that the efforts of the council were "the only practical way to deliver the profession from one of the greatest curses that ever came on it."[70]

The Council on Pharmacy and Chemistry was established by George H. Simmons, a physician who had joined the AMA in 1899 and become both the editor of the *Journal of the American Medical Association* and the general manager of the organization itself. However, the origins of the idea for such an organization lay with Francis Stewart and his efforts to reform the drug market along scientific lines. Stewart had argued for the need for an independent body of scientific experts to evaluate new drugs as early as 1881 as part of his efforts to overcome the medical community's skepticism toward commercially introduced products.[71] By the turn of the century, Stewart had come to the conclusion that the federal government was the proper place to house such an institution, and in 1901 he published an article in the *Journal of the American Medical Association* in which he proposed establishing a national bureau made up of experts from various fields that would investigate new drugs submitted to it by manufacturers and then publish its findings as working bulletins. Moreover, the bureau would actively recommend that physicians only prescribe remedies that had gone through this process and that manufacturers advertise the fact that their products had been evaluated. The intention behind the bureau was thus to simultaneously "collect knowledge of material medica products . . . and publish it for the benefit of science" and to "aid the manufacturers . . . in the introduction of their brands to commerce by advocating that the medical profession in prescribing shall specify only those brands which comply with scientific and professional requirements."[72]

Simmons was deeply impressed by the idea. Stewart's proposed bureau seemed a reasonable solution to the difficult question of how to suppress quackery and reform the therapeutic market along scientific lines. "I have just read your excellent article, and I must say that I am surprised and delighted at the way in which you have handled the subject," Simmons wrote to Stewart in 1901. "It seems to me also that you have solved the problem and all that is necessary now is for the scientific men who are working in the pharmaceutical field, and honorable physicians, to unite and ask for the creation of such a bureau."[73] For Simmons, such a bureau also promised to help editors like himself determine which products should be allowed to advertise in the medical press; this would in turn help suppress the trade in nostrums and elevate the practice of medicine. Beginning around 1901 Stewart tried to organize the bureau but without much success, and in 1903 he suggested that a joint committee of the AMA and the American Pharmaceutical Association be established to perform the functions of his proposed bureau. This committee only existed for a brief time before itself being abolished, at which point Simmons established the Council on Pharmacy and Chemistry under the sole auspices of the AMA.[74] The goal of the council, like Stewart's proposed national bureau, was to act as a mediating force between pharmaceutical manufacturers and physicians, rationalizing the commercial introduction of new drugs along scientific lines.

After the Council on Pharmacy and Chemistry was established, Simmons gave it a regular column in the pages of the *Journal* to describe drugs that merited the "patronage" of the profession.[75] The council began printing notices of new remedies that met its approval in the pages of the *Journal*, and in 1907 it also began annually issuing *New and Nonofficial Remedies*, which listed all accepted products.[76] Any manufacturer that sought recognition for a product by the council was required to submit it for examination, along with the product's formula and information regarding tests that could be used to identify its identity, purity, and strength. No product would be accepted—with a few exceptions, such as mineral water—that was advertised to the public, that was labeled or advertised with the names of diseases, or whose manufacturer made "unwarranted, exaggerated, or misleading statements" about the product in its advertising.[77] Notably, the council allowed manufacturers to submit drugs for consideration that were protected by both trademarks and patents— including product patents. The Council on Pharmacy and Chemistry was intended to distinguish between products developed along scientific lines and quackish nostrums that gained market share through deceptive advertising. The rules the council used to determine whether or not a product would be accepted thus focused on the validation and promulgation of what the council considered scientific facts; products that were advertised with unwarranted therapeutic claims, for example, were prohibited because such claims ran contrary to the facts about the product that had been established through the scientific process.[78] Whether or not a product was patented meant little from this perspective. Once the basic ethical legitimacy of using remedies protected by patents had been accepted, the patent status of the product had little to do with the question of whether or not the drug had been manufactured along scientific lines.

Trademarks were a more complicated matter. Therapeutic reformers in the first decade of the twentieth century continued to have mixed feelings about trademarked names. "Catchy names," as one critical editorial called them, often sounded very similar to one another, and the rapidly growing number of trademarked remedies on the market made it increasingly difficult for physicians to keep the names of various drugs straight: by 1908, more than eight thousand trademarks had been issued for medicinal compounds, more than two thousand of which had been issued in the past eight years alone.[79] What's more, some of these names evoked the drugs' curative properties or problems they could be used for—a practice reformers considered an unethical form of therapeutic advertising— and the tendency to use trademarked names in medical and scientific literature seemed to distort scientific communication toward commercial ends. And, of course, to many critics it seemed unfair that products sold under trademarked names often cost significantly more than what were presumably the same substances sold under official or other common names, especially when the culprit was a German manufacturer.[80]

Yet there were also benefits to trademarked names. Long and complex chemical names were unwieldy at best, and many products simply did not have adequately short and usable nonproprietary names, especially if they had not yet been included in the *USP*. Many physicians also preferred to use trademarked names in their prescriptions because they considered some manufacturers more trustworthy than others, and if a physician prescribed a drug using a nonproprietary name, then a pharmacist could ethically dispense *any* manufacturer's version of that substance, including versions that the physician might consider inferior or untrustworthy. Prescribing according to trademarked name was thus a means for physicians to assert a degree of control over what product their patients actually received. A growing number of observers also recognized that trademarks played an important role in promoting the fortunes of a firm, and some argued that manufacturers had what Francis Stewart called a "natural right" to protect their reputations.[81] This argument focused on trademarks that were taken out at the level of the company and covered multiple goods made by the same firm; from this perspective, trademarks protected the legitimate interests of the manufacturer while leaving the names of products themselves free for the use of the medical and scientific communities. Stewart made this argument repeatedly in the first decade of the twentieth century, as he long had, but a growing number of other physicians did so as well. As one physician noted in 1907, "In this way we give the manufacturer his legitimate protection; we concede and make valuable his property right to his brand. We also protect the medical profession and science in their right to what is the general property—the descriptive name of the product."[82] Of course, this position failed to recognize the fact that the arbitrary names that manufacturers frequently used for their goods—names that operated at the level of product *species* rather than product *genus*—increasingly acted as a type of brand themselves, in that they pointed to a particular line of products made by specific manufacturers and thereby contributed to the overall reputation of the firm in question. Indeed, the arguments made by physicians such as Stewart contributed to this transformation, since they were part of the process through which trademarks on the names of products lost their power to monopolize the sale of those goods by operating as the common name of things themselves.

The position of the Council on Pharmacy and Chemistry on trademarked names reflected this complex situation. In the first years of its existence, the council accepted products under their trademark names, followed by whatever scientific name they might have been given. Hoechst, for example, patented the chemical *1-para-aminobenzoyl-2-diethyl-aminoethane hydrochloride* in 1906, which it sold under the trademarked name "Novocain."[83] The drug was accepted under that name in the first edition of the *NNR*—after all, the scientific name was clearly too long for therapeutic use, and there was no other name by which it could be called. However, the council also believed that such names could undermine scientific medicine, and as a result it adopted a number of rules to guide the naming of submitted products. In general, the council discouraged submitting products under "more or less arbitrarily selected or 'coined'" names because, as they put it, such names were "intimately associated" with many of the "abuses connected with proprietary medicines." Coined names, whether protected or not, were therefore not allowed for nonproprietary goods that already had well-established names (such as an official name in the *USP*). However, the council did recognize the "right" of manufacturers to give names to new synthetic chemicals and other inventions. The council preferred that such products be given truly scientific names rather than overtly commercial ones and that these names not be trademarked; as the council noted, "The protection of the manufacturer can be amply secured by appending the firm or 'brand' name to the official name, and to this there can be no objection." Still, the council allowed manufacturers to submit new drugs under trademarked names that operated at the level of the product as long as these names did not violate its other rules. The council also reserved the right to coin what it called "generic" names for new products if it decided that the names the manufacturers had chosen were unsatisfactory.[84] The council appears to have used this power only sparingly in the years before World War I, but as I describe below, it did so in at least one high-profile case when it began to use the term "epinephrin" to refer to Adrenalin and other suprarenal products then on the market.

The power of the council was initially conceptualized as a form of moral exhortation: physicians would be encouraged to adopt recommended remedies in lieu of other products, and this would encourage the reform of the market. Soon after the council was established, however, the delegates to the 1905 annual convention of the AMA adopted a resolution that requested "the purification of the *Journal's* advertising pages" by excluding "nostrums." The Board of Trustees then asked Simmons to refuse advertising space to those that did not pass muster.[85] The AMA also adopted this policy for the other medical journals it published, and over the course of the next decade a significant number of independent journals followed suit and rejected products for advertising based on the judgments of the council.[86] The result was that the failure to gain council approval meant that a product could not be advertised in a large portion of the medical press, significantly impacting the ability of the manufacturer to develop a market for the product. Moreover, manufacturers that refused to submit their products to the council opened themselves to suspicion and criticism for their refusal. "No honest firm will hesitate for a moment to have their products examined by the Council," noted the *Journal of the Indiana Medical Association* in 1908. "The truth of the matter is that a firm taking any such stand fears the results of an examination of their products."[87] For manufacturers that hoped to advertise to the orthodox medical community, the question of whether they could gain the council's approval rapidly became an important consideration in their efforts to introduce new drugs.[88]

This gave the Council on Pharmacy and Chemistry a significant amount of power to shape the therapeutic market. A conflict between Simmons and the Abbott Alkaloidal Company illustrates the point. The company had been founded in 1888 by a physician named Wallace C. Abbott, who had begun producing pills from what he claimed were the alkaloids of various plants, and within a decade the company was manufacturing more than seven hundred items and had opened branches in New York, San Francisco, and several other cities. Following the establishment of the council, Abbott began to submit his products for evaluation and by 1908 had gained acceptance for at least one of his products. Abbott published widely in the medical press—very widely—and heavily promoted his goods.[89] These promotional efforts raised difficult questions about the relationship between science and advertising, and some critics decried them as little more than quackery. In 1906, for example, Abbott introduced a preparation made from hyoscin and morphine for use in childbirth and published numerous articles in the medical press about the supposed wonders of the preparation, suggesting that "nothing like it has ever appeared" and that it was "extinguishing the fear of child-birth."[90] Reports about the product causing infant mortality soon appeared, however, and Abbott's articles began to attract criticism as a type of "free advertising" that corrupted medical science.[91] As one editorial published in the *Journal of the American Medical Association* put it, Abbot's efforts presented "an interesting example of the subordination of science to commercialism."[92]

In 1908 the council rejected a number of Abbott's products for the "wildness and unreliability" of the claims made about them. Simmons then wrote a highly critical article that accused Abbott of "flood[ing] the reading pages of medical journals with so-called original articles which are but thinly veiled advertisements" for his products. According to Simmons, Abbott and his "literary fecundity" were corrupting the practice of scientific medicine toward commercial ends by convincing physicians to purchase nostrums that had been unable to gain the approval of the council.[93] The combined weight of the council's rejection of his products and Simmons's blistering attack on his reputation appears to have convinced Abbott to change his ways. By the outbreak of the war he had had changed his promotional efforts so that they no longer attracted criticism, and his company had more than fifteen products accepted by the council, including a digitalis preparation sold under the name "Digipoten," a product made from the "essential salts of the bile," and various serums, vaccines, and antitoxins.[94] Following the war, the Abbott Alkaloidal Company would go on to become one of the most successful ethical manufacturers in the country.

Simmons's efforts to suppress quackery by establishing the Council on Pharmacy and Chemistry is an important example of how therapeutic reformers in the early twentieth century sought to promote a rational therapeutics in which both laboratory and clinical science would serve as the basis for establishing and validating scientific knowledge about pharmaceuticals. The AMA worked to impose this therapeutic framework on an unruly market, using the power of the *Journal* to shape the overlapping scientific and promotional strategies of firms intent on introducing new products. The result was that, at times, the emerging framework—in which manufacturers developed new products, generated evidence about the utility of these products, and then submitted them for evaluation by other experts—sometimes led to significant battles among various parties involved in the process. These conflicts grew out of difficult questions about what constituted adequate evidence to justify therapeutic claims, how research was conducted, the appropriate boundaries on advertising, and other issues. Yet noticeably absent, at least from a historical perspective, was any significant concern about the patent status of the drugs involved. Whether or not a drug was patented had little to do with such issues. Instead, what mattered was that the claims made about products were true, that products were prescribed and used according to rational principles, and that the commercial motives of manufacturers did not overwhelm the commitment to good science.

Therapeutic reformers thus accepted the role of private interest in the scientific process, but skepticism about its impact remained. Outright quackery could be attacked, of course, but even highly reputable firms sometimes used promotional methods that blended the goals of science and commerce in ways that reformers found troubling. Sometime around 1907, for example, the New York branch of Bayer sent a memo to its sales representatives detailing how the firm prepared a "general article" about each of its products once a month. "These articles are sent to the editors of the different journals in which the announcements or advertisements of the Farbenfabriken product appear," noted the memo, "and these articles are very often accepted by the editors who assume responsibility for them and publish them as efforts of their own."[95] Critics denounced this type of promotion as a form of commercial exploitation, arguing that it subordinated the goals of medical science to the pursuit of individual profit. More broadly, reformers worried about the growing impact of advertisers on editorial decisions and the publication of supposedly impartial journals, textbooks, and other scientific literature by drug manufacturers. Francis Stewart, for example, was deeply concerned about the issue, which is not surprising given his sometimes contentious history of working with George S. Davis. Noting that the Hippocratic oath "imposes the obligation upon each member of the medical profession to report the results of his experience and observations in the practice of the healing art to the common fund of knowledge," in 1911 Stewart argued that it was "essential" that the medical profession should "have control" of the "educational machinery" of the profession to "prevent commercial exploitation and the teaching of error." Stewart still believed that physicians and ethical manufacturers should report the truth, the whole truth, and nothing but the truth. Yet how to ensure that manufacturers followed this dictate was increasingly unclear.[96]

Federal Regulation and the Quest for Therapeutic Equivalence

In 1906 reformers secured a major victory with the passage of the Pure Food and Drug Act. The law grew out of two decades of efforts to pass federal food and drug legislation in the face of what reformers took to be a deeply irrational market. The result of political organizing by business groups, women's clubs, grocers and pharmacists, and many others, the law was in part a response to the sometimes conflicting food and drug laws that operated at the state and local levels, in part a response to intense and seemingly unfair competition in the food and drug markets, and in part a response to the fear of adulterated, poisonous, and otherwise dangerous goods. It was a tremendously important piece of legislation. Historians should not ignore its significance simply because, in retrospect, we realize that it was inadequate to meet the challenges of the time.[97]

The 1906 law reflected the shifting attitudes toward patents and trademarks among therapeutic reformers. The views of Harvey Wiley, the head of the Bureau of Chemistry (BOC) and the single most important figure responsible for the passage of the law, serve as an important example. After receiving his medical degree in 1871 from Indiana Medical College, Wiley had become chief chemist of the Department of Agriculture in 1882 and from there rose to the head of the BOC when it was established in 1901. Like other chemists of his time—and similar to a growing number of physicians—he considered patenting to be an important spur to scientific innovation; he also secured a number of patents himself, including a patent on smokeless gunpowder.[98] In a paper he presented at the 1904 annual conference of the American Medical Association, Wiley laid out the case for federal control of drugs based on a distinction between unscientific nostrums with secret ingredients and legitimate medicines manufactured according to scientific methods. Wiley's argument hinged on the idea that patents are only given for true inventions, that patenting requires the disclosure of ingredients, and that secrecy was not necessary to protect truly novel goods and was instead indicative of quackery.[99] Despite this example, however, Wiley did not spend much time making these types of arguments. He was not particularly interested in patents from a regulatory perspective, and he appears to have believed that reforming the patent law held little promise for addressing the problems of the drug market.[100] Wiley believed that patenting played an important role in promoting scientific innovation, and he occasionally dabbled in thinking about patents in broader terms, but in general he was focused on other problems and had little interest in the topic.

Trademarks were a different matter. From Wiley's perspective, food and drug purity was fundamentally a question of the relationship of names to things. "By the word 'purity,'" he noted in 1905 in reference to distilled spirits, "I mean that they are true to name and are exactly what they are represented to be or what the consumer believes them to be."[101] From this perspective, drugs should only be sold under their own "proper names," and official drugs should be sold under official names and matched to the standards set out in the *USP* in order to ensure that they are, in fact, what they are claimed to be. At the same time, combinations of drugs should not be dispensed under "fanciful or assumed" names that concealed their true nature. This did not mean that such names needed to be abandoned or that trademarks had no place in commerce. Wiley believed that preassembled formulas served a legitimate role in medicine when made according to reputable formulas, and he believed that such products could legitimately be sold under trademarked names—but only if their ingredients were listed on their packaging, and only if those ingredients were listed using their true names.[102] For Wiley, as for other reformers at the time, the effort to rationalize the drug market could not be separated from complex questions about the relationship between names and things. Correspondence between the two was a central goal of therapeutic reform.

The passage of the 1906 law was a bruising political fight, and the final law did not reflect all of Wiley's wishes. In general, however, the law corresponded to his views about the importance of the names of things representing the true nature of the things themselves. The law defined a drug as being adulterated if (1) it was sold under a name included in the *USP* and violated the standards imposed by that text for the drug in question or (2) its strength or purity fell below "the professed standard or quality under which it is sold."[103] The second clause meant that any drug that the BOC determined to be of sufficiently poor quality could be considered adulterated, since no one advertised their goods as filthy, putrid, or otherwise of substandard quality. Outside of cases covered by the second clause, however, the possibility of adulteration only applied to goods that were sold under official names. However, there was a very important exception to this known as the "variation clause," which stated that a drug would not be deemed to be adulterated if the "standard of strength, quality, or purity" under which it was manufactured was plainly stated on the container it was sold in, even if this standard differed from the official one laid out for the good in the *USP*.[104] Thus, for all products, gross violations of quality—as determined by the BOC—could lead to a determination of adulteration. For goods that were sold under official names, the 1906 law linked the designation of a product as "adulterated" or not—as either being the thing it was claimed to be or being something else—to either the standards for that type of good promulgated by the *USP* or to the standards promulgated for the product by the manufacturer himself. The key relationship was thus between the name of the good and the underlying thing; as long as the properties of the product conformed to the standards promulgated for it, the product was assumed to actually be what it was claimed to be. If they did not, then the name of the product did not reflect what it actually was—adulterated or putrid opium, in a sense, was not really opium at all.

The 1906 law thus gave the Bureau of Chemistry the authority to police the relationship between names and things by determining what was and was not adulterated. A similar process was at work with misbranding. The 1906 law defined a drug as misbranded if its package or label bore "false" or "misleading" statements, if the product sold "be an imitation of or offered for sale under the name of another article," of if products made with alcohol, opium, cocaine, acetanilide, or a number of other drugs or their derivatives failed to list the amount or proportion of each of these ingredients on its label. Shortly after the passage of the law, the Department of Agriculture also adopted a list of forty rules and regulations for its enforcement.[105] A number of these related to misbranding, some of which were modified in subsequent years. Regulation 19, as modified in 1908, required that a food or drug product not bearing a "distinctive name" should "be designated by its common name in the English language"—or, in the case of drugs, any name recognized in the *USP*. Regulation 20 defined a "distinctive name" as a "trade, arbitrary, or fancy name" that "clearly distinguishes" one product from another and declared that "a distinctive name shall give no false indication of origin, character, or place of manufacture, nor lead the purchaser to suppose that it is any other food or drug product."[106] And finally, Regulation 28 listed the various derivatives and preparations of the drugs listed under section 8 of the original law that needed to be included on product labels; for acetanilide, these included ditrophen, diacetanilide, and *both* acetphenetidin and phenacetin, which were two different names for the same chemical.

The ability to designate a good as "misbranded" enabled the bureau to police the relationship between the name of a product and the underlying good. This power intersected with trademark law in complex ways. For example, section 8 of the law, which prohibited "false or misleading" statements on labels, gave the Department of Agriculture the right to prosecute manufacturers who used counterfeit trademarks or manu-facturers who designed their trademarks in a way that would cause customers to confuse their product with that of another manufacturer.[107] Regulation 20, which stated that "a distinctive name shall give no false indication of origin, character, or place of manufacture," strongly implied that any use of geographic names by manufacturers that did not manufacture their goods in the area named would be prohibited. This caused a significant amount of concern among many companies, since it meant that they might lose the considerable investment they had already made in their advertising. "If the new National Food and Drug Act is literally enforced," noted one observer, "it is feared that millions of dollars' worth of trade-marks will no longer be permissible."[108]

The prohibition on "false and misleading claims" was also interpreted by the government to mean that inaccurate claims about the effectiveness of a drug on its label was a form of misbranding. In the 1911 case *United States v. Johnson*, however, the Supreme Court ruled that the "false or misleading" clause law did not pertain to claims about efficacy but only to claims about the "identity of the article." According to Oliver Wendell Holmes's decision, the law as written regulated drugs as objects of commerce and nothing more. The label on the drug, as well as the claims made on it, "forms a part of the commerce in the article only in so far as it deals with the identity of the commodity contained in the package." This meant that "a statement which gives no information concerning the commodity itself, its physical constituents, or its chemical ingredients is not so related to the commodity as to form a part of the commerce in the article and is not, therefore, a part and parcel of the commerce within the regulating power contemplated by this statute." In other words, claims made about the effects of a drug on people should not be considered claims about the drug *itself*. Moreover, as the court noted, different schools of medicine have different "opinions" concerning "the curative properties of drugs." "No method has yet been devised by finite man to harmonize these warring factions," the court noted, "and indeed, it cannot be said that the truth lies entirely with any one of them. Congress cannot under the circumstances be deemed to have intended by this legislation to invade a field so speculative and conjectural."[109]

In response to the decision, in 1912 Congress passed the Sherley Amendment, which explicitly prohibited misleading statements about a product's therapeutic effects. Debate about the amendment centered on a number of difficult questions about how to regulate "unjust, unfair, and fraudulent" claims made by manufacturers.[110] One of the most significant problems was that in many cases deceptive claims about products were not made on labels at all but rather in advertising and through other promotional strategies. Another problem was the fact that the names of goods themselves were often deceptive, in that they suggested the supposed therapeutic properties of the product. The most difficult issue, however, was related to the variation clause. Critics of the clause argued that it should be removed because it allowed manufacturers to vary their products as they saw fit and thereby to make claims for their goods that did not conform to accepted medical opinion without fear of sanction. Others, however, argued that removing the clause would mean that improvements to pharmacopoeial products could not be made without having to invent a new name for the product in question—after all, if a product no longer conformed to the standards of the *USP*, it could no longer be called by an official name without being considered adulterated.[111]

In the end, Congress passed a narrow amendment that defined a product as misbranded if its package or label bore any "statement, design, or device regarding the curative or therapeutic effect" that is "false and fraudulent."[112] The amendment also explicitly stated that any product "offered for sale under the name of another article" was misbranded. However, the amendment failed to state whether or not trademarked names that were therapeutically suggestive in deceptive ways were a form of misbranding—a concession that probably grew out of concerns over granting the BOC unilateral authority to declare legally adopted trademarks violations of the 1906 law, thereby destroying their value to the firms in question. The inclusion of the term "fraudulent" in the definition of misbranding was also extremely significant. It meant that in order to secure a conviction under the law, the government was required to prove not only that the therapeutic claims made by an accused manufacturer were false but also that they were intended to deceive. This was a difficult bar because it required not just a demonstration that the products in question did not in fact have the therapeutic effects that the manufacturer claimed for them but also that the manufacturers *knew* that this was the case and sought to defraud the public. Combined with the fact that the government had no authority to force manufacturers to reveal their ingredients, it was exceedingly difficult to prove this type of intentional deception. Harvey Wiley thus denounced the clause as a "joker" that "practically nullifies [the law's] intended effects."[113]

In some ways the 1906 Pure Food and Drug Act was a weak law that did little to accomplish its goals: cosmetics and medical devices were not included under its provisions, penalties were relatively small, there was no requirement for establishing drug safety, and the scope of the law did not extend beyond regulating claims made on the labels of products. Furthermore, the Sherley Amendment made it difficult to prove misbranding based on efficacy claims. Yet despite these limitations the 1906 law represented a fundamental transformation in the relationship between manufacturers, the goods they produced, and the federal government. The ability to define drugs as both adulterated and misbranded inserted the power of the state in the relationship between a manufacturer and the thing that he produced. It allowed the federal government to enforce what it considered the proper relationship between names and things and to enforce a relative degree of equivalence among products. Since all goods sold under official names were supposed to conform to the same standards, they could be assumed to be roughly equivalent to one another. A similar effect can be seen even in official products made under the variation clause—since the manufacturer promulgated its own standards, each instance of a product should be at least roughly equivalent to every other instance made by the same manufacturer under the same standard. And finally, even though the ability to enforce honesty in therapeutic claims was not as strong as many proponents would have liked, under the Sherley Amendment the Bureau of Chemistry acquired at least limited power to force manufacturers to conform their claims for their products to the emerging norms of rational science. The power of the state and the emergent epistemological frameworks of both laboratory and clinical science were thus combined into a new means of structuring the market in therapeutic goods.

Efforts to reform the therapeutic market continued to be troubled by difficult questions related to patents and trademarks. The 1906 law delegated a significant amount of authority to the *USP* because the linking of official standards to official names served as the basis upon which drugs were defined as adulterated or not. Yet long-standing difficulties about including monopolized products remained. The revision convention met in 1910, and the ninth revision was finally published in 1916 after a long and protracted series of debates. Participants in the revision process were well aware that the results of their deliberations would now carry the weight of federal law, and much of the strife had to do with the difficult issue of whether or not to include patented and trademarked goods. From the perspective of reformers in the medical community, the refusal to include useful patented goods was untenable in the face of rational therapeutics. In 1912, for example, the executive committee voted against including acetylsalicylic acid because of its patented status. George Simmons, perhaps the most influential advocate for admitting patented goods, found the vote baffling.[114] Ethical manufacturers, on the other hand, were generally opposed to the inclusion of patented goods and strongly lobbied the revision committee against their inclusion. Manufacturers believed that the inclusion of goods patented by their competitors would give those companies an unfair advantage in the market, since inclusion would in effect give the *USP*'s stamp of approval to a product that only the patent holder could manufacture. To most manufacturers this seemed deeply unfair. "We believe that no patented or proprietary products should be recognized in the U.S.P., for obvious reasons," noted representatives of the Mallinckrodt Chemical Works. "Such recognition would be misconstrued to mean Government approval of the preparations and the advertising value of such recognition would be enormous."[115]

The question of names was also complex. Members of the revision committee generally assumed that drugs could not be included in the *USP* under trademarked names, and in cases where scientific names were overly cumbersome the revision committee used shortened names instead. The committee tried to adopt either names that were already being used or to coin new names that were convenient enough that they might reasonably be expected to be adopted into general use. However, significant problems about this issue remained: the revision committee did not really have an effective way to popularize the use of these names and it was not at all clear that they would actually be adopted in medical practice. There was also a significant amount of confusion about whether to adopt names that had once been monopolized by a single firm but had since moved into common use. Under the *Singer* decision, of course, such names were supposedly available to all, but in practice manufacturers tried to retain control over these types of names for as long as possible. Furthermore, a close association between names that had moved into the common domain and the product's original manufacturer often persisted for an extended amount of time. Even after Bayer lost control of the name "phenacetin," for example, many physicians continued to assume that if they prescribed the drug under that name, pharmacists would dispense the Bayer product. Given this, it is not surprising that manufacturers opposed the inclusion of names that were closely associated with the manufacturer of the original product in the *USP*. On the one hand, inclusion of these types of names furthered the interests of the original manufacturer, since it worked to promote a name with which the company was closely associated. On the other hand, it also worked to dilute this association by popularizing the name as a generic term. When the revision committee had originally included acetphenetidin in the eighth revision (which became official in 1905), it had done so under the name "acetphenetidinum" and listed *acetphenetidin* as its scientific name. In the ninth revision, the committee added "phenacetin" as one of its syno-nyms.[116] This probably seemed reasonable, since by this point multiple companies sold the product under this name. Bayer, however, was prob-ably not pleased.

One suggestion for addressing this difficult situation came from Fran-cis Stewart. In 1906 Stewart had joined H. K. Mulford & Company as its scientific director. Stewart retired in 1920, but while with the company he formulated the firm's scientific policy and implemented a working bul-letin system similar to the one he had devised for Parke-Davis.[117] He also continued to critique the use of product patents and trademarks on the names of goods themselves. The adoption of common names for all phar-maceuticals was central to Stewart's vision of a rationally operating drug market, and he argued against the revision committee's including prod-ucts under trademarked names—although he did not do so strenuously, given that opposition to including products under trademarked names was the generally accepted position at the time. Stewart also suggested that the revision committee should feel free to make use of formerly mo-nopolized names that had moved into common use, even if doing so was against the wishes of the original holders of the trademarks. Although, as he noted, manufacturers "are strongly opposed to relinquishing the con-trol over the names of their products and are doing all in their power to circumvent the law," he pointed out that under the *Singer* decision such names properly belonged to all. Still, Stewart recognized the difficulties in using formerly monopolized names, and in order to avoid such prob-lems in the future he recommended that the *USP* revision committee be empowered to issue an annual list of all new products introduced to the market and that it assign "generic" names "comfortable with scien-tific nomenclature" to each new drug upon its introduction. Doing so would allow manufacturers to sell these products under their own trade names while simultaneously allowing them to be standardized through inclusion in the *Pharmacopeia*. It would also curtail the ability of manu-facturers to monopolize the use of these drugs following the expiration of their patents, and it would preserve the integrity of both scientific nomenclature and scientific literature against the dangerous threat of exploitation by unethical firms.[118]

Stewart's perspective on the importance of the *Pharmacopeia* to the rationalization of the drug market points to the way in which therapeutic reformers both conformed to and helped create the developing logic of the corporate state, a state in which authority over the market is partially delegated to the interests of private organizations. From Stewart's perspec-tive, unethical advertising, secret ingredients, trademarks on the names of things themselves, and other forms of what he called "commercialism" undermined rational therapeutics and robbed legitimate manufacturers of their just rewards. The effort to rationalize the drug market was thus an effort to promote a rational therapeutics and simultaneously to promote the commercial interests of legitimate firms. "The remedy for unfair competition is to be found in drug standardization," Stewart argued in 1913. "Demand created by false advertising claims represents unfair competition. Business taken away from competitors in this way is stolen."[119] Stewart did not dwell on the fact, but the linkage of federal authority to the deliberations of the revision committee meant that a significant amount of authority over the therapeutic market was delegated to private interests. Reformers who participated in the revision process, whether directly or indirectly, were empowered to shape the market along what they considered rational lines. In doing so, they worked toward both the health of the public and the accumulation of corporate profit. These were not opposing goals. In many cases they were deeply intertwined with one another, despite the anxieties that the blending of the two also sometimes raised.

Adrenalin and the Embrace of Product Patents

I now turn to a discussion of Adrenalin, the single most important drug introduced by an American firm in the early twentieth century. Scientific interest in the adrenal gland dates to the late 1850s but began in earnest following George Oliver and Edward Schäfer's 1895 discovery that an extract made from the gland increased blood pressure when injected intravenously in animals. Researchers in both the United States and Europe quickly set about trying to discover the active principle of the gland. In 1896 Sigmund Frankel of Vienna isolated a "syrup-like body" which he named "sphygmogenin," and around the same time Otto von Fürth of Strasbourgh derived a substance he named "suprarenin." The following year the pharmacologist John J. Abel of Johns Hopkins isolated a highly purified crystalline form of the active principle of the gland, which he named "epinephrin." Pharmaceutical manufacturers also began to be interested in suprarenal products around this time: Armour & Company, for example, introduced a crystalline product under the name Suprarenalin in 1900.[120]

Adrenalin was developed by Jokichi Takamine and commercially introduced by Parke-Davis. Following the successful development of Taka-Diastase, Takamine established a laboratory in New York City in

1897 with considerable financial support from the company. He then began searching for a way to purify the "active principle" of the adrenal gland. This was a logical choice for his research efforts, given the developing interest in the topic among both academic scientists and manufacturers. In 1900 Takamine isolated what he considered to be a pure version of the principle.[121] Takamine was initially unsure whether his substance was the same as Abel's, so he coined a new name for it—as he put it in a letter to Abel, "Not being sure whether or not this substance is identical either with your Epinephrin or Furth's Suprarenin, I have for the sake of convenience, named it 'Adrenalin.' "[122] Parke-Davis then began to conduct animal experiments with the drug and to distribute it to physicians for clinical testing.[123] The company recognized the importance of Adrenalin almost immediately and began commercially distributing the product as early as April 1901.[124]

After some internal debate at the company, Takamine also applied for a patent on both the product and the process used to manufacture Adrenalin.[125] Patenting the manufacturing process was surprisingly complicated because of questions about the relationship of different processes to one another and the scope of Takamine's various applications; after more than two years of trying he ended up securing four patents on his manufacturing process after having separated the process and product claims into different applications.[126] Obtaining a patent on the product itself was even more difficult. James B. Littlewood, the examiner in charge of the case, rejected the initial application because he considered "Adrenalin" a "coined" name that could not be used in an application.[127] More seriously, he also rejected Takamine's product claims on the basis that they were little more than descriptions of the "natural principle" itself, and therefore not patentable. Citing *Cochrane v. Badische Anilin* and *Ex Parte Latimer*, Littlewood ruled that one claim "is drawn to a product of nature, merely isolated by applicant, and hence is not drawn to such patentable invention as required by statute," while the second "discloses nothing regarding the properties of the substance to be covered . . . except that it has the same properties as the natural principle. The natural principle not being patentable, neither is this."[128] From Littlewood's perspective, the substance that Takamine described in his application was essentially the same thing as the substance in its natural state and therefore could not be patented. Takamine's lawyers responded in a subsequent application, arguing that "the product as it exists in nature is certainly not a white, solid, crystalline body as defined in claim 7."[129] Littlewood was not convinced and in rejecting the application cited the *American Wood Paper* case, noting that "a process to obtain an extract from a subject from which it has never been taken may be the creature of invention, but the thing itself when obtained, can not be called a new manufacture."[130] Littlewood was making a basic point here, though somewhat cryptically: changes in nonessential characteristics, including the degree of purification, do not justify a claim to novelty. Just because something has a new color, consistency, or degree of purification does not mean it has become a new thing.

Faced with Littlewood's decisions, Takamine's attorneys changed gear. In addition to describing a variety of different physical characteristics, they also began to emphasize, as Christopher Beauchamp puts it, "the relationship between purity and function" in an effort to demonstrate the novelty of Takamine's product. They also argued that *Ex Parte Latimer* supported their claim in this regard, since the commissioner in that case had suggested that if Latimer's fiber had been curled, it might have been patentable. Takamine's lawyers emphasized "definite properties and characteristics which [the substance] does not possess in nature," arguing that the substance had undergone a "complete transformation." Importantly, this hinged on the assertion that the product had not just new physical characteristics but also new utility. As they noted:

There is a much greater distinction between a mere curling of a natural fiber, which the Commissioner intimated would have made Latimer's claim patentable, and the complete transformation which applicant has accomplished and defined in his claims. The article set forth is not anticipated and has never before been produced. It is therefore new. It is a useful product. Having invented and produced a new and useful article applicant is entitled to a patent.

They were now on the right track. After three more unsuccessful applications, in which Takamine's lawyers made similar arguments, Littlewood was either convinced or exhausted. In June 1903 a patent on the product was granted.[131]

Adrenalin was a remarkably successful product. By 1904 Parke-Davis already sold more than $200,000 worth of the drug per year, and the company considered it to be the most successful product it had ever introduced.[132] Other companies quickly introduced competing products, and by late 1904 there were at least five additional suprarenal preparations on the market that were roughly equivalent to Adrenalin.[133] The most successful of these was H. K. Mulford's Adrin, which was initially manufactured using Abel's method. Because Takamine had used a different method of isolating the substance than Abel, and because he had identified his substance by both a different chemical formula and a different name, H. K. Mulford considered its product to be a different substance than Adrenalin. Parke-Davis, however, considered Adrin and the other similar adrenal preparations on the market "imitations" of Adrenalin and violations of its patent rights. The firm hired the well-known patent attorney Livingston Gifford and sent notices to the other manufacturers telling them to desist selling their products. The company also retained Charles F. Chandler of Columbia University to serve as an expert witness in preparation for a suit.[134] H. K. Mulford refused to comply, and in 1905 Parke-Davis began the legal process against the firm. The two companies spent the next five years gathering expert testimony about research on the adrenal gland, the chemistry of the two substances involved, and numerous other issues.

When Stewart became the head of H. K. Mulford's scientific department in 1906, he began to watch the case closely. He was not pleased. Stewart believed that his former firm had behaved unethically by securing a patent on the substance rather than limiting itself to a patent on its manufacturing process; as he noted in 1910, "The evils of our product patent system are well illustrated by the 'Adrenalin' patent now under litigation."[135] Reflecting these views, the company made a concerted effort to position itself in the medical press as defending the ethics of the profession, arguing that product patents "work an injustice on the medical and pharmaceutical professions" and are "inimical to the public good."[136] Parke-Davis ignored these criticisms. By this point the medical community had come to accept the basic legitimacy of product patents, the company had an excellent reputation, and few paid attention to H. K. Mulford's complaints.

Stewart also criticized Parke-Davis for popularizing the term "Adrenalin" within the medical community as if it were the name of the product itself. One of the key difficulties in the suprarenal situation was the multiplicity of names for what might or might not have been the same substance. According to Stewart, the name "epinephrin" operated as a generic term that referred to all the various adrenal products on the market, while names such as "Adrenalin" and "Adrin" were brand names that manufacturers applied to their own products. The name "adrenaline"—with an "e"—was also sometimes used as a generic term for the substance.[137] From the perspective of Parke-Davis, however, competing products were either inferior substances that were not equivalent to Adrenalin or they actually *were* Adrenalin and therefore infringed on the company's patent rights; from this perspective, the term "epinephrin" was a misnomer, since Parke-Davis's name, "Adrenalin," applied to both their own product and to all equivalent products, including the product made according to Abel's method.[138] Parke-Davis had been granted a trademark on the name of its product in early 1906, and the company undoubtedly felt that it had every right to monopolize the name of the substance, since it had a patent on it.[139] Stewart, on the other hand, considered this argument to be little more than an unethical effort to substitute the company's brand name for the product's generic name, thereby distorting the scientific process to its own ends. As he noted, the strategy allowed the company to "use the educational machinery of the medical profession for commercial exploitation" and to "convert every textbook in which the name appears into a permanent advertisement for which the manufacturer pays nothing."[140]

Stewart was not the only one to consider "epinephrin" the generic name for Adrenalin and other adrenal products.[141] When the Council on Chemistry and Pharmacy issued its first list of accepted new remedies in 1906, it included Adrenalin, under that name, and described it as the "active alkaloid of the suprarenal gland."[142] Over the next several years the council accepted several different suprarenal products, including H. K. Mulford's Adrin and Armour & Company's Suprarenalin.[143] However, in the 1909 edition of *New and Nonofficial Remedies* the Council also adopted the term "epinephrin" as a "non-proprietary" name and listed the various companies' products as proprietary versions of the substance. The *Journal of the American Medical Association* also began to use the name "epinephrin" to refer to the principle when discussing it in general terms and to occasionally substitute it for trademarked names in the abstracts of articles. "The fact that 'adrenalin' is regarded by many, both here and abroad, as a common, generic name does not alter the fact that it is claimed as a trade name by a commercial house," noted the *Journal* in 1911. "It cannot be too strongly emphasized that 'epinephrin' is a true scientific name for the active principle of the suprarenal gland, [and] that it should be used on all occasions when the active principle and not some particular firm's make is referred to."[144]

This prompted an angry response from Parke-Davis. In 1909 E. M. Swift, the general manager of the firm, accused the *Journal* of both discriminating against the company and undermining the practice of rational medicine. Swift argued that because of the company's investment in Takamine's research, it had a right to commercialize the product and recoup its investment. He also argued that competing suprarenal products were little more than "imitations" of Adrenalin and suggested that substituting the term "epinephrin" for "Adrenalin" undermined the practice of scientific medicine by encouraging the substitution of untested products. Swift pointed out that the utility of the drug had been clinically established using the Parke-Davis product and argued that competing products varied so dramatically in strength and other characteristics that scientific knowledge about Adrenalin did not apply to them. As he rather pointedly noted, a recent report by the Hygienic Laboratory of the Bureau of Public Health and Marine-Hospital Service (soon to be renamed the Public Health Service) had shown a tremendous amount of variability among adrenal products on the market. From the perspective of Parke-Davis, this was a question both of justice to the firm and of rational medicine. The company clearly believed that the effort to promote "epinephrin" as a generic name threatened its ability to profit from its investment in Takamine's research. Yet it also believed that the various suprarenal products on the market were not equivalent to one another and that the effort to use a new name to refer to them all, even in an abstract sense, undermined the practice of scientific medicine by encouraging the substitution of one for the other. If a physician were to prescribe "epinephrin," who knows what product might actually be dispensed? This seemed both irrational and dangerous. As Swift put it, "The more you encourage the use of the blanket name 'epinephrin,' the more you put the physician at the mercy of 'thirty or forty' different manufacturers whose products range in value from the worthless to the best."[145]

Ironically, even as the company made these arguments, it also pursued a legal strategy based on the supposed equivalence between Adrenalin and Adrin. Gifford's legal strategy for the firm depended on two main tactics: first, showing that the product manufactured by H. K. Mulford & Company was the same as the product covered by Takamine's patent and second, proving that Takamine's patent was in fact valid. The first claim was established through scientific testing of the two products—a complicated process that involved a variety of disagreements about technical issues in chemistry. The second part of the strategy was even more complex. It depended on showing that Adrenalin was a new substance and therefore patentable. The problem for Parke-Davis was that simply arguing that Adrenalin was a more highly purified form of the suprarenal principle was not a viable legal strategy. As Gifford told Chandler,

> The grand difficulty of this case arises from the fact that the product is only new in the sense of being isolated from the bodies with which it was associated in the glands and, therefore, is open to all of the degree arguments. One position of our opponents may therefore be that the invention was not really a new product but a new process by which the old product was purified to a greater degree and that, therefore, the product claims are invalid. . . . It will be necessary for you to maintain on the other hand that Takamine's product is substantially a new body and that a new body may

be just as well created for practical purposes by isolating it from other bodies whereby its utility was dominated, throttled or suppressed, as by building up of a new body.[146]

Gifford's argument was thus based on the idea that Adrenalin was an essentially new substance, both because it had physical characteristics different from those of the suprarenal principle in its natural state and because it had been created "for practical purposes" and had a substantively new type of utility not available to other forms of the substance.

The case finally came to trial in early February 1911. Judge Learned Hand's decision was based on the idea that Adrenalin is a fundamentally different substance from the suprarenal principle in an unpurified state because of its therapeutic utility. In a much-quoted passage, Hand ruled that even if Adrenalin had been "merely an extracted product without change"—by which he meant changes in physical characteristics—the patent would still have been valid because Takamine was the first to have made the substance available for practical use; it was, Hand ruled, "a new thing commercially and therapeutically." This was, he noted, "a good ground for a patent."[147] As Christopher Beauchamp has noted, Hand's decision followed the pragmatic rationale of the *Kuehmsted* decision and similar recent cases.[148] As such, it was also a continuation of the long-standing legal doctrine that the application of human ingenuity to previously known things could render them patentable inventions if their essential characteristics were substantively changed. Purification alone was not enough to produce new substances, but substantively new utility might be; therapeutic—and commercial—value could therefore be a determining factor in whether or not something was considered novel enough to be patentable. Adrenalin was something *new* not because it was more purified but because it was useful in a categorically different way than the suprarenal principle in an unpurified state. "Everyone, not already saturated with scholastic distinctions," Hand noted, "would recognize that Takamine's crystals were not merely the old dried glands in a purer state. . . . The line between different substances and degrees of the same substance is to be drawn rather from the common usages of men than from nice considerations of dialectic."[149]

Following the decision, H. K. Mulford had little choice but to stop manufacturing Adrin. The company also sent out a bitter statement about the decision—probably written by Stewart—in which it reiterated its belief that product patents "are a hindrance to, rather than a means of promoting, progress in the practice of medicine" and are "inimical to the public good."[150] A number of medical journals reprinted selections from the company's statement verbatim, including its critique of product patents, but in general the medical press ignored the decision. Although such arguments could still find a sympathetic ear, for the most part they held little sway in the medical community. The decision was also noted with interest in the pharmaceutical press but with little concern about the ethics of monopolizing the sale of what was one of the most important drugs introduced in the early twentieth century.[151]

The development of Adrenalin was a tremendously important turning point in the history of the American pharmaceutical industry: the result of significant financial investment on the part of Parke-Davis, it was an important therapeutic advance at a time when the United States was significantly behind Europe in its scientific and technical capacities to develop new drugs. As Brian Hoffman has recently noted, later research on Adrenalin also led to a wide number of important discoveries that underlay our current understanding of hormones and other drugs.[152] Yet the importance of Adrenalin goes beyond even this. The willingness of Parke-Davis to acquire and then publicly defend its patent on the drug marked the end of the traditional prohibition on monopoly in the American pharmaceutical industry. In 1913, for example, the company filed for a patent on the active extract of the posterior lobe of the pituitary gland, which the company had recently begun to market under the name Pituitrin.[153] Other ethical manufacturers cautiously moved into patenting products as well, and by the outbreak of World War I a small handful of firms had applied for or secured patents on pharmaceutical products.[154] Monopoly was no longer excluded from the republic of medical science.

Conclusion:
The Promise of Reform

In February 1913, Carl Alsberg gave a speech to the National Association of Manufacturers of Medicinal Products, a newly organized trade group for manufacturers in the ethical wing of the drug industry. Alsberg had succeeded Harvey Wiley as chief of the Bureau of Chemistry (BOC) in 1912, and as a physician with extensive research experience in biochemistry he brought a deep commitment to scientific rigor to his dealings with the companies that his agency regulated. Yet this does not mean that he was hostile to the commercial interests of his audience. Quite the contrary. "Many of you realize that the time has come for the co-operation of the medical investigator and the manufacturer of remedial agents," Alsberg told the group. "It may be that there are some who fear that such co-operation will commercialize the medical profession. I am not one of them." The reason for this, he suggested, was that the great therapeutic advances of recent years had resulted from manufacturers and physicians working together to develop and introduce powerful new drugs such as Adrenalin and Salvarsan. The willingness of drug manufacturers to invest significant resources in this process had proved an important component of its success. "It has been amply shown that the solution of many therapeutic problems may be largely a matter of money," Alsberg noted. "The fact that the manufacturer offers his help as a speculation by which he hopes to gain does not alter the fact that the net result may be of immense benefit to mankind."[1] For Alsberg, as for other therapeutic reformers in the early twentieth century, the promotion of medical science and the pursuit of corporate profits were deeply intertwined projects. Reputable pharmaceutical manufacturers had a central role to play in solving the therapeutic problems of the day.

Alsberg's faith in the power "speculation" to improve the condition of humanity through the development of new pharmaceuticals was based on his recognition that harnessing the power of modern science to the pursuit of private profit had already delivered immense therapeutic dividends. Yet Alsberg also believed that the boundary between science and the drive for profit needed to be patrolled in order for this promise to be fulfilled. Outright quackery needed to be suppressed, of course, but even well-intentioned firms needed to be regulated to ensure that they conformed to the developing norms of rational therapeutics and did not inappropriately prioritize their own commercial interests. Under his direction, the BOC thus worked to monitor the drug market and to enforce an emergent epistemological regime in which laboratory and clinical science served as the basis for evaluating pharmaceuticals. Toward the end of 1913, for example, Alsberg met with the chief chemist of Rumford Chemical Works about the company's labeling of acid phosphate. He informed the firm that the therapeutic claims made on its labels were not supported by "medical authority based on scientific research" and asserted that the BOC would not tolerate therapeutic statements not based on "modern research work." Rumford responded by grudgingly adjusting its labels and removing the most egregious claims.[2]

The BOC was not the only government agency concerned about the drug market in the early twentieth century. In 1918, shortly after the United States entered the Great War, Congress authorized the newly formed Office of Alien Property to seize patents and trademarks held by enemy nationals.[3] The agency confiscated a large number of pharmaceutical patents and trademarks held by German firms, and the Federal Trade Commission then began licensing domestic companies to manufacture Salvarsan and other drugs under the seized patents. However, the FTC refrained from issuing licenses to use the seized trademarks. Instead, the FTC coined new names for the drugs and required that licensed manufacturers sell them under these names; Salvarsan, for example, was renamed "arsphenamine," Veronal was given the name "barbital," and so on. As Julius Stieglitz explained, the decision to precede this way was based on the recognition that the trademarked names for these drugs actually acted as their common names and if German firms reentered the market following the war, the association between the trademarked name and the original manufacturer would put American companies at a disadvantage. Stieglitz was both a former employee of Parke-Davis and a professor of chemistry at the University of Chicago; he was as familiar with the issues facing the domestic pharmaceutical industry as anyone, and he was deeply involved in the decision to coin the new names. As Stieglitz explained it, the FTC was "inspired by the idea of encouraging the establishment of a permanent American industry in these important articles" and "wisely decided that American houses should be put on the same footing as competing foreign houses for the after-the-war competition, by imposing on all licenses the obligation to use *new official names* for the article, names which after the war will be open to all competitors, domestic and foreign."[4]

The management of the seized patents and trademarks was soon engulfed in controversy.[5] The details of this process are beyond the scope of this volume, as is a detailed discussion of the rapid maturation of the industry following the war. For my purposes, it is enough to point out that the ethical wing of the industry expanded rapidly during the 1920s, in part as a result of the ability to exploit the seized patents, and as it expanded, the industry embraced the cooperative ideal that Alsberg had laid out. In 1921, for example, H. K. Mulford & Company finally revised its patenting policy to allow product patents. A central part of the new policy was a commitment to issue licenses to noncommercial institutions at no cost, while those whose purpose was to sell the goods for profit would be licensed under royalty. The company formulated its policy in this way as a means of reconciling the need to protect its products with its desire to promote scientific cooperation with the medical community, academic scientists, and other manufacturers. As the president of the company noted, the new policy was designed to ensure protection "against applications or patents by others," to "control or prohibit competition by incompetents," and "to permit co-operation and study by scientific and medical bodies."[6]

As H. K. Mulford's new policy makes clear, in the decade following the war patenting was understood as a means toward cooperation among reputable parties rather than a barrier to it. Patenting prevented unethical and disreputable manufacturers from entering the market in a given product, thereby ensuring the safety of the public. It also allowed manufacturers to promote cooperation among different parties through licensing, to enforce high manufacturing standards and thereby protect their reputation as reputable firms dedicated to the public good, and, of course, to profit from their investment in the scientific process. Not surprisingly, even as ethical manufacturers worked to promote industrial cooperation, they also defended what they saw as their rightful interests through increasingly aggressive patenting strategies. Companies struggled to monitor patenting activity at rival firms, worked closely with patent lawyers to acquire and enforce their patents, and otherwise sought to maximize the impact of patenting on their commercial success. In this process, they contended with shifting notions of patent law. Patents on chemical compounds, biological products, and other compositions of matter raised extremely complicated legal, scientific, and epistemological issues, including questions about the relationship between novelty and utility and questions about how to distinguish between essential and nonessential characteristics.[7] As ethical manufacturers fully embraced product patents, in other words, scientific innovation, technical issues in patent law, the creation and maintenance of pharmaceutical markets, and efforts by manufacturers to maintain their reputations as public servants were increasingly intertwined in complex and sometimes contradictory ways.

A growing tolerance of medical patenting within the orthodox medical community was an important part of all this. In the years immediately before World War I it still seemed troubling that physicians might personally benefit from acquiring patents on drugs. Yet the medical community also began to consider whether or not physicians should patent their own discoveries in order to keep unscrupulous or unreliable manufacturers from exploiting their work and manufacturing unsafe products. As a result, in 1912 the Principles of Ethics (formerly the Code of Ethics) of the American Medical Association were once again revised. The prohibition on physicians holding patents was dropped. However, the revised principles declared it "unprofessional to receive remuneration from patents for surgical instruments or medicines."[8] Two years later, the delegates to the annual convention also voted to give the Board of Trustees of the AMA the authority to hold medical patents "as trustees for the benefit of the profession and the public, provided that neither the American Medical Association nor the patentee shall receive remuneration from these patents."[9] The idea was short lived—in 1916 the organization decided not to manage patents in this way—but it points to both the ongoing concern that commercialism might corrupt the scientific process and the developing idea that patents could be used to enforce the norms of rational therapeutics through the practice of selective licensing. Patenting could be used to promote the public interest by ensuring that only ethical and reliable manufacturers had the right to commercially introduce new discoveries.

This idea became increasingly influential in the years following the war as academic researchers began to patent new drugs that they had discovered and then assign their patents to the institutions at which they worked; licenses were then issued to ethical manufacturers for the manufacture and distribution of the products in question. The two best-known examples of this dynamic are the discovery and patenting of insulin by Frederick Banting and his colleagues at the University of Toronto and Harry Steenbock's discovery and patenting of a method for producing vitamin D in food using irradiation at the University of Wisconsin. Banting and his colleagues turned their patent for insulin over to the University of Toronto, which then licensed Eli Lilly and Company to manufacture the drug, while Steenbock turned the patent on his irradiation process over to the newly formed Wisconsin Alumni Research Foundation, which then licensed the rights to his discovery to a small number of manufacturers—and, in the process, earned the University of Wisconsin a significant amount of money. Other discoveries in the 1920s and early 1930s followed this pattern as both academic scientists and physicians in clinical practice worked increasingly closely with manufacturers.[10] Neither physicians nor academic scientists were supposed to benefit personally from such arrangements, so royalties were typically directed back into the research process in order to prevent the desire for personal gain from unduly influencing the scientific process. Nor were they supposed to personally benefit from the growing number of research grants, fellowships, and other types of financial support that increasingly came from the industry. Still, the trend is unmistakable: over the course of the interwar period both academic scientists and practicing physicians increasingly pursued financial relationships with pharmaceutical manufacturers to advance both the cause of science and their own personal careers. The pursuit of individual interest, the creation of corporate profit, and the battle against disease were increasingly intertwined.

There was an important tension at the heart of this process. By the 1930s it had become common for scientists and physicians at universities, medical schools, and other institutions to work closely with pharmaceutical manufacturers to investigate new therapeutic substances. It had also become common for them to patent their discoveries and for their institutions to license the manufacturers who financially supported their work. Doing so was generally understood as a legitimate part of the effort to ensure that only high quality products reached the market; it was also assumed that drug manufacturers needed to be able to recoup their investment in the scientific process, and granting the company that had funded the research that led to a patentable discovery an exclusive licenses to commercially introduce the resulting product seemed both reasonable and fair. Yet the royalties that accrued from this process also raised the unseemly possibility that efforts to enforce patent rights might be about more than simply protecting the public. During the 1930s, for example, the Wisconsin Alumni Research Foundation assertively enforced its patent rights to Steenbock's irradiation technology.[11] From the perspective of some critics, the organization seemed to be acting contrary to the public good by prioritizing its own financial interests.[12] Thus, even as physicians and other academic scientists embraced patenting, purportedly for the public good, they also exhibited a significant amount of concern that doing so might distort the practice of science away from its noble ends. Despite Carl Alsberg's confident assertion two decades earlier that cooperation with the industry would not "commercialize" the medical profession, by the 1930s the embrace of patenting seemed to threaten this very danger. Notably, however, the question at hand was no longer about the legitimacy of acquiring patents per se. The legitimacy of physicians acquiring patents, and even receiving modest royalties, was no longer in doubt. The question was about how to best manage the potentially deleterious effects of patenting on the scientific process.

The attitudes of Morris Fishbein illustrate this dynamic. The editor of the *Journal of the American Medial Association* from 1924 to 1950, Fishbein was a ferocious and exceedingly influential critic of the drug market. Like other therapeutic reformers of his day, he believed that the traditional prohibition on medical patenting no longer suited the times. He argued that patents were a normal part of commercial drug development and that all investigators, including physicians, should receive a "reasonable royalty on a medical discovery."[13] On the other hand, Fishbein also believed that patenting had the potential to undermine the scientific process by fomenting competition, shaping the direction of medical research, and leading scientists and universities to prioritize profit over the public good.[14] "Steenbock has shown how remarkably remunerative a patent may be for a fund for a university," he noted in 1933 with concern. "The word has gone around that the Wisconsin University has gone royalty crazy."[15] The following year, Fishbein suggested that the AMA develop a committee to study the difficult questions involved in academic patenting.[16] The committee met periodically over the next several years, but was unable to come up with a solution that acknowledged the importance of patenting to the development and introduction of new drugs in an ethical manner yet also kept science free from distortion by the commercial impulse. It was not at all clear how to balance the need for patenting with its potentially corrosive effects on the scientific process, yet the risks of ignoring the problem were great. As Fishbein noted, "In the scramble for patent rights, it is no profit to a university to gain the whole world and lose its own soul."[17]

These types of anxieties deeply informed the ongoing project of therapeutic reform. By the early 1920s reformers had succeeded in erecting a complex, overlapping set of regulations at the municipal, state, and federal levels intended to control the manufacture, buying, and selling of drugs. Deceptive advertising and other unfair promotional strategies remained a central area of concern: from this perspective, unethical manufacturers made dishonest claims for their products, thereby artificially inflating demand, promoting the dangerous practice of self-medication, and duping the public into spending money on worthless or even dangerous goods. Many such problems were ascribed to patent medicines, but both proprietary and ethical manufacturers were also implicated because of their increasingly sophisticated promotional strategies. Reformers thus sought to distinguish between responsible and scientific advertising conducted by reputable firms and unethical advertising that exploited the public. The goal was to both promote what reformers considered rational therapeutics and to suppress what drug wholesaler William J. Schieffelin called the "inflated demand which we have so often experienced for drugs with very moderate merits and often very dangerous properties, a demand created by skillful propaganda and advertising on a huge scale."[18] As a result, even as they worked increasingly closely with industry, academic researchers were quite careful to protect their reputations for impartiality. Linking the practice of science too closely to the commercial interests of individual firms—even highly reputable firms—risked prioritizing the pursuit of profit over the practice of dispassionate science. Thus, even as academic scientists and physicians worked increasingly closely with drug manufacturers they maintained a distinct skepticism toward the motives of industry as one part of the broader project of therapeutic reform. This skepticism continues to characterize the relationship between the two groups to this day, albeit in a more muted form.

The federal government also played an increasingly important role in the effort to rationalize the pharmaceutical market. Following World War I the BOC worked closely with ethical manufacturers, industry groups such as the National Association of Manufacturers of Medicinal Products, and professional organizations such as the American Pharmaceutical Association to resolve the numerous difficulties that ethical drug firms faced in manufacturing their goods according to official standards, accurately labeling their products, and otherwise conforming to the government's regulatory requirements. The BOC also brought significant numbers of manufacturers to court for violations of the 1906 law in its efforts to protect the health of the public. In 1927 the regulatory functions of the BOC were housed under the newly formed Food, Drug, and Insecticide Administration, and in 1930 the name of this agency was shortened to the Food and Drug Administration (FDA). Still, despite what were at times aggressive enforcement efforts by the BOC—and later the FDA—the 1906 law was widely considered insufficient to meet the needs of the day. Following the economic collapse in 1929, efforts to improve the law accelerated, and after a drug sold under the name Elixir Sulfanilamide was linked to more than a hundred deaths in the late 1930s, reformers managed to pass a major revision of the law. The 1938 Food, Drug, and Cosmetic Act transformed the regulatory relationship between the federal government and pharmaceutical manufacturers, instituting a premarket review requirement for drug safety (though not effectiveness), eliminating the use of secret ingredients in drug manufacturing, and making other significant changes. A series of important amendments followed, including the 1962 Kefauver-Harris Amendment, which required that manufacturers demonstrate clinical effectiveness as a part of the premarket approval process.[19]

Not surprisingly, therapeutic reformers continued to be deeply concerned about the relationship between names and things. By the early 1920s it was widely understood in advertising circles that trademarks played an important role in creating distinctive brand identities for consumer products, and ethical manufacturers fully turned to the use of marks that operated at the level of the product in their efforts to create and maintain markets for their goods.[20] These efforts intersected with trademark law, the broader law of unfair competition, and therapeutic reform in complex ways, such as by provoking significant concern about whether or not similar products sold under different names were functionally equivalent to one another. At the same time, there was a great deal of confusion about the proliferation of drug names and no clear system for assigning what were increasingly called "generic" names to new products. Short nonproprietary names sometimes emerged out of the scientific community as convenient replacements for long and complex chemical names. Manufacturers also increasingly realized that giving their products distinct generic names might prevent their trademarked names from moving into the common domain at the expiration of their patents rights on the underlying goods, thus protecting their investment in developing a brand identity for their products. A variety of other interested parties also sometimes gave generic names to new products, including the members of various committees involved in revising the *USP* and the Council on Pharmacy and Chemistry. The result was that a single chemical could have multiple scientific, generic, and—once it went off patent—commercial names. This unruly proliferation of names continued to be a source of profound concern for reformers until the early 1960s, when a special committee was finally established by the American Medical Association, the United States Pharmacopeial Convention, and the American Pharmaceutical Association to systematically give all new pharmaceuticals generic names. Still, to this day the relationship among products that are made by different manufacturers yet sold under the same generic name is not always clear. The *generic* remains an unstable and contested category.[21]

Historians of the pharmaceutical industry tend to ignore the history of the Federal Trade Commission, but over the course of the twentieth century the FTC has also played a tremendously important role in shaping the drug market. During the interwar period the agency actively worked to suppress what it considered unfair forms of competition, including resale price maintenance, the use of "loss leaders"—trademarked products sold under market value in order to attract customers to stores—and other efforts to manipulate the price of goods in ways that seemed both monopolistic and unfair.[22] The FTC also focused on suppressing unfair advertising techniques, such as the use of trademarked names that were therapeutically suggestive in inaccurate ways.[23] At the same time, however, the authority of the agency to regulate fraudulent claims was limited to preventing practices that harmed other competitors by a series of court decisions that ruled that injury to consumers did not comprise a restraint on trade.[24] In response, newly formed consumerist groups organized around the issue as a part of their broader efforts to reform the market. These efforts culminated in the 1938 Wheeler-Lea Act, which expanded the authority of the FTC to restrict practices that harmed consumers. Ironically, one of the most significant obstacles to passing the law was a conflict with the effort to reform the 1906 Pure Food and Drug Act. The question at hand was a surprisingly difficult one: what agency should be given the authority to regulate advertising for drugs? The final version of the Wheeler-Lea Act was dramatically shaped by this conflict, in that it granted the FTC authority over drug advertising but excluded authority over both drug labels and advertising directed at the medical profession.[25]

The new authority of the FTC over drug advertising was matched by efforts to suppress other types of anticompetitive behavior. Early in his administration, for example, Franklin D. Roosevelt strengthened the FTC and brought the composition of its members in line with his ideals of strong regulatory government. Following the sharp economic downturn of 1937, his Justice Department launched what Alan Brinkley has called an "anti-monopoly crusade," much of which focused on the use of licensing arrangements to establish monopolistic cartels.[26] Not surprisingly, the pharmaceutical industry was one important target of this effort—in early 1941, the department instituted actions against Alba Pharmaceutical Company, Sterling Products, Winthrop Chemicals, and the American Bayer Company, alleging restraint of trade through illegal patent arrangements with the German corporation I. G. Farben.[27] In the decades following World War II regulators continued to investigate and at least try to suppress collusive arrangements, what seemed to be unfairly high prices, and other monopolistic patenting practices in the industry. From late 1959 to 1963, for example, Senator Estes Kefauver's Antitrust and Monopoly Subcommittee called widespread attention to what many reformers saw as grossly unfair profit margins, dishonest advertising, abuse of the patent law, and other problems in the industry.[28] Efforts to restrain monopolistic and anticompetitive behavior continue to this day. To take just one example, the Federal Trade Commission recently won a significant victory when the Supreme Court ruled that so-called pay-for-delay cases—in which drug manufacturers pay other companies not to introduce generic versions of products that have gone off patent—are appropriately subject to the rule-of-reason doctrine, which remains the standard for evaluating anticompetitive market behavior.[29]

The pharmaceutical industry is now truly massive—in 2013 Pfizer, the pharmaceutical giant that absorbed Parke-Davis in 2000, earned $59 billion alone—and patenting is at the heart of how the industry operates. Patents are used to establish a temporary monopoly over the results of scientific and technological innovation, thereby enhancing—and many would say making possible—the ability of corporations and other actors to invest and then recoup resources in the development of new products. Although outside the scope of this volume, the 1952 Patent Act established the basic framework for current patent law, while the 1984 Drug Price Competition and Patent Term Restoration Act (better known as the Hatch-Waxman Act) established a regulatory pathway for an expedited approval process of generic drugs, thereby laying the framework for the emergence of the generic industry in the following decade.[30] Since 1984 generic drugs have occupied a growing share of the drug market—in 2012 about 84 percent of all prescriptions were written for generics—and patenting strategies, the management of competition with generic manufacturers, and similar issues are now central to how the industry functions. Yet all is not well in the industry. As competition from generics has increased, other segments of the industry have faced declining profits as more and more of their products go off patent. The year 2012 was particularly bad. More than a dozen major drugs went off patent, sometimes leading to a drop of 90 percent or more in sales for the original manufacturer within a matter of weeks; as one industry newsletter put it, "For sheer terror, nothing matched the patent-cliff headlines as the year unfolded."[31] A lack of promising products in the development pipeline only adds to the current woes of the industry. How successfully manufacturers respond to these problems remains to be seen, but clearly legal, scientific, and business strategies related to patenting will remain at the heart of how the industry is organized and conducts itself.

Despite its centrality to the industry, pharmaceutical patenting and related practices continue to stir controversy. Questions swirl around the appropriate extent of patent protection, the relationship between patenting and drug prices, the role of intellectual property regimes in creating global health inequalities, and numerous other issues.[32] To take just one example, during the height of the health care reform battle in 2009, Representatives Anna Eshoo (D-CA), Jay Inslee (D-WA), and Joe Barton (R-TX) introduced an amendment to the larger health care bill intended to expedite the approval process for biosimilars, new versions of biopharmaceutical products that have passed out of patent protection. In order to encourage innovation in the field, and to address what some observers saw as the weakness of patent protections for these highly complex drugs, the amendment also established "data exclusivity" for the original makers of biopharmaceutical products for twelve years. Biopharmaceuticals can be incredibly expensive, and the amendment was strongly criticized for delaying the manufacture of biosimilars longer than necessary and thereby unfairly restricting access to relatively affordable products; the amendment, noted one critic, "will cost many of my fellow breast cancer survivors everything they own, and quite possibly their lives."[33] Such arguments were significantly bolstered when the Federal Trade Commission issued a report suggesting that twelve years of data exclusivity was unnecessarily long to encourage scientific innovation.[34] Biopharmaceutical manufacturers, however, successfully lobbied for inclusion of the amendment in the Patient Protection and Affordable Care Act of 2010, and these products are now protected by a full twelve years of data exclusivity. The Obama administration has also begun to push for twelve years of data exclusivity for biopharmaceuticals in multilateral trade talks under the assumption that the ability of corporate bodies to earn a return on their investment is a necessary component of scientific drug development.[35] Given that other countries offer significantly less time of data protection, with some offering none at all, the effort is sure to provoke significant debate.

Controversy also continues to swirl around the promotional efforts of the industry. Cooperation among pharmaceutical manufacturers, academic scientists, and physicians in clinical practice has been central to the development and popularization of numerous effective drugs over the course of the past century. We should not forget this. Yet the blurring of commerce and science at the heart of this cooperative arrangement continues to provoke anxieties and concern. Few critics worry about trademarks anymore, but aggressive and sometimes illegal efforts by pharmaceutical manufacturers to brand and promote their goods raise difficult questions about how health care decisions are made and whose interests these decisions ultimately advance.[36] Pharmaceutical companies use a variety of promotional strategies that strike many critics as deeply troubling, including ghostwriting (in which companies write or help to write scientific articles about their products and then conceal their role in the process) and the use of so-called seeding trials—clinical trials designed primarily to raise interest in a drug among physicians rather than to produce scientifically useful data. As Sergio Sismondo has argued, these and other practices are part of an entire "ghost management" process in which the industry shapes the production of biomedical knowledge toward the goal of corporate profit.[37] Over the past decade critics have increasingly decried this type of management as unethical, corrosive to the scientific process, and dangerous to public health.[38] Efforts have thus been made to reduce undue commercial influence on the conduct of both medical practice and biomedical research; some medical journals have adopted editorial policies intended to reduce ghostwriting, for example, and industry-physician payments are now subject to mandatory disclosure under the Physician Payments Sunshine Act, which was passed into law as part of the Patient Protection and Affordable Care Act and recently went into effect.

These types of reforms should be applauded. They point to the fact that we continue to believe in the possibility of a rationally operating therapeutic market, one in which markets are fair, science and commerce work together to benefit the public, and biomedicine is an undisputed good. We continue to believe in the promise of pharmaceuticals to heal the world. Yet if we are the inheritors of this optimistic vision, we also struggle with the fact that the world in which we live is much different than the one we believe to be possible. Indeed, it sometimes seems as if reality falls short of our aspirations precisely because of the fact that science and commercial interest are now so deeply intertwined. Pharmaceuticals play a tremendously important role in determining how medicine is practiced, in how disease is distributed, and in the definition and meaning of health itself. Increasingly, the scientific, patenting, and promotional strategies of the pharmaceutical industry shape the most intimate aspects of our lives—not just our experience of physical health and illness, but also how we think and feel, how we function on a daily basis, and how we organize our social worlds.[39] At the same time, it seems increasingly difficult to think about the pressing problem of health inequalities—whether at the local, national, or global levels—except through the lens of promoting access to drugs. Even as we debate the proper limits of patents, data exclusivity, and other forms of intellectual property, we confront the fact that it now appears virtually impossible to think seriously about the problems of health and disease without also thinking about the promotion of corporate profit. The two seem almost inextricably linked.

It is at this moment that history is of use to us: in our glimpses of a world that has gone by we can see the possibilities of something different than what we now know. The past teaches us that the world can be other than it is, that we have arrived at the current moment through contingent processes, and that the future is up for grabs. The therapeutic market might be organized differently; science and profit might be delinked from one another or rearranged in some unforeseeable way; patenting, trademarks, and other forms of intellectual property might be replaced by some other type of social arrangement. Health itself might mean something completely different. Who knows? Yet whatever the future holds, for the moment the goals of medical science and commercial profit are deeply intertwined. For better or for worse, monopoly has entered the house of medicine. And so the project of therapeutic reform continues, one part of our broader efforts to remake the world.

Acknowledgments

Any scholarly project incurs more debts than can be repaid. Scholarship, like science, is a cooperative endeavor built on the labor of many. I would like to thank the following people for their encouragement, support, and intellectual engagement as this project has taken shape: Keith Wailoo, Paul Israel, James Livingston, Jackson Lears, Stephen Pemberton, Nathan Crick, Curt Cardwell, Trysh Travis, Fritz Davis, Richard Mizelle, Jennifer Koslow, Judith Rushin, Meegan Kennedy, Edward Gray, Daniel Goldberg, Melissa Schroeder-Stein, Abena Dove Osseo-Asare, Rob Alegre, Matthew Osborn, Matthew Crawford, Scott Podolsky, Will Hanley, Jonathan Metzl, Erika Dyck, David Courtwright, Elizabeth Watkins, Christopher Crenner, and Howard Kushner.

I would also like to thank Jeremy A. Greene and David Herzberg for their friendship and critical engagement with pharmaceutical history. I have learned a tremendous amount from both of them, and this book would simply not be the same without the benefit of the many conversations that we have had. Nor would it be the same without the efforts of Susan Strasser, Anthony Palmieri, and Gregory Higby, each of whom read versions of the manuscript and offered helpful comments. A special word of thanks also goes to Neil Jumonville for his friendship and support. It has meant the world to me. And, of course, my gratitude and love to Richard Aviles. I am very fortunate to have met you those many years ago.

Lauren Gray and Adam Park both helped with research and critically engaged with my work in very thoughtful ways. Additional research assistance was provided by Kory

Gallagher, Anna Bessler, Sandra Holyoak, Jessica Nickrand, Laura Ann Twagira, and Edra Sorrentino. Jon Harkness offered important insights on the history of Adrenalin, and Arthur Daemmrich provided an extensive bibliography on the history of chemistry for which I am thankful. Numerous archivists provided expert and courteous assistance, including Christopher Magee at the National Archives in Kansas City and Gene Morris at the National Archives in College Park. Early research for this project was conducted in 2006–2007 when I held a National Science Foundation postdoctoral fellowship in the Science Studies Program at the University of California, San Diego. In 2007 a Reynolds Associates Research Fellowship from the Lister Hill Library of Health Sciences at the University of Alabama at Birmingham also provided financial support and encouragement. I am grateful to Steven Epstein and my other former colleagues at UCSD and to Michael Flannery for their early support of this project.

Portions of this book appeared in different form in two essays, "Restricting the Sale of 'Deadly Poisons': Pharmacists, Drug Regulation, and Narratives of Suffering in the Gilded Age," *Journal of the Gilded Age and Progressive Era* (July 2010): 145–169; and "A Thing Patented Is a Thing Divulged: Francis E. Stewart, George S. Davis, and the Legitimization of Intellectual Property Rights in Pharmaceutical Manufacturing, 1879–1911," *Journal of the History of Medicine and Allied Sciences* (April 2009): 135–172. Alan Lessoff and Margaret Humphreys have my thanks for their editorial guidance on these two papers, and I am grateful for the valuable feedback provided by the anonymous referees during the review process. A small amount of the material on ether also appeared in my essay "Anesthetics and the Chemical Sublime," *Raritan: A Quarterly Review* (Summer 2010): 69–74. I would also like to thank Jackson Lears, mentioned above, for inviting me to write about the fascinating history of anesthesia for *Raritan*—and, of course, for serving as my dissertation advisor, which is where I first began thinking about both Francis Stewart and anesthetics. Although only a small portion of the essay appears here, the opportunity to expand my work on ether provided an important stimulus for my ideas about patenting and medical ethics more broadly.

I would also like to thank Karen Merikangas Darling at the University of Chicago Press and Robert Devens, now with the University of Texas Press, for their enthusiasm for this project. Working with both of them has been a true pleasure. I am also grateful to the two anonymous referees of the manuscript, to George Roupe for his expert copyediting, and to Sophie Wereley for her editorial assistance. I would also like thank both the editorial board of the Synthesis series at the University of Chicago

Press and the Chemical Heritage Foundation for their support of this project.

Since 2007 I have had the honor of being on the faculty of the Department of Medical Humanities and Social Science at the Florida State University College of Medicine. Georgia Arnold, Leslie Beitsch, Debra Bernat, Kimberly Driscoll, Gareth Dutton, Heather Flynn, Mary Gerend, Suzanne Johnson, Gorana Knezevic-Zec, Chris Linton, Elena Reyes, Jeffrey Spike, Carmen Sualdea, and Angelina Sutin have all made this department a remarkably warm and supportive environment. In particular, I would like to thank my colleagues Robert Glueckauf and Michael Nair-Collins for their friendship, support, and intellectual companionship. It has been tremendously important to me. I would also like to thank Michelle Carter and Amelia Jones for their outstanding office support, their warmth and professionalism, and their tolerance of my idiosyncrasies. I am also grateful to my colleagues in the College of Medicine more broadly. The sense of community and shared dedication to the mission of the institution has made this an exceptionally rewarding place to work.

Finally, my deepest love and gratitude to Anne Viets, Laurie Ransom, and Claudia Sperber, my beloved wife and companion through this journey we call life. And to Jacob, my son. You mean more to me than I can ever express.

Archival Collections Consulted

Bayer Company Archives, Leverkusen, Germany
College of Physicians of Philadelphia
    Joseph McFarland Papers, 1900–1943
    Medical Trade Ephemera Collection
    National Convention for Revising the Pharmacopoeia, Committee of Revision and Publication, Records, 1858–1864
    Papers Concerning the Pharmacopoeia of the United States, 1818–1821
    George Bacon Wood Papers, 1793–1921
Columbia University, Rare Book and Manuscript Library
    Charles Frederick Chandler Papers, 1847–1937
Johns Hopkins University, Alfred Mason Chesney Medical Archives
    John Jacob Abel Collection
Library of Congress
    Harvey Washington Wiley Papers
Lloyd Library, Cincinnati
    John Uri Lloyd Papers, 1849–1936
    Lloyd Brothers Pharmacists, Inc., Collection, 1870–1938
Massachusetts Historical Society
    William T. G. Morton Papers, 1846–1876
    John Collins Warren Papers, 1738–1926
Missouri Historical Society, Saint Louis Branch
    Sappington-Marmaduke Family Papers, 1810–1941
National Archives and Records Administration, Washington, DC
    RG 46: Records of the US Senate Committee on Patents
National Archives and Records Administration, College Park, MD (NARA II)
    Papers Relating to the Administration of the US Patent Office

during the Superintendency of William Thornton, 1802–1828, Microfilm collection

RG 88: Food and Drug Administration

Records of the Board of Food and Drug Inspection, Seizure Cases, 1908–1912

Records of the Bureau of Chemistry, Miscellaneous Records, 1877–1910

RG 122: Federal Trade Commission

Trading with the Enemy Files, 1916–1924

RG 131: Records of the Office of Alien Property

Seized Records of the Farbenfabriken Co. Regarding Patent Interference Suits

Records of the Bureau of Investigations. Records of Investigations, 1917–1921

RG 241: Records of the Patent and Trademark Office

Files of the Office of the Commissioner of Patents, 1850–1949

Interference Case Files, 1836–1905

Patent Wrapper Files

National Archives and Records Administration, Kansas City, MO

RG 241: Patent Wrapper Files

National Archives and Records Administration, Philadelphia

RG 21: Records of the District Courts of the United States

RG 276: Records of the United States Courts of Appeals

National Archives and Records Administration, New York City Branch

RG 21: Records of the District Courts of the United States

RG 276: Records of the United States Courts of Appeals

Rhode Island Historical Society

Rumford Chemical Company Records, 1853–1951

Rutgers University Special Collections

Lucius F. Wood Papers, 1882–1889

Rutgers Medical College Records, 1792–1973

Rumford Chemical Works Records, 1848–1948

William Waldron Papers

Schlesinger Library, Radcliff Institute for Advanced Study

Lydia E. Pinkham Company Records, 1776–1968

Smithsonian Institution, National Museum of American History, Archives Center

Parke, Davis Research Laboratory Records, 1902–1950. Additional company records, AC001.

South Carolina Department of Archives and History

Vertical file collection

University of Chicago, Special Collections Research Center

Morris Fishbein Papers, 1912–1976

University of Michigan, Bentley Historical Library

Upjohn Family Papers, 1795–1974

University of Missouri–St. Louis, Western History Manuscript Collection
  Edward J. Mallinckrodt Jr. Papers, 1798–1981
University of Wisconsin, American Institute for the History of Pharmacy, Kremers
    Reference Files
  Parke, Davis & Company records
  Frederick Stearns & Company records
  McKesson & Robbins records
  H. K. Mulford & Company records
Western Michigan University, Special Collections
  Pfizer, Inc., Collection, 1888–1995
Wisconsin Historical Society
  Francis Edward Stewart Papers, 1853–1941
  United States Pharmacopeial Convention Records, 1819–2000
  O. N. Falk & Sons Druggists Records, 1904–1946
  Robert P. Fischelis Papers, 1821–1981

Notes

ADPR	*American Druggist and Pharmaceutical Record*
FESP	Francis Edward Stewart Papers, 1853–1941, Wisconsin Historical Society
JAMA	*Journal of the American Medical Association*
NARA	National Archives and Records Administration
NARA II	National Archives and Records Administration, College Park, MD, Branch
OGPO	*Official Gazette of the United States Patent Office*
PDC	Parke, Davis & Company (as author or recipient of correspondence)
PDCR	Parke, Davis Research Laboratory Records, 1902–1950, Additional Company Records, AC001, Smithsonian Institution, National Museum of American History, Archives Center
PWFKC	Patent Wrapper Files, RG 241, NARA, Kansas City Branch, followed by patent number
USPN	US patent number
USTN	US trademark number

INTRODUCTION

1. For example, Joseph M. Gabriel, "Restricting the Sale of 'Deadly Poisons': Pharmacists, Drug Regulation, and Narratives of Suffering in the Gilded Age," *Journal of the Gilded Age and Progressive Era* 9, no. 3 (July 2010):145–169; Nancy Tomes, "The Great American Medicine Show Revisited," *Bulletin of the History of Medicine* 79, no. 4 (Winter 2005): 627–663; Tomes, "Merchants of Health: Medicine and Consumer Culture in the United States, 1900–1940," *Journal of American History* 88, no.

2 (Sept. 2001): 519–547; Keith Wailoo, *Drawing Blood: Technology and Disease Identity in Twentieth-Century America* (Baltimore: Johns Hopkins University Press, 1997), 99–133. This argument stands in contrast to Paul Starr's claim that medicine "escaped" the corporation during this period. See Paul Starr, *The Social Transformation of American Medicine: The Rise of a Sovereign Profession and the Making of a Vast Industry* (New York: Basic Books, 1982), 198–232.

2. Martin J. Sklar, *The Corporate Reconstruction of American Capitalism, 1890–1916: The Market, the Law, and Politics* (New York: Cambridge University Press, 1988), 12–13.

3. On this topic, see Jeremy A. Greene, *Generic: The Unbranding of Modern Medicine* (Baltimore: Johns Hopkins University Press, 2014).

4. Harry Marks, *The Progress of Experiment: Science and Therapeutic Reform in the United States, 1900–1990* (New York: Cambridge University Press, 1997), 3. Other studies of therapeutic reform in the twentieth century include Scott H. Podolsky and Jeremy A. Greene, "Combination Drugs—Hype, Harm, and Hope," *New England Journal of Medicine* 365 (Aug. 11, 2011): 488–491; Scott H. Podolsky, "Antibiotics and the Social History of the Controlled Clinical Trial, 1950–1970," *Journal of the History of Medicine and Allied Sciences* 65, no. 3 (July 2010): 327–367; George Weisz, "From Clinical Counting to Evidence-Based Medicine," in *Body Counts: Medical Quantification in Historical and Sociological Perspective*, ed. Gérard Jorland, Annick Opinel, and George Weisz (Montreal: McGill-Queen's University Press, 2005), 377–393.

5. Over the past decade historians, sociologists, and other scholars have increasingly turned their attention to the industry, examining the relationship between the introduction of new pharmaceuticals and shifting definitions of disease and other important topics. Examples of recent work include Jeremy A. Greene and Elizabeth Siegel Watkins, *Prescribed: Writing, Filling, Using, and Abusing the Prescription in Modern America* (Baltimore: Johns Hopkins University Press, 2012); Joseph Dumit, *Drugs for Life: How Pharmaceutical Companies Define Our Health* (Durham, NC: Duke University Press, 2012); Dominique A. Tobbell, *Pills, Power, and Policy: The Struggle for Drug Reform in Cold War America and Its Consequences* (Berkeley: University of California Press, 2011); David Herzberg, *Happy Pills in America: From Miltown to Prozac* (Baltimore: Johns Hopkins University Press, 2010); Elizabeth Siegel Watkins, *The Estrogen Elixir: A History of Hormone Replacement Therapy in America* (Baltimore: Johns Hopkins University Press, 2010); Jeremy A. Greene, *Prescribing by Numbers: Drugs and the Definition of Disease* (Baltimore: Johns Hopkins University Press, 2008); Andrew Lakoff, *Pharmaceutical Reason: Knowledge and Value in Global Psychiatry* (New York: Cambridge University Press, 2006); Jonathan Michel Metzl, *Prozac on the Couch: Prescribing Gender in the Era of Wonder Drugs* (Durham, NC: Duke University Press, 2003).

6. Examples include David Healy, *Pharmageddon* (Berkeley: University of California Press, 2013); Carl Elliot, *White Coat, Black Hat: Adventures on the Dark Side of Medicine* (Boston: Beacon, 2010); John Abramson, *Overdosed America:*

The Broken Promise of American Medicine (New York: HarperCollins, 2004); Howard Brody, *Hooked: Ethics, the Medical Profession, and the Pharmaceutical Industry* (Lanham, MD: Rowman and Littlefield, 2007); Jerry Avorn, *Powerful Medicines: The Benefits, Risks, and Costs of Prescription Drugs* (New York: Alfred A. Knopf, 2004); Marcia Angell, *The Truth about the Drug Companies: How They Deceive Us and What to Do about It* (New York: Random House, 2004).

CHAPTER 1

1. Caspar Winstar Eddy, "Plantæ Plandomenses, or a Catalogue of the Plants Growing Spontaneously in the Neighborhood of Plandome, the Country Residence of Samuel L. Mitchill," *Medical Repository of Original Essays and Intelligence* (Aug.–Oct. 1807): 123.

2. Joseph M. Gabriel, "Anesthetics and the Chemical Sublime," *Raritan* 30, no. 1 (June 2010): 68–93; Matthew Turner, *An Account of the Extraordinary Medicinal Fluid, Called Aether* (1761).

3. Sharla M. Fett, *Working Cures: Healing, Health, and Power on Southern Slave Plantations* (Chapel Hill: University of North Carolina Press, 2002); Wonda L. Fontenot, *Secret Doctors: Ethnomedicine of African Americans* (New York: Praeger, 1994). On slavery, health, and medicine in early America see also Steven M. Stowe, *Doctoring the South: Southern Physicians and Everyday Medicine in the Mid-Nineteenth Century* (Chapel Hill: University of North Carolina Press, 2004); W. Michael Byrd and Linda A. Clayton, *An American Health Dilemma*, vol. 1, *A Medical History of African Americans and the Problem of Race: Beginnings to 1900* (New York: Routledge, 2000), esp. 151–321; Kenneth F. Kiple and Virginia Himmelsteib King, *Another Dimension to the Black Diaspora: Diet, Disease, and Racism* (New York: Cambridge University Press, 1981); Todd L. Savitt, *Medicine and Slavery: The Diseases and Health Care of Blacks in Antebellum Virginia* (Urbana: University of Illinois Press, 1981). On Indian practices, see Virgil J. Vogel, *American Indian Medicine* (Norman: University of Oklahoma Press, 1970).

4. *New-York Commercial Advertiser*, Jan. 7, 1806, 3.

5. For example, "To the Public," *New York Journal*, Nov. 5, 1767, 2.

6. Quoted in Diana Ross McCain, *It Happened in Connecticut* (Guilford, CT: Globe Pequot, 2008), 65.

7. "Interesting to All Sea-Faring People. Doctor LEE's Patent New-London Bilious Pills," in Michael Walsh, *A New System of Mercantile Arithmetic* (1803), 267. At least three other manufacturers in Connecticut secured patents for bilious or antibilious pills, including Samuel Cooley (1798), John Hawkes (1799), and Thomas Rauson (1802 and 1803). Edmund Burke, *List of Patents for Inventions and Designs, Issued by the United States, from 1790 to 1847, with Patent Laws and Notes of Decisions of the Courts of the United States for the Same Period* (1847), 8. For examples of advertisements, see "Dr. Hawks' Pills or American Extract," *Norwich Courier*, Sept. 1, 1802, 4; "Dr. Cooley's genuine

Bilious Pills," *Springer's Weekly Oracle*, Aug. 19, 1799, 4. The recipe for Samuel H .P. Lee's pills can be found as an enclosure in William Thornton to James Monroe, Dec. 20, 1813, Papers Relating to the Administration of the US Patent Office during the Superintendency of William Thornton, 1802–1828, NARA II, Microfilm collection.

8. Catherine L. Fisk, *Working Knowledge: Employee Innovation and the Rise of Corporate Intellectual Property, 1800–1930* (Chapel Hill: University of North Carolina Press, 2009), 31.

9. Edward C. Walterscheid, *To Promote the Progress of Useful Arts: American Patent Law and Administration, 1798–1836* (Littleton, CO: Fred B. Rothman, 1998); Steven Lubar, "The Transformation of Antebellum Patent Law," *Technology and Culture* 32, no. 4 (Oct. 1991): 934–942; B. Zorina Kahn, *The Democratization of Invention: Patents and Copyrights in American Economic Development, 1790–1920* (New York: Cambridge University Press, 2005), 49–65; Edward C. Walterscheid, "Patents and the Jeffersonian Mythology," *John Marshall Law Review* 29, no. 1 (Fall 1995): 269–314.

10. William Thornton, *Patents* (Patent Office, 1811), 1.

11. Christine MacLeod, *Inventing the Industrial Revolution: The English Patent System, 1660–1800* (New York: Cambridge University Press, 1988), 75–96.

12. Patrick N. I. Elisha, *Patent Right Oppression Exposed; or, Knavery Detected* (1813). On antimonopoly sentiment in the nineteenth century, see Richard R. John, "Robber Barons Redux: Antimonopoly Reconsidered," *Enterprise and Society* 13, no. 1 (Mar. 2012): 1–38.

13. M. F. Bailey, "History of Classification of Patents," *Journal of the Patent Office Society* 28, no. 7 (July 1946): 468. See also Lubar, "Transformation of Antebellum Patent Law," 934–942.

14. *New-York Commercial Advertiser*, Nov. 11, 1820, 4. On patent medicines in the colonial period and early republic, see Ann Anderson, *Snake Oil, Hustlers and Hambones: The American Medicine Show* (Jefferson, NC: McFarland, 2000), 1–30; George B. Griffenhagen and James Harvey Young, "Old English Patent Medicines in America," *Pharmacy in History* 34, no. 4 (1992): 199–229; James Harvey Young, *The Toadstool Millionaires: A Social History of Patent Medicines in America before Federal Legislation* (Princeton, NJ: Princeton University Press, 1961). For an interesting set of essays examining secrecy in early modern European science and medicine, see Elaine Leong and Alisha Rankin, eds., *Secrets and Knowledge in Medicine and Science, 1500–1800* (Burlington, VT: Ashgate, 2011).

15. *National Gazette and Literary Journal*, Jan. 8, 1825, 4; *Daily National Journal*, Sept. 26, 1826, 2; *Virginia Herald*, Oct. 7, 1826, 1.

16. Glenn Sonnedecker, reviser, *Kremers and Urdang's History of Pharmacy* (Madison, WI: American Institute of the History of Pharmacy, 1976), 326–327; "Powers-Weightman-Rosengarten Co.," *Pharmaceutical Era*, Jan. 30, 1908, 134–135; "Carpenter's Chemical Warehouse," *American Journal of Science and Arts*, Jan. 1829, 200.

17. This figure is based on an analysis of *A List of Patents Granted by the United States, for the Encouragement of Arts and Sciences: Alphabetically Arranged from 1790 to 1820* (1820) and Burke, *List of Patents for Inventions and Designs*. Evidence of a small number of additional patents that are not recorded in these sources appears in the *Journal of the Franklin Institute* and other publications of the time, including occasional notices for patents on medicinal goods. For one example, see "Patent Granted to Richard Burgess Doctor of Medicine for the [indecipherable] Invented a Drink for the Cure, Prevention, or Relief of Gout," *Journal of the Franklin Institute*, Mar. 1, 1833, 186.

18. Willard Phillips, *The Law of Patents for Inventions, Including the Remedies and Legal Proceedings in Relation to Patent Rights* (1837), 115.

19. For example, the manufacturer of "Audler's Patent Columbian Opodeldoc" threatened that "all persons are forewarned that any infringement on said patent will be prosecuted according to law." *New-York Commercial Advertiser*, Apr. 13, 1820, 3. I have been unable to find any material documenting a patent for this product, either in the two lists of patents cited above or in other sources. However, despite the difference in the spelling of the last names, this may have been a product manufactured by a man named Ezra Aulder, who two years earlier had acquired a patent for his "Asiatic lenitive for pain." Burke, *List of Patents for Inventions*, 99. Perhaps he realized that acquiring a patent was not worth the expense and trouble.

20. This claim is provisional. The failure to find cases of drug manufacturers enforcing their patent claims may have to do with the fact that equity law was in the process of being established at the time and disputes may have ended up in local criminal courts. As I describe below, for example, Thomson arranged to have a competitor arrested for violating his patent system and held on an extraordinarily high bail before a judge dismissed the case. Other disputes over patents on medicines may also have been treated as criminal proceedings. Further research on this topic is needed to better understand the history of patenting in the early drug industry and the extent of legal protections that such patents offered.

21. As one manufacturer noted in 1740, "King *George*, in Consideration of the rare and uncommon Cures performed by the Drops on Thousands of his Subjects, hath granted unto *Benjamin Okel*, the sole Inventor thereof . . . his Letters Patent under the Great Seal of *Great Britain*." *Pennsylvania Gazette*, June 12, 1740, 3.

22. Samuel Lee of Windham, Connecticut, for example, prominently mentioned in his advertising that his "True & Genuine Bilious Pills" had been "secured to him by patent by the President of the United States, dated April 30, 1796, agreeable to an act of Congress." "Bilious Pills," *Albany Centinel*, Apr. 26, 1799, 1. For other examples, see "Cooley's Pills," *United States Chronicle*, Dec. 26, 1799, 2; "Fresh Drugs and Medicines," *Columbian Centinel*, Aug. 27, 1800, 4; "Wheaton's Genuine Jaundice Bitters," *Norwich Courier*, Sept. 9, 1801, 3. As late as 1827, Ezra Aulder was prominently advertising his "Asiatic Lenitive

for Pain," which he had patented in 1818, as a "new and valuable medicine, patented by the United States of America." *Augusta Chronicle and Georgia Advertiser*, Aug. 8, 1827, 4.

23. "No Relief No Pay," *Portsmouth Oracle*, Oct. 7, 1809, 4. Of course, records documenting a patent may not have survived.

24. Elisha, *Patent Right Oppression Exposed*, 139.

25. Phillips, *Law of Patents for Inventions*, 113–114.

26. Richard L. Hills, *Power from Steam: A History of the Stationary Steam Engine* (New York: Cambridge University Press, 1989); John Kanefsky and John Robey, "Steam Engines in 18th-Century Britain: A Quantitative Assessment," *Technology and Culture* 21, no. 2 (Apr. 1980): 161–186; F. M. Scherer, "Invention and Innovation in the Watt-Boulton Steam-Engine Venture," *Technology and Culture* 6, no. 2 (Spring 1965): 165–187.

27. *Boulton v. Bull* (1795), reprinted in Charles R. Brodix, ed., *Decisions on the Law of Patents for Inventions Rendered by English Courts since the Beginning of the Seventeenth Century*, vol. 1 (Washington, DC: C. R. Brodix, 1887), 59–97.

28. Phillips, *Law of Patents for Inventions*, 95–101.

29. Robley Dunglison, *New Remedies: The Method of Preparing and Administering Them; Their Effects on the Healthy and Diseased Economy, & c.* (1839), 137; John W. Webster, *A Manual of Chemistry, on the Basis of Professor Brande's* (1828), 506.

30. *Boulton v. Bull* (1795), reprinted in Brodix, *Decisions on the Law of Patents for Inventions*.

31. John Kunitz acquired a patent on "bleeding, with leeches" in 1805. Burke, *List of Patents for Inventions*, 341.

32. "Specification of the patent granted to Joseph Pelletier and Jean Adrien Desprez, for improvements in making, or manufacturing, sulphate of quinine. Dated July 25, 1833," *Journal of the Franklin Institute*, July 1834, 47–48. This patent does not appear in Burke, *List of Patents for Inventions*.

33. *Woodcock v. Parker*, 30 F. 491 (1813).

34. *Earle v. Sawyer*, 8 F. 254 (1825).

35. *Woodcock v. Parker*, 30 F. 491 (1813).

36. William Thornton to James Monroe, Dec. 20 1813, Papers Relating to the Administration of the US Patent Office during the Superintendency of William Thornton, 1802–1828, NARA II, microfilm collection..

37. *Thompson et al. v. Haight et al.*, 23 F. 1040 (1826).

38. On Dyott, see Young, *Toadstool Millionaires*, 31–37.

39. "Catalogue of Drugs and Medicines, Patent Medicines, Dye Stuffs," *Franklin Gazette*, Oct. 14, 1820, 4.

40. "Medicines," *New-York Columbian*, June 21, 1819, 4.

41. On the rapid growth of newspapers in the early republic, see Paul Starr, *The Creation of the Media: Political Origins of Modern Communication* (New York: Basic Books, 2004), 84–94.

42. "Patent Medicine," in *The Spirit of the Public Journals; or, Beauties of the American Newspapers, for 1805* (Baltimore, 1806), 158.

43. "Be Cautious," *Franklin Gazette*, June 25, 1819, 4.

44. Historians have long assumed that consumers purchased patent medicines primarily because they were duped into doing so. James Harvey Young, in his classic book *The Toadstool Millionaires* (1961), described manufacturers of patent medicines as dangerous quacks who preyed on a gullible public, unscrupulous hucksters who sold nostrums with outlandish names to unwitting consumers. Young's book is still cited as the standard work on the topic, and this framework is common enough among historians that it is rarely questioned. However, the problem with this view is that it uncritically adopts the position of therapeutic reformers in the orthodox medical community without substantively interrogating the variety of reasons that people may have taken patent medicines, including the possibility that they were, at times, effective. There are complex methodological problems involved in making claims about the effectiveness of products used centuries ago that are beyond the scope of this volume. However, it is perhaps worth pointing out that patent medicines were—as far as can be determined—often made from many of the same ingredients that orthodox physicians prescribed and that many of the same plants are used in herbal medicine today. Fraud and hucksterism were undoubtedly important forces at the time, but to understand patent medicines solely through this lens is to miss much of the complexity of the early American drug market. On this issue, see Susan Strasser, "Sponsorship and Snake Oil: Medicine Shows and Contemporary Public Culture," in *Public Culture: Diversity, Democracy, and Community in the United States*, ed. Marguerite S. Shaffer (Philadelphia: University of Pennsylvania Press, 2008), 91–113.

45. *Singleton v. Bolton* (Nov. 13, 1783), in *Reports of Cases Argued and Determined in the English Courts of Common Law* (1853), 196–197.

46. "Be Cautious," *Franklin Gazette*, June 25, 1819, 4.

47. "John C. Morrison,"" *New-York Columbian*, Nov. 8, 1820, 4.

48. For example, "Caution," *New-York Daily Advertiser*, Sept. 8, 1819, 3.

49. "Medicines," *Connecticut Centinel*, Aug. 7, 1804, 3.

50. Philadelphia College of Pharmacy, *Formulae for the Preparation of Eight Patent Medicines* (1824).

51. "Reports of the Medical Society of the City of New York, on Nostrums, or Secret Medicines," *Journal of the Philadelphia College of Pharmacy*, Nov. 1827, 114–115.

52. John Harley Warner, *Against the Spirit of System: The French Impulse in Nineteenth-Century American Medicine* (Baltimore: Johns Hopkins University Press, 1998), 223–252; Warner, *The Therapeutic Perspective: Medical Practice, Knowledge, and Identity in America, 1820–1885* (Princeton, NJ: Princeton University Press, 1997), 37–57; Charles Rosenberg, "The Therapeutic Revolution: Medicine, Meaning, and Social Change in Nineteenth-Century America,"

in *Explaining Epidemics and Other Studies in the History of Medicine* (New York: Cambridge University Press, 1992), 9–31.

53. Preface to *Medical Repository of Original Essays and Intelligence* (1803), iii–ix. On early American medical publishing, see James H. Cassedy, "The Flourishing and Character of Early American Medical Journalism, 1797–1860," *Journal of the History of Medicine and Allied Sciences* 38, no. 2 (Apr. 1983): 135–150.

54. Stephanie B. Browner, *Profound Science and Elegant Literature: Imagining Doctors in Nineteenth-Century America* (Philadelphia: University of Pennsylvania Press, 2005), 15–38; Warner, *Therapeutic Perspective*, 11–80.

55. "Physiology of Circulation," *Medico-Chirurgical Review, and Journal of Medical Science,* June 1, 1823, 38.

56. Complex negotiations took place between physicians and the patients that they experimented on, but these negotiations took place in the context of hierarchal social relations that operated very differently than they do today. As a result, in my view, it is difficult to apply modern notions of consent to the interaction between physicians and patients at the time with any real rigor. This was not a matter of physicians obtaining consent from their patients to conduct their experiments in the way that we mean it today, even if, as Martin Pernick and others have argued, "truth-telling and consent seeking" were part of "an indigenous medical tradition" among orthodox physicians (quoted in Sydney A. Halpern, *Lesser Harms: The Morality of Risk in Medical Research* [Chicago: University of Chicago Press, 2004], 3). On medical experimentation before the Civil War, see Alexa Green, "Working Ethics: William Beaumont, Alexis St. Martin, and Medical Research in Antebellum America," *Bulletin of the History of Medicine* 84, no. 2 (Summer 2010): 193–216; Todd Savitt, "The Use of Blacks for Medical Experimentation and Demonstration in the Old South," *Journal of Southern History* 48, no. 3 (Aug. 1982): 331–348. On the debate about the comparability of modern notions of consent and past practices among physicians, see Halpern, *Lesser Harms*, 3–9; John C. Fletcher, "A Case Study in Historical Relativism: The Tuskegee (Public Health Service) Syphilis Study," in *Tuskegee's Truths: Rethinking the Tuskegee Syphilis Study*, ed. Susan Reverby (Chapel Hill: University of North Carolina Press, 2000), 276–298; Ruth R. Faden and Tom Beauchamp, *A History and Theory of Informed Consent* (New York: Oxford University Press, 1986), 53–101; Martin S. Pernick, "The Patient's Role in Medical Decisionmaking: A Social History of Informed Consent in Medical Therapy,' in *Making Health Care Decisions: A Report on the Ethical and Legal Implications of Informed Consent in the Patient-Practitioner Relationship*, vol. 3 (Washington, DC: President's Commission for the Study of Ethical Problems in Medicine and Biomedical and Behavioral Research, 1982).

57. "Patent for Bleeding Leeches!!!," *The Philadelphia Medical Museum, Conducted by John Redman Coxe, M.D.* (1806), 90.

58. Anderson, *Snake Oil, Hustlers and Hambones*, 7–11.

59. In 1790, for example, an English physician named James Adair published an

exposé of quack remedies and other topics with a long but revealing title: *Essays on Fashionable Diseases; the dangerous Effects of hot and crowded Rooms, the Clothing of Invalids, Lady and Gentlemen Doctors, and on Quacks and Quackery, with the genuine Patent Prescriptions of Dr. James's Fever Powder, Tickell's Ethereal Spirit, and Godbold's Balsam, taken from the Rolls in Chancery, and also the Composition of many celebrated Quack Nostrums, & c. & c.* (1790).

60. "Quack and Patent Medicines," *Medical and Agricultural Register* (1806–1807), 69.

61. Thomas Percival, *Medical Ethics; or, A Code of Institutes and Precepts, Adapted to the Professional Interests of Physicians and Surgeons* (1803), 44.

62. "Present Study of Medical Science," *American Medical and Philosophical Register*, July 1812, 81, 84.

63. Ebenezer Alden, *Historical Sketch of the Origin and Progress of the Massachusetts Medical Society* (1838), 36.

64. *Boston Medical Police* (Boston, 1820), 7. See also Robert Baker, "An Introduction to the Boston Medical Police of 1808," in *The Codification of Medical Morality: Historical and Philosophical Studies of the Formalization of Western Medical Morality in the Eighteenth and Nineteenth Centuries*, ed. Robert Baker, vol. 2, *Anglo-American Medical Ethics and Medical Jurisprudence in the Nineteenth Century* (Dordecht, the Netherlands: Kluwer, 1995), 25–39.

65. Examples include Georgia Medical Society, *A System of Medical Ethics, Adopted by the Georgia Medical Society, on the Fourth Day of February, 1821: For Regulating the Conduct of the Members of That Society* (1822); Central Medical Society of Pennsylvania, *The Constitution and Medical Ethics of the Central Medical Society of Pennsylvania* (1825); Medico-Chirurgical Society of Baltimore, *A System of Medical Ethics* (1832); Medical Society of the District of Columbia, *Regulations and System of Ethics of the Medical Association of Washington* (1833); Newark Medical Association (NJ), *Regulations and System of Ethics of the Medical Association of Newark* (1835).

66. *A System of Medical Ethics, Published by Order of the State Medical Society of New-York* (1823), 8–9.

67. J. Stuart, *A Popular Essay on the Disorder Familiarly Termed a Cold* (1808), 195.

68. "Are Inventions in Surgery and in Chemistry Legitimate Subjects for Patents?," *Boston Medical and Surgical Journal*, Dec. 23, 1836, 436.

69. Alexander Coventry, "Observations on Endemic Fever," *New York Medical and Physical Journal*, Mar. 1825, 10–11.

70. Rosemary Stevens, *American Medicine and the Public Interest: A History of Specialization* (Berkeley: University of California Press, 1998), 9–33; William C. Rothstein, "The Botanical Movements and Orthodox Medicine," in *Other Healers: Unorthodox Medicine in America*, ed. Norman Gevitz (Baltimore: Johns Hopkins University Press, 1988), 37–38; Paul Starr, *The Social Transformation of American Medicine: The Rise of a Sovereign Profession and the Making of a Vast Industry* (New York: Basic Books, 1984), 37–59.

71. John C. Whotron, *Nature Cures: The History of Alternative Medicine in America*

(New York: Oxford University Press, 2002), 25–48; Alex Berman and Michael Flannery, *America's Botanico-Medical Movements Vox Populi* (Binghamton, NY: Pharmaceutical Products Press, 2001); John S. Haller Jr., *The People's Doctors: Samuel Thomson and the American Botanical Movement, 1790–1860* (Carbondale: Southern Illinois University Press, 2000); Rothstein, "The Botanical Movements and Orthodox Medicine," 29–51. See also Samuel Thomson, *The Constitution, Rules and Regulations to Be Adopted and Practiced by the Members of the Friendly Botanic Society in New Hampshire and Massachusetts* (1812); Thomson, *Address to the People of the United States* (1817); Thomson, *New Guide to Health; or, Botanic Family Physician* (1822 and subsequent editions). The breadth and complexity of the botanic medical movement is demonstrated by the fact that more than eighty botanic medical journals were founded between 1822 and the beginning of the Civil War. For a list, see Haller, *The People's Doctors*, appendix H.

72. For details of Thomson's original patent, see "Specification of a Patent Granted for 'Fever Medicine.' To Samuel Thomson, of Surrey, County of Cheshire, New Hampshire, March 2, 1813," *Journal of the Franklin Institute of the State of Pennsylvania*, Feb. 1829, 130–131.

73. "Public Notice," and "Dr. John Locke," *Columbian Centinel*, Aug. 6, 1825, 3.

74. "Just Received," *(Charleston, SC) City Gazette*, Nov. 3, 1820, 4.

75. "To the Public," *Medical News-Paper; or, The Doctor and the Physician*, Jan. 1, 1822, 1.

76. "To the Public Again—or the Doctor's Trial," *Medical News-Paper; or, The Doctor and the Physician*, May 28, 1822, 34.

77. "Medical Pocket Book, & c.," *Medical News-Paper; or, The Doctor and the Physician*, May 28, 1822, 36.

78. For details of Thomson's 1823 patent, see "Specification of a Patent Granted for a Mode of Preparing, Mixing, Compounding, Administering, and Using, the Medicine Therein Described, to Samuel Thomson, of Boston, Suffolk County, Massachusetts, January 28, 1823," *Journal of the Franklin Institute of the State of Pennsylvania*, Feb. 1829, 212–215.

79. For example, John J. Waldron of New Brunswick, New Jersey, manufactured "rheumatic drops," "hot drops," and other products that he sold as "Genuine Thomsonian Botanic Medicines, Essential Oils, Essences, & c.," despite the fact that many of them used ingredients that had no place in Thomson's original system, including opium and rum. Waldron's advertisements and recipe book can be found in the William Waldron Papers, Rutgers University Special Collections, b. 2, f. 10.

80. Samuel Thomson, quoted in "Essay III," *Thomsonian Recorder*, Dec. 1, 1832, 97–98.

81. E. G. House, *The Botanic Family Friend: Being a Complete Guide to the New System of Thomsonian Medical Practice* (1844), 131–132.

82. "Thomsonism—the Term Reform, &c.," *Western Medical Reformer*, Mar. 1838, 43.

83. Wooster Beach, *The American Practice Condensed, or the Family Physician* (1848), 82, ix.

84. Haller, *The People's Doctors*, 128. Thomson himself was put on trial again in 1822 in Massachusetts for supposedly causing the death of a man named Ezra Lovett. For a description of the trial, see "Medical," *Independent Chronicle & Boston Patriot*, (Nov. 16, 1825), 4.

85. "Report of the New Haven Co. Medical Society," *Boston Medical and Surgical Journal* (Aug. 2, 1837), 26.

86. Haller, *The People's Doctors*, 130; Caleb Ticknor, *An Exposition of Quackery and Imposture in Medicine* (1839), 22. Useful acounts of trials include "Circuit Court," *New York Spectator*, Nov. 10, 1826, 1; "Miscellany," *New York Telescope*, Dec. 15, 1827, 114; *Trial of Dr. Frost, before the Court of Sessions for the City and County of New York, for Manslaughter, Alleged to Have Been Committed on Tiberius G. French by the Administration of Certain Thomsonian Remedies* (1838).

87. "A System of Medical Ethics," reprinted in *Transactions of the Medical Society of the State of New York* 1 (1833): 53.

88. Felix Pascalis et al., "To the Honourable Legislature of the State of New-York, in Senate and Assembly Convened," n.d., Rutgers Medical College Records, 1792–1973, Rutgers University Special Collections, b. 1, f. 4.

89. "An Address to the State's Rights Party," *Southern Botanic Journal*, Nov. 11, 1837, 326, 331.

90. *A Memorial to the Legislature of South-Carolina, Praying for a Repeal of the Medical Law* (1837), 12. A copy can be found in South Carolina Department of Archives and History, vertical file collection, item no. 63. A large number of other petitions are also available at the South Carolina Department of Archives and History.

91. Thomson, *New Guide to Health* (1832),132.

92. Thomson, *New Guide to Health* (1825), 163, 167.

93. *Thompson [sic] v. Staats* 15 Wend. 395 (1836).

94. Stephen D. Law, *Copyright and Patent Laws of the United States, 1790 to 1866* (New York, 1866), 10. See also "Interesting Case Law," *Portsmouth Journal of Literature and Politics*, Aug. 21, 1824, 2; *Daniel Jordan v. The Overseers of Dayton*, 4 Ohio 294 (1831); *Bryant v. the State*, 2 Miss. 351 (1836).

95. Committee of Correspondence, Medical Society of New Haven County, to John Redman Coxe, 18 July 1790, Papers Concerning the Pharmacopoeia of the United States, 1818–1821, College of Physicians of Philadelphia.

96. *Pharmacopoeia nosocomii New-Eboracensis; or the Pharmacopoeia of the New-York Hospital* (1816).

97. *Pharmacopoeia nosocomii New-Eboracensis*, vii.

98. On the establishment of the national pharmacopoeia, see Sonnedecker, *Kremers and Urdang's History of Pharmacy*, 255–263; Gregory J. Higby, "The Early History of the USP," in Lee Anderson and Higby, *The Spirit of Volunteerism: A Legacy of Commitment and Contribution; The United States Pharma-*

copeia 1820–1995 (Rockville, MD: United States Pharmacopeial Convention, 1995), 3–39; Glenn Sonnedecker, "The Founding Period of the U.S. Pharmacopeia I. European Antecedents," *Pharmacy in History* 35, no. 4 (1993): 152–161; Sonnedecker, "The Founding Period of the U.S. Pharmacopeia II. A National Movement Emerges," *Pharmacy in History* 36, no. 1 (1994): 3–25; Sonnedecker, "The Founding Period of the U.S. Pharmacopeia III. The First Edition," *Pharmacy in History* 36 no. 3 (1994): 103–121.

99. On Spalding, see James Alfred Spalding, *Dr. Lyman Spalding, the Originator of the United States Pharmacopeia* (Boston: W. M. Leonard, 1916).

100. *The Pharmacopeia of the United States of America* (1820), 17–18.

101. Ibid., 3.

102. Ibid., 23–25.

103. "Art. XX. The Pharmacopoeia of the United States," *Philadelphia Journal of the Medical and Physical Sciences* 2 (1821): 383.

104. Lewis Condict to Franklin Bache, June 29, 1832, MSS 2/0188—01 Acc. 1991—022-04, College of Physicians of Philadelphia.

105. Higby, " Early History of the USP," 55.

106. "Art. XIII. The Pharmacopoeia of the United States of America," *American Journal of the Medical Sciences*, May 1831, 142.

107. George B. Wood and Franklin Bache, *The Dispensatory of the United States of America* (1834), 214.

108. Thomas Spencer Baynes, "Cinchona," *Encylopaedia Britannica*, vol. 5 (1833), 780–781.

109. "Pharmacopoeia of the United States," *North American Medical and Surgical Journal*, Jan. 1831, 442.

110. *Pharmacopoeia of the United States* (1831), 9.

111. Wood and Bache, *Dispensatory* (1834), 201–235. The *Dispensatory* continued to play this role until the 1880 revision of the *USP*, at which point it began to be transformed into what Gregory Higby has called "a repository for obscure drug lore unavailable elsewhere." Higby, "Early History of the USP," 55.

112. On Linnaeus and his taxonomic system, see Paul Lawrence Farber, *Finding Order in Nature: The Naturalist Tradition from Linnaeus to E. O. Wilson* (Baltimore: Johns Hopkins University Press, 2000); Donald Worster, *Nature's Economy: A History of Ecological Ideas* (New York: Cambridge University Press, 1994), 26–56.

113. For example, Hugo Reid, *Outlines of Medical Botany* (1839), 181.

114. Tomas Miner, *Typhus Syncopalis, Sinking Typhus, or the Spotted-Fever of New-England, as It Appeared in the Epidemic of 1823, in Middletown, Connecticut* (1825), 29.

115. Robert Christison, *A Dispensatory, or Commentary on the Pharmacopoeias of Great Britain* (1842), 665.

116. John Eberle, *A Treatise of the Materia Medica and Therapeutics*, vol. 1 (1834), 140.

117. *Pharmacopoeia of the United States of America* (1830), 69; Edward Polehampton

and John M. Good, *The Gallery of Nature and Art; or, A Tour through Creation and Science* (1821), 127–130.

CHAPTER 2

1. Quoted in Jan R. McTavish, *Pain & Profits: The History of Headache and Its Remedies in America* (New Brunswick, NJ: Rutgers University Press, 2004), 27.
2. Robley Dunglison, *The Medical Student; or, Aids to the Study of Medicine* (1844), 232.
3. Steven Lubar, "The Transformation of Antebellum Patent Law," *Technology and Culture* 32, no. 4 (Oct. 1991): 942.
4. M. F. Bailey, "History of Classification of Patents," *Journal of the Patent Office Society* 28, no. 7 (July 1946): 467–468.
5. Quoted in Lubar, "Transformation," 957.
6. Lubar, "Transformation," 958.
7. Kara W. Swanson, "The Emergence of the Professional Patent Practitioner," *Technology and Culture* 50, no. 3 (July 2009): 525–526.
8. *Annual Report of the Commissioner of Patents, for the Year 1843* (1844), 316.
9. *Extracts from the Report of the Commissioner of Patents for the Year 1845* (1846), 31.
10. On Booth, see Eric Paul Wittkopf, "James Curtis Booth: Chemistry in Antebellum Philadelphia" (PhD diss., University of Delaware, 1994). On Philadelphia as a center of scientific and technological innovation, see Thomas C. Cochran, "Philadelphia: The American Industrial Center, 1750–1850," *Pennsylvania Magazine of History and Biography* 106, no. 3 (July 1982): 323–340; Bruce Sinclair, *Philadelphia's Philosopher Mechanics: A History of the Franklin Institute, 1824–1865* (Baltimore: Johns Hopkins University Press, 1974).
11. *Annual Report of the Commissioner of Patents for the Year 1843* (1844), 314.
12. USPN 3,633 (June 15, 1844).
13. My account of Goodyear and the patenting of vulcanized rubber is based on Cai Guise-Richardson, "Redefining Vulcanization: Charles Goodyear, Patents, and Industrial Control, 1834–1865," *Technology and Culture* 50, no. 2 (2010): 357–387, quotes from 377 and 364.
14. Ibid., 378.
15. Ibid., 379.
16. Willard Phillips, *Law of Patents and Invention; Including the Remedies and Legal Proceedings in Relation to Patent Rights* (1837), 95–115.
17. As Joseph Story put it in 1841, "In a race of diligence between two independent inventors, he, who first reduces his invention to a fixed, positive, and practical form, would seem to be entitled to a priority of right to a patent therefor." *Reed v. Cutter* 20 F. 435 (1841).
18. As George Ticknor Curtis put it, neither a principle nor a "method merely as such" could be patented. "If, when a patent is obtained for a method, it is in fact granted for *tangible means* of carrying that method into practice." Curtis,

Treatise on the Law of Patents for Useful Inventions in the United States of America (1849), 204.

19. For an interesting discussion of many of these issues, see *In re Kemper* 14 F. 286 (1841), in which the Circuit Court of the District of Columbia pondered the difficult question of whether John F. Kemper could obtain a patent for packing blocks of ice on their edge, rather than flat, since doing so reduced the rate of melting. The judge in the case, after much deliberation, decided that he could not.

20. Quoted in Guise-Richardson, "Redefining Vulcanization," 381.

21. *Goodyear et al. v. Central Railroad Co. of N.J.*, 10 F. 664 (1853).

22. *Hotchkiss v. Greenwood*, 52 US 248 (1850).

23. *Thomas Otis Le Roy et al. v. Benjamin Tatham et al.*, 55 US 156 (1852).

24. On patent medicines during this period see Susan Strasser, "Sponsorship and Snake Oil: Medicine Shows and Contemporary Public Culture," in *Public Culture: Diversity, Democracy, and Community in the United States*, ed. Marguerite S. Shaffer (Philadelphia: University of Pennsylvania Press, 2008), 91–113; McTavish, *Pain & Profits*, 27–42; Ann Anderson, *Snake Oil, Hustlers and Hambones: The American Medicine Show* (Jefferson, NC: McFarland, 2000); T. J. Jackson Lears, *Fables of Abundance: A Cultural History of Advertising in America* (New York: Basic Books, 1994), 103–104, 142–158; J. Worth Estes, "The Pharmacology of 19th-Century Patent Medicines," *Pharmacy in History* 30, no. 1 (1988): 3–18; James Harvey Young, *The Toadstool Millionaires: A Social History of Patent Medicines in America before Federal Legislation* (Princeton, NJ: Princeton University Press, 1961).

25. For example, *The Story of Sara Goodwin and Her Boys, to Which Is Added a Guide to Health, Wealth, and Happiness* (Dr. Herrick and Brother, Chemists, 1856).

26. On Sappington, see Lynn Morrow, "Dr. John Sappington: Southern Patriarch in the New West," *Missouri Historical Review* 90, no. 1 (Oct. 1995): 38–60. On malaria in the antebellum South, see Margaret Humphreys, *Malaria: Poverty, Race, and Public Health in the Untied States* (Baltimore: Johns Hopkins University Press, 2001).

27. Loren Humphrey, *Quinine and Quarantine: Missouri Medicine through the Years* (Columbia: University of Missouri Press, 2000), 18. Numerous historians have claimed that Sappington patented his pills. For example, Perry McCandless, *A History of Missouri*, vol. 2, *1820 to 1860* (Columbia: University of Missouri Press, 2000), 217. However, I have been unable to find any direct evidence for this claim. For an example of the charge of quackery leveled against Sappington, see "Travelling Letters from the Senior Editor," *Western Journal of Medicine and Surgery*, Oct. 1, 1844, 356.

28. W. Spillman, "The Truth, the Whole Truth, and Nothing but the Truth," newspaper clipping, Sept. 6, 1842, Sappington-Marmaduke Family Papers, 1810–1941, Missouri Historical Society, Saint Louis, b. 3, f. 2.

29. G. Hill, "To the Public," circular, Aug. 30, 1842, ibid, b. 3, f. 2.

30. Ibid.

31. State governments also began to pass laws against trademark infringement around this time. According to one source, New York passed the first trademark law in 1845, with eleven more states following suit during the next two decades, including Connecticut (1847), Pennsylvania (1847), Massachusetts (1850), and Ohio (1859). Arthur P. Greeley, "Dissenting Report of Mr. Greeley with Reference to the Revision of the Trademark Law," in *Report of the Commissioners Appointed to Revise the Statues Relating to Patents, Trade and Other Marks, and Trade and Commercial Names under Act of Congress Approved June 4, 1898* (Washington, DC: Government Printing Office, 1900), 91–92. Antebellum state trademark law is beyond the scope of this volume, but additional scholarly work is needed to understand the relationships between state trademark law, the developing common law tradition, and the antebellum drug trade.

32. As Mark McKenna argues, early trademark law was oriented toward protecting manufacturers from unfair competition. It was not primarily intended to protect purchasers. Mark P. McKenna, "Normative Foundation of Trademark Law," *Notre Dame Law Review* 82, no. 5 (June 2007): 1858–1871.

33. Francis H. Upton, *A Treatise on the Law of Trade Marks, with a Digest and Review of the English and American Authorities* (1860), 99.

34. *Amoskeag Manufacturing Co v. Spear & Ripley*, 2 Sandf. 599 (1849).

35. Francis Hillard, *The Law of Torts or Private Wrongs* (1859), 209.

36. For example, *Taylor v. Carpenter*, 23 F. 744 (1846); *Coffeen v. Brunton*, 5 F. 1184 (1849).

37. *Samuel Thomson v. Hosea Winchester*, 36 Mass. 214 (1837).

38. For example, Ross Petty, "The Codevelopment of Trademark Law and the Concept of Brand Marketing in the United States before 1946," *Journal of Macromarketing* 31, no. 1 (2011): 87.

39. Upton, *Treatise*, 24–26.

40. "Fetridge v. Wells. New-York Superior Court; Special Term, January, 1857," in Benjamin Vaughn Abbott and Austin Abbott, *Reports of Practice Cases, Determined in the Courts of the State of New York Practice Reports* (1857), 146.

41. Manufacturers sometimes tried to assert property rights in their names, marks, and labels in other ways. For example, manufacturers occasionally tried to copyright their labels under an 1831 law that had consolidated the various laws on copyrights and argue that their labels were the equivalent of books. For an example of an unsuccessful effort of this type, see *Scoville v. Toland et al.*, 21 F. 863 (1848).

42. "Self-Reformation of the Medical Profession," *Buffalo Medical Journal and Monthly Review*, Mar. 1850, 575.

43. William H. Hefland, "Advertising Health to the People," in *"Every Man His Own Doctor": Popular Medicine in America* (Philadelphia: Library Company of Philadelphia, 1998), 34; Young, *Toadstool Millionaires*; William D. Murphy, "Benjamin Brandreth," *Biographical Sketches of the State Officers and Members of the Legislature of the State of New York* (Albany, 1859), 38–40.

44. John Sappington, *The Theory and Treatment of Fevers* (1844), 198.
45. George Bicknell to M. M. Marmaduke, May 19, 1846, Sappington-Marmaduke Family Papers, 1810–1941, Missouri Historical Society, Saint Louis, b. 4, f. 1.
46. Ibid.
47. Charles Caldwell, "A Valedictory Address, by Charles Caldwell M.D. Delivered, by Appointment, to the Lexington Medical Society, on the 10th Day of March 1831," *Transylvania Journal of Medicine and the Associate Sciences*, Jan.-Mar. 1832, 39.
48. Charles Caldwell, *Phrenology Vindicated, and Antiphrenology Unmasked* (New York, 1838), 132.
49. "Code of Medical Ethics," *Proceedings of the National Medical Conventions, Held in New York, May, 1846, and in Philadelphia, May, 1847* (Philadelphia, 1847), 91–106. On the 1847 code and its implications, see Robert B. Baker, Arthur L. Caplan, Linda L. Emanuel, and Stephen R. Latham, eds., *The American Medical Ethics Revolution: How the AMA's Code of Ethics Has Transformed Physicians' Relationships to Patients, Professionals, and Society* (Baltimore: Johns Hopkins University Press, 1999); Robert Baker, Dorothy Porter, and Roy Porter, eds., *The Codification of Medical Morality*, vol. 2, *Anglo-American Medical Ethics and Medical Jurisprudence in the Nineteenth Century* (Boston: Kluwer Academic, 1993); John S. Haller Jr., *American Medicine in Transition, 1840–1910* (Urbana: University of Illinois Press, 1981), 234–279.
50. *Proceedings of the State Medical Convention of the Medical Society of the State of North Carolina* (1849), 9, 12. Other examples include Georgia Medical Society, *Medical Ethics, Being the Introduction and Code Adopted by the National Medical Convention, Held at Philadelphia, May 1847: also Adopted by the Georgia Medical Society of Savannah* (1847); *Proceedings of the Sate Medical Convention, Held in Lancaster, April 1848, and Constitution of the Medical Society of the State of Pennsylvania* (1848); "American Medical Intelligence: Proceedings of the Medical Convention of the State of Alabama, Held in Mobile, December, 1847," *New Orleans Medical and Surgical Journal*, Mar. 1848, 674–675; "Code of Medical Ethics of the American Medical Association, Adopted by the Kentucky State Medical Society, October 1, 1851," *Transactions of the First Annual Meeting of the Kentucky State Medical Society* (1851), 24; *Proceedings of the Fifth Annual Meeting of the Indian State Medical Society* (1854), 9.
51. *Transactions of the American Medical Association*, vol. 5 (1852), 35–36.
52. For an example of a physician being expelled for making and vending a patent medicine, see "Report of the State Medical Society of Indiana," in *Ohio Medical and Surgical Journal*, Mar. 1857, 271–272.
53. At times, orthodox physicians debated the limits of the prohibition on medical patenting. According to some physicians at the time, for example, inventors of new surgical devices should be able to patent their inventions, just as any other inventor, in order to guarantee that they earn a profit off of their labor. In 1855 a brief controversy over this issue broke out when the

State Medical Society of Ohio adopted a resolution allowing the patenting of surgical instruments; in response, the AMA voted to expel the Ohio medical society from membership for this "obnoxious action" unless it rescinded the resolution, which it did in short order. *Transactions of the American Medical Association*, vol. 8 (1855), 56–57. On the debate over patents for surgical equipment, see also Haller, *American Medicine in Transition*, 240–242. I have decided to ignore the issue of medical patenting on surgical instruments, largely because of space constraints, but in general physicians were more tolerant of patenting surgical devices than they were of patenting medicines. Nevertheless, patents on surgical instruments were frequently denounced as a form of quackery.

54. Medicus, "Patent Surgical and Medical Instruments," *Western Lancelet*, May 1855, 273–275.

55. Nathan Rice, *Trials of a Public Benefactor, as Illustrated in the Discovery of Etherization* (1859), 92.

56. Theo Rinney [Binney?] to John C. Warren, Jan. 18, 1848, John Collins Warren Papers, 1738–1926, Massachusetts Historical Society, b. 11, f. 3. The discovery of general anesthesia has long been a topic of considerable historical interest. Recent examples include Jacqueline H. Wolf, *Deliver Me from Pain: Anesthesia and Birth in America* (Baltimore: Johns Hopkins University Press, 2009); Stephanie J. Snow, *Blessed Days of Anaesthesia: How Anaesthetics Changed the World* (New York: Oxford University Press, 2008); Thomas Dormandy, *The Worst of Evils: The Fight against Pain* (New Haven, CT: Yale University Press, 2006), 208–226; Julie M. Fenster, *Ether Day: The Strange Tale of America's Greatest Medical Discovery and the Haunted Men Who Made It* (New York: HarperCollins, 2002); Richard J. Wolfe, *Tarnished Idol: William Thomas Green Morton and the Introduction of Surgical Anesthesia* (San Anselmo, CA: Jeremy Norman, 2001).

57. USPN 4,848 (Nov. 12, 1846).

58. "Insensibility during Surgical Operations Produced by Inhalation," *British American Medical and Physical Journal*, Jan. 1847, 248.

59. "The Inhalation of an Ethereal Vapor to Prevent Sensibility to Pain During Surgical Operations," *Boston Medical and Surgical Journal*, Dec. 2, 1846, 357.

60. Stephanie P. Browner, *Profound Science and Elegant Literature: Imagining Doctors in Nineteenth-Century America* (Philadelphia: University of Pennsylvania Press, 2005), 15–38.

61. Henry J. Bowditch [?], n.d., William T. G. Morton Papers, 1846–1876, Massachusetts Historical Society, b. 1, loose leaf, 21–22.

62. Joshua B. Flint, "Report on the Best Mode of Rendering the Patronage of the National Government Tributary to the Honor and Improvement of the Profession," in *Transactions of the American Medical Association*, vol. 9 (1856), 533–539.

63. For example, Alexander H. Stevens, "Annual Address, Delivered before the New-York State Medical Society," in *Transactions of the Medical Society of the*

State of New York, during Its Annual Session, Held at Albany, February 6, 1849 (Albany, 1849), 4–11.

64. Henry Jacob Bigelow, "Insensibility during Surgical Operations Produced by Inhalation," *Boston Medical and Surgical Journal*, Nov. 18, 1846, 316–317. In response to the controversy, the Massachusetts Medical Society briefly declared it ethically legitimate to acquire patents on medical discoveries. "Massachusetts Medical Society—Patent Medicines," *Boston Medical and Surgical Journal*, Feb. 17, 1847, 63–64. However, the society quickly came under blistering criticism and soon reversed its position; as one critic of the society noted, "we can only deplore a state of things so humiliating and inexcusable" as their position on the ether patent. "Varia," *Annalist*, Feb. 1, 1847, 262.

65. Henry Jacob Bigelow, *Ether and Chloroform: A Compendium of Their History, Surgical Use, Dangers, and Discovery* (1848), 9, 14.

66. One notable defense was physician Nathan P. Rice's *Trial of a Public Benefactor as Illustrated in the Discovery of Etherization* (1859), esp. 128–141. Rice's book was a spirited defense of Morton. It included a chapter on the controversy in which Rice explicitly defended Morton's patent and reproduced many of the arguments that both Morton and Bigelow had made, such as the idea that Morton had pursued a patent in order to be able to prevent the use of ether for "nefarious purposes" or "indiscriminate and careless use." Reviews of Rice's book also occasionally described Morton's patent in a sympathetic light. For example, see the lengthy review in *New York Journal of Medicine*, Mar. 1859, 265–274.

67. *Morton v. New York Eye Infirmary Circuit Court*, 17 F. 879 (1862). See also Martin A. Ryan, "Patentability of a New Use for an Old Composition of Matter," *Journal of the Patent Office Society*, Nov. 1947, 791–792.

68. "A Code of Ethics Adopted by the Philadelphia College of Pharmacy," *American Journal of Pharmacy*, Apr. 1848, 149–151. On pharmacy and ethics in the nineteenth century, see Glenn Sonnedecker, reviser, *Kremers and Urdang's History of Pharmacy* (Madison, WI: American Institute of the History of Pharmacy, 1976), 200–202; Robert A. Buerki, "The Historical Development of an Ethic for American Pharmacy," *Pharmacy in History* 39, no. 2 (1997): 54–72.

69. "Code of Ethics of the American Pharmaceutical Association," *Proceedings of the National Pharmaceutical Convention Held at Philadelphia, October 6th, 1852* (1852), 45.

70. Edward Parrish, "American Pharmacy," *American Journal of Pharmacy*, July 1854, 290.

71. "Constitution of the American Pharmaceutical Association," *Proceedings of the American Pharmaceutical Association* (1856), 79–80. Note that here and in subsequent citations the year given for the *Proceedings of the American Pharmaceutical Association* refers to the year the meeting took place, not the year the proceedings were published.

72. Examples include "An Act to Regulate the Preparation and Dispensing of Medicines in the City of New-York," *Laws of the State of New-York Passed at*

the Sixty-Second Session of the Legislature (1839), 57–58; "An Act Regulating the Sale of Poisons," *The Public Statutes at Large of the State of Ohio*, vol. 3 (1854), 1791 (prohibiting the sale of articles of medicine "belonging in the class usually known as poisons" without a prescription from a physician); "Offenses against Public Health," *The Code of Iowa Passed at the Session of the General Assembly of 1850–1* (1851), 377 (prohibiting the sale of certain drugs or poisons without labels bearing the word "poison" and the true name of the substance).

73. "Patent Medicine Law in Maine," *New York Journal of Medicine*, Mar. 1847, 252.

74. "Report of the Committee on Quack Medicines," *Proceedings of the American Pharmaceutical Association* (1854), 36.

75. Jan R. McTavish, "What Did Bayer Do before Aspirin? Early Pharmaceutical Marketing Practices in America," *Pharmacy in History* 41, no. 1 (1999): 3–15. As McTavish points out, ethical manufacturers sometimes ignored professional idealism by including therapeutic claims in their advertisements, using fancy graphics, and so on. In doing so they pushed the boundaries of what was considered acceptable behavior and risked the wrath of elite opinion.

76. Untitled editorial, *American Journal of Pharmacy*, July 1849, 287–288.

77. For example, Edward Parrish, "Patents in Their Relation to Pharmacy," *Proceedings of the American Pharmaceutical Association* (1860), 173–176; untitled editorial, *American Journal of Pharmacy*, July 1849, 287–288; "Report of the Committee on Patent Medicines to the Board of Trustees of the College of Pharmacy of the City of New York," *Journal of the Philadelphia College of Pharmacy*, Apr. 1834, 61.

78. Alex Berman and Michael A. Flannery, *America's Botanico-Medical Movements: Vox Populi* (New York: Pharmaceutical Products Press, 2001), 110–111.

79. Edwin T. Freedley, *Leading Pursuits and Leading Men: A Treatise on the Principal Trades and Manufactures of the United States* (1856), 179–180. See also Tilden & Co., *A Catalogue of Pure Medicinal Extracts* (New York, 1852).

80. Edward Parrish, *An Introduction to Practical Pharmacy* (Philadelphia, 1859), 179.

81. "Tilden & Company's Extracts," *Medical and Surgical Reporter*, Jan. 1856, 34.

82. "The Purity of Medicine," *Boston Medical and Surgical Journal*, Oct. 28, 1858, 262–64.

83. "Tilden & Co. Again," *New Orleans Medical News and Hospital Gazette*, Mar. 1860, 39–40.

84. "George W. Carpenter," in Stephen N. Winslow, *Biographies of Successful Philadelphia Merchants* (1864), 124–129.

85. "Geo. W. Carpenter's Compound Fluid Extract of Wahoo," *American Medical Intelligencer*, June 1842, 261–262.

86. "Euonymus Americanus," *New York Journal of Medicine*, Sept. 1847, 165.

87. "Euonymus Atropurpureus," in George B. Wood and Franklin Bache, *The Dispensatory of the United States of America* (1854), 1409–1410.

88. "Fluid Extract of Wahoo," *Tilden & Co.s Medical Advertiser*, (n.d.), 4, reprinted in *Western Journal of Medicine and Surgery*, Dec. 1855, back advertising material.

89. "American Pharmaceutical Association," *New York Journal of Pharmacy* 3 (1854): 382.

90. Daniel Henchman to George B. Wood, March [n.d.] 1840, in "Minutes of the Committee for revising and publishing the U.S. Pharmacopoeia," George Bacon Wood Papers, 1793–1921, College of Physicians of Philadelphia, b. 3, f. 20.

91. *Pharmacopoeia of the United States of America*, 4th revision (1863), 46.

92. "Art. XIII—The Pharmacopoeia of the United States of America," *American Journal of Medical Sciences*, Oct. 1863, 434.

93. *Pharmacopoeia of the United States of America*, 4th revision (1863), 41.

94. "Report of the Committee on Quack Medicines," *Proceedings of the American Pharmaceutical Association* (1852), 36.

95. "Report of the Select Committee of the House of Representatives of the United States; to Whom Was Referred the Subject of Imported Adulterated Drugs, Medicines, and Chemical Preparations," *New Jersey Medical Reporter*, July 1848, 305.

96. United States Department of the Treasury, *General Regulations under the Revenue and Collection Laws of the United States*, sec. X, art. 249 (Washington, 1857), 156; "Effects of the New Drug Law," *New Jersey Medical and Surgical Reporter*, Jan. 1849, 166; James Harvey Young, *Pure Food: Securing the Federal Food and Drugs Act of 1906* (Princeton, NJ: Princeton University Press, 1989), 6–17.

97. "Patent Medicine Tax," *American Journal of Pharmacy*, Apr. 1851, 187.

98. Edward Parish, William Procter, and Ambrose Smith, "Report for the City of Philadelphia," *Proceedings of the American Pharmaceutical Association* (1852), 33. See also "Patent Medicine Tax," *American Journal of Pharmacy*, Apr. 1851, 186.

99. [Indecipherable], written response to "The Medical Society of the State of New-York Requests the Medical Societies Represented in its Organization to Reply to the Following Inquiries Before the 20th of April, 1860," National Convention for Revising the Pharmacopoeia, Committee of Revision and Publication, Records, 1858–1864, College of Physicians of Philadelphia, b. 2, f. "Medical Society of the State of New York, Circular."

100. On Squibb, see Lawrence Goldtree Blochman, *Doctor Squibb: The Life and Times of a Rugged Idealist* (New York: Simon and Schuster, 1958).

101. E. R. Squibb, "Materia Medica and Pharmacy," *American Medical Times*, Dec. 8, 1860, 408–409.

CHAPTER 3

1. David Prince and Thomas Antisell, "Patent Rights among Medical Men," *Transactions of the American Medical Association* (1866), 521–528. On Antisell,

see "Dr. Thomas Antisell and His Associates in the Founding of the Chemical Society of Washington," *Journal of the Patent Office Society* 20, no. 8 (Aug. 1938): 651–671. On Prince, see Frank B. Norbury, *David Prince: A Pioneer in Surgical Therapeutics in Central Illinois* (Springfield: Southern Illinois University School of Medicine, 1981).

2. "Patent Rights," *Medical and Surgical Reporter*, Oct. 12, 1867, 320–321.

3. Other statements include "Patents in Medicine and Surgery," *Medical and Surgical Reporter*, Aug. 31, 1867, 190–191.

4. "The American Patent System," *OGPO*, Feb. 12, 1901, 1383. In 1865, 6,616 patents were issued; in 1868, 13,378 patents were issued. Between 1870 and 1879 the number ranged from a low of 12,864 (1873) to a high of 15,595 (1876). Data based on "Comparative Statement of the Business of the Office from 1837 to 1921, Inclusive," in *Annual Report of the Commissioner of Patents for the Year 1921* (1922), viii.

5. Steven W. Usselman and Richard R. John, "Patent Politics: Intellectual Property, the Railroad Industry, and the Problem of Monopoly," *Journal of Policy History* 18, no. 1 (2006), 96–125.

6. Between 1867 and 1886, for example, imports of quinine coming through New York increased from about 22,100 ounces to over 1,122,100 ounces, licorice root increased from just over 3 million pounds to over 48 million, and asafetida from about 34,750 pounds to 254,050 pounds. Figures based on "Appendix to the Report of the Committee on the Drug Market," *Proceedings of the American Pharmaceutical Association* (1867), 290–293; "Report on the Dug Market," *Proceedings of the American Pharmaceutical Association* (1886), 2–3. There are a variety of problems in using import statistics from individual cities to estimate overall trends, but in general it appears that the amount of imported pharmaceuticals increased substantially during this period. Note that here and in subsequent citations the year given for the *Proceedings of the American Pharmaceutical Association* refers to the year the meeting took place, not the year the proceedings were published.

7. "Report of the Committee on the Drug Market," *Proceedings of the American Pharmaceutical Association* (1867), 274.

8. C. Lewis Diehl, "Indigenous Drugs," *Proceedings of the American Pharmaceutical Association* (1870), 137–139.

9. Edward R. Squibb, "Circular" (Feb. 1, 1863), National Convention for Revising the Pharmacopoeia, Committee of Revision and Publication, Records, 1858–1864, College of Physicians of Physicians of Philadelphia, b. 2, f. "Correspondence (4)."

10. The growth of the pharmaceutical industry during this period is an understudied area, but see William Haynes, *American Chemical Industry*, vol. 1, *Background and Beginnings, 1609–1911* (New York: Van Nostrand, 1945), 319–334; Haynes, *American Chemical Industry*, vol. 6, *The Chemical Companies* (New York: Van Nostrand, 1954). The December 31, 1896, issue of the *Pharmaceutical Era* also provides useful histories of many pharmaceutical

companies, including E. R. Squibb & Sons, Frederick Stearns & Company, and McKesson & Robbins. For a helpful discussion of drug manufacturing and supply during the Civil War, see Michael A. Flannery, *Civil War Pharmacy* (Binghamton, NY: Pharmaceutical Products Press, 2004), 91–114.

11. Milton L. Hoefle, "The Early History of Parke-Davis and Company," *Bulletin for the History of Chemistry* 25, no. 1 (2000): 30.

12. "Historical Financial Data," in PDCR, b. 5, f. 32.

13. Additional complications include the fact that the ethical segment of the manufacturing industry was itself divided between those that marketed primarily to orthodox physicians and those that marketed their goods primarily to eclectics and other sects; manufacturers sometimes sold to both groups, but some firms were identified primarily with one market. At the same time, the distinction between manufacturers and wholesalers was not as firm as might be assumed. McKesson & Robbins, for example, is sometimes described as a wholesale firm by historians, but by the 1880s it had also begun to manufacture a wide variety of products. McKesson & Robbins, *McKesson & Robbins, New York 1885* in McKesson & Robbins records, Kremers Reference Files, American Institute for the History of Pharmacy, University of Wisconsin, Madison, WI; "McKesson & Robbins," *Pharmaceutical Era*, Dec. 31, 1896, 948–954.

14. Based on patents listed in *Annual Report of the Commissioner of Patents* for those years.

15. "Physicians' Prescriptions and Patent Medicines," *American Journal of Pharmacy*, Jan. 1, 1874, 42.

16. For example, USPN 72,705 (Jan 25, 1876) and USPN 231,236 (Aug. 17, 1880); USPN 156,398 (Oct. 8, 1874) and USPN 206,536 (July 30, 1878); USPN 176,890 (May 2, 1876).

17. For example, "Tasteless Compounds of Iodide of Iron," *Proceedings of the American Pharmaceutical Association* (1873), 143; "Chlorine for Chlorinated Lime," *Proceedings of the American Pharmaceutical Association* (1873), 275; "Maryland College of Pharmacy," *American Journal of Pharmacy*, Aug. 1, 1974, 394; E. T. Ellis, "A New Method of Making Suppositories," *American Journal of Pharmacy*, Apr. 1879, 184–186.

18. *Proceedings of the American Pharmaceutical Association* (1874), 565–566.

19. On the relationship between employers and employees in terms of patent rights, see Catherine L. Fisk, *Working Knowledge: Employee Innovation and the Rise of Corporate Intellectual Property, 1800–1930* (Chapel Hill: University of North Carolina Press, 2009). Although this important topic is beyond the scope of this volume, trends in the pharmaceutical industry appear to have been delayed somewhat beyond the pattern Fisk describes simply because patenting played such a small role in the industry during the nineteenth century.

20. The earliest evidence I have found for a major ethical firm acquiring a patent on a manufacturing process is an 1880 patent licensed to Parke-Davis

for a process of coating pills, USPN 231,236 (Aug. 17, 1880). Whether or not the company used this process is not clear. However, the claim that ethical manufacturers refrained from using patented machinery or methods during the 1870s is tentative. It is certainly possible that I have missed some examples of the early use of process patents during this period.

21. During the 1874 debate about including descriptions of patented machinery in the American Pharmaceutical Association's *Proceedings*, for example, there was some discussion about whether or not E. R. Squibb had patented a retort stand. Joseph Remington made clear that he had not and that Squibb had "brought that stand here, and made it for the benefit of the whole profession, with no view to reaping pecuniary benefit at all, and that any insinuation to that effect is not right." *Proceedings of the American Pharmaceutical Association* (1874), 566.

22. For examples of the different attitudes toward fluid extracts, see "Editorial," *Western Lancet*, Sept. 1872, 551; "Selected Articles," *Chicago Medical Journal*, Oct. 1866, 471.

23. B. Senn, "Report of the Committee on New Remedies," in *Transactions of the State Medical Society of Wisconsin for the Year 1869* (1870), 9.

24. "Inhalation of Calomel," *New Remedies* (1872), 21; "Prescriptions and Formulas," ibid., 74; "Elixir of Calisaya Bark," ibid., 75; "Action of Chloroform," ibid., 147–148.

25. *Pharmacopoeia of the United States of America*, 6th revision (1882), 21.

26. Jan R. McTavish, *Pain & Profits: The History of the Headache and Its Remedies in America* (New Brunswick, NJ: Rutgers University Press, 2004), 69–71.

27. "Introduction of Salicylic Acid into the Pharmacopeia," *Cincinnati Lancet and Clinic*, Aug. 23, 1879, 176.

28. "Pamphlet of E. R. Squibb," in *A Reprint of the Pamphlets of Dr. H. C. Wood, Mr. Alfred B. Taylor, the Philadelphia County Medical Society, and the National College of Pharmacy, with a Rejoinder Addressed to the Professions of Medicine and Pharmacy by Edward R. Squibb, M.D.* (1877), 12.

29. *Proceedings of the Philadelphia County Medical Society in Reference to the Proposed Plan for the Management of the U.S. Pharmacopoeia, as Presented by Dr. Squibb to the American Medical Association for their Action in 1877* (1877), 7.

30. On Stearns, see "Frederick Stearns & Company," in James J. Mitchell, *Detroit in History and Commerce* (1891), 33–34.

31. The 1869 edition of Wood and Bache's *Dispensatory* noted that cinchonia and quinine are therapeutically identical but that the taste of cinchonia "is very bitter." Franklin Wood and George Bache, *Dispensatory of the United States of America* (1869), 1045.

32. *Proceedings of the American Pharmaceutical Association* (1869), 99–102, 112.

33. The Editor [William Procter Jr.], " 'Sweet Quinine': What Is it?," *American Journal of Pharmacy*, July 1869, 302–304.

34. Mitchell, *Detroit in History and Commerce*, 33–34.

35. For example, Charles Wright & Company of Detroit (founded 1880) and

Greensfelder & Company in San Francisco (1881) both adopted his methods. See Mitchell, *Detroit in History and Commerce*, 35; Frederick H. Hackett, *The Industries of San Francisco* (1884), 117.

36. For example, *Proceedings of the Illinois Pharmaceutical Association at Its Sixth Annual Meeting* (1885), 89.

37. On Parke, Davis & Company, see Joseph M. Gabriel, "A Thing Patented Is a Thing Divulged: Francis E. Stewart, George S. Davis, and the Legitimization of Intellectual Property Rights in Pharmaceutical Manufacturing," *Journal of the History of Medicine and Allied Sciences* 64, no. 2 (Apr. 2009): 135–172; Hoefle, "Early History of Parke-Davis"; Joseph F. Spillane, *Cocaine: From Medical Marvel to Modern Menace in the United States, 1884–1920* (Baltimore: Johns Hopkins University Press, 2000), 58–73, 132–134; Haynes, *American Chemical Industry*, 6:320–324. A catalog from 1863, when the firm was still known as "Duffield, Parke & Co.," can be found in Parke, Davis & Company records, Kremers Reference Files, American Institute for the History of Pharmacy, University of Wisconsin, Madison, f. "1868–1885."

38. For example, an early trade catalog of the company includes an "appeal to physicians" in which the company notes that it adheres to the standards of the USP. *Duffield's Medicinal Fluid Extracts* (1868), PDCR, b. 1, f. 1.

39. Spillane, *Cocaine*, 69.

40. Plants introduced in the 1870s included black haw, boldo, and jaborandi. "Black Haw," *New Preparations*, Oct. 1877, 21; "Jaborandi," *New Preparations*, Jan., 1877, 14–17; "New Preparations for 1878 and 1879," *New Remedies*, Oct. 1878, 88. See also Hoefle, "Early History of Parke-Davis," 29.

41. "The New California Remedies," *Medical and Surgical Reporter*, Dec. 21, 1878, 547; J. H. Bundy, "Cascara Sagrado [*sic*]," *New Preparations*, Jan. 15, 1878, 1–2; Bundy, "Yerba Reuma," ibid., 2–3; Bundy, "Berberis Aquifolium," ibid., 3; "Cascara Cordial" in Parke, Davis & Company, *Epitome of the Newer Materia Medica, Standard Medicinal Products and Fine Pharmaceutical Specialties* (1880), 53-54.

42. "Cascara Sagrada," *Louisville Medical News*, Mar. 22, 1879, 136. See also "Doubtful Novelties," *Medical and Surgical Reporter*, Nov. 23, 1878, 455.

43. Horatio R. Bigelow, "Physicians, Pharmacists and the Therapeutic Gazette," *New England Medical Monthly*, Feb. 1882, 204–205. For a defense of Parke-Davis, see "New Remedies and Journalism of California," *Michigan Medical News*, Mar. 25, 1879, 1–2. It is worth noting that in 1882 Davis purchased the *Michigan Medical News* and combined it with the *Detroit Clinic* to form the *Medical Age*. John J. Mulheron, the editor of the *Michigan Medical News*, stayed on as editor of the *Medical Age*. See "Adieu the 'Michigan Medical News,'" *Michigan Medical News*, Dec. 26, 1882, 369.

44. Francis E. Stewart, "Parke, Davis & Co.'s Work," (1879), FESP, b. 6, f. 8.

45. L. Wolff, "Cascara Sagrada," *Monthly Review of Medicine and Pharmacy*, Nov. 1881, 349–350; Lewis Lehn, "Report of the Committee on the Drug Market," *Proceedings of the American Pharmaceutical Association* (1880): 369–370; Vic-

tor C. Vaughan, "A Case of Insufficient Formation of Gastric Juice," *Physician and Surgeon*, Jan. 1880, 13. For a history of the introduction of cascara sagrada, see "How Indigenous Drugs Become Officinal: A Drug with a Interesting History," *Medical Times and Register*, Mar. 15, 1889, 423–425.

46. John Harley Warner, "The Fall and Rise of Professional Mystery: Epistemology, Authority, and the Emergence of Laboratory Medicine in Nineteenth-Century America," in *The Laboratory Revolution in Medicine*, ed. Andrew Cunningham and Perry Williams (New York: Cambridge University Press, 1992), 110–141; Warner, "Ideals of Science and Their Discontents in Late Nineteenth-Century American Medicine," *Isis* 82, no. 3 (Sept.1991): 454–478; Warner, *The Therapeutic Perspective: Medical Practice, Knowledge, and Identity in America, 1820–1885* (Princeton, NJ: Princeton University Press, 1997), 235–283; Gerald L. Geison, " 'Divided We Stand': Physiologists and Clinicians in the American Context," in *The Therapeutic Revolution: Essays in the Social History of American Medicine*, ed. Morris J. Vogel and Charles E. Rosenberg (Philadelphia: University of Pennsylvania Press, 1979), 67–90.

47. Warner, *The Therapeutic Perspective*, 7.

48. Albert B. Prescott, "Chemical and Microscopical Analysis of Rhamnus Puurshiana (Cascara Sagrada)," *American Journal of Pharmacy*, Apr. 1879, 165.

49. On Lyons, see "Albert Brown Lyons," *Journal of the American Pharmaceutical Association*, May 1926, 411–412. On Rusby, see Susan M. Rossi-Wilcox, "Henry Hurd Rusby: A Biographical Sketch and Selectively Annotated Bibliography," *Harvard Papers in Botany* 4 (1993): 1–30; David E. Williams and Susan M. Fraser, "Henry Hurd Rusby: The Father of Economic Botany at the New York Botanical Garden," *Brittonia* 44, no. 3 (1992): 273–279; George A. Bender, "Henry Hurd Rusby: Scientific Explorer, Societal Crusader, Scholastic Innovator," *Pharmacy in History* 23, no. 2 (1981): 71–85.

50. *New Preparations*, Jan. 1, 1877. The name of the journal was changed in 1880. Other journals included the *Detroit Medical Journal* (first published in 1877 and later renamed the *Detroit Lancet* and then the *American Lancet*) and the *Medical Age* (first published in 1883), which was established after Davis purchased the *Michigan Medical News* and combined it with another of his journals, the *Detroit Clinic*. Davis clearly understood that this type of publishing would advance the interests of his firm, by disseminating information about its new products to the medical community and by enhancing the overall reputation of the company. However, he also genuinely believed that he was promoting the cause of scientific medicine. His letters to Stewart, for example, are filled with complaints that the medical profession was not sufficiently appreciative of his efforts. For Davis, there was no contradiction between the promotion of medical science and the creation of markets through the promotion of his company's products.

51. Francis E. Stewart, "Defibrinated Blood as a Substitute for Extract of Beef," *Medical Record*, Mar. 6, 1880, 285; PDC to Francis E. Stewart, n.d., FESP, b. 6, f. 8. For a detailed overview of Stewart's development of desiccated bullock's

blood, see Stewart, "Hæmoglobin Compound, or 'Bullock's Blood in Therapeutics,'" *Proceedings of the Philadelphia County Medical Society* (1890), 115–123.

52. Francis E. Stewart to PDC, Dec. 1, 1879, FESP, b. 6, f. 8.

53. Francis E. Stewart, "The 'Therapeutic Boom' and Unscientific Advertising," *Medical and Surgical Reporter*, Nov. 27, 1880, 465.

54. Francis E. Stewart, typed notes, n.d., FESP, b. 6, f. 8.

55. Francis E. Stewart, "The Relation of Pharmacists to Physicians and the Relation of Pharmacy to Materia Medica and Drug Therapeutics," *Transactions of the American Therapeutic Society* (1903), 30.

56. "Plan Originated by F. E. Stewart, Ph.G., M.D. for Securing Cooperation between Professional and Commercial Interests to Promote Original Materia Medica Research," unpublished notes, FESP, b. 17, f. 4. A working bulletin for *Berberis aquifolium* (ca. 1880) can be found in PDCR, b. 21, f. 3. Working bulletins for *Erythroxylum coca* (ca. 1880) and *Aspidosperma quebracho* (ca. 1880) can be found in Medical Trade Ephemera Collection, College of Physicians of Philadelphia, b. "Parke Davis." Working Bulletins for *Franciscea uniflora* (1884), *Grindelia robusta* (1884), *Cascara cordial* (1884), *Viburnum prunifolium* (1887), *Liquid Ergot, Normal* (n.d.), and *Fabiana imbricata* (1886) can be found in Parke, Davis & Company, *The Ethical Relations Existing between Medicine and Pharmacy, with Illustrations of an Improved Method for the Collective and Scientific Investigation of New Drugs* (1887).

57. Horace Greeley, *The Great Industries of the United States* (1872), 1105–1115.

58. USPN 75,271 (Mar. 10, 1868).

59. "Horsford's Acid Phosphate," *New York Medical Journal* (June 1870), advertisement.

60. "Horsford's Acid Phosphate," *Boston Morning Journal*, May 31, 1871, 3; "Horsford's Acid Phosphate," *New Haven Register*, May 20, 1881, 4.

61. "To Farmers and Others," *Easton Gazette*, Sept. 24, 1864, 3; "Pacific Guang Company's Compound Acid Phosphate of Lime," *Columbus Daily Enquirer*, Apr. 18, 1871, 4; "Wilson's Ammoniated Super Phospate of Lime Has No Superior," *New Hampshire Sentinel*, May 25, 1871, 4.

62. Quoted in "The Acid Phosphate in Bread-Making," *Medical and Surgical Reporter*, Aug. 14, 1869, 146.

63. *Rumford Chemical Works v. Lauer*, 10 Blatch. 122 (1872).

64. For example, *Tilghman v. Proctor*, 102 US 707 (1881).

65. *Rumford Chemical Works v. Lauer*, 10 Blatch. 122 (1872). Horsford unsuccessfully tried to suppress competition using his process patent in other cases as well. For example, *Rumford Chemical Works v. Hecker*, 20 F. 1342 (1876).

66. USPN 49,230 (Aug. 8, 1865); 49,502 (Aug. 22, 1865); 56,179 (July 10, 1866).

67. USPN 99,500 (Feb. 1, 1870).

68. Robert Chesebrough to Commissioner of Patents, Apr. 18, 1870, Patent Wrapper Files, RG 241, NARA II, USPN 127,568; USPN 127,568 (June 4, 1872); *R. A. Chesebrough v. C. Toppan* in *Decisions of the Commissioner of Patents for the Year*

1872 (1873), 100–102. Documents related to the interference case between Chesebrough and Toppan, including a hearing transcript, can be found in Records of the Patent and Trademark Office, Interference Case Files, 1836–1905, NARA II, RG 241, b. 580.

69. *The American Wood-Paper Co. v. The Fibre Disintegrating Co.*, 90 US 566 (1874). On *Young v. Fernie*, see Tal Golan, *Laws of Men and Laws of Nature: The History of Scientific Expert Testimony* (Cambridge, MA: Harvard University Press, 2007), 93–96.

70. *Smith v. Nichols*, 88 US 112 (1874). For another example, see *Roberts v. Ryer*, 91 US 150 (1875).

71. *Smith v. Goodyear*, 93 US 486 (1877). For another example of the Supreme Court deliberating on the relationship between novelty and utility, see *Wicke v. Ostrum*, 103 US 461 (1881).

72. USPN 193,476 (July 7, 1877).

73. *Bowker v. Dows*, 3 F. 1070 (1878).

74. USPN 156,802 (Nov. 10, 1874) and 190,801 (May 15, 1877); USPN 229,130 (June 22, 1880); USPN 163,493 (May 18, 1875); USPN 169,818, (Nov. 9, 1875); USPN 171,662 (Jan. 4, 1876); USPN 177,534 (Dec. 29, 1875).

75. "An Act to Revise, Consolidate, and Amend the Statutes Relating to Patent and Copyright," H.R. 1714, 41st Cong. (1870). On the international context of the 1870 law, see Paul Duguid, "French Connections: The International Propagation of Trademarks in the Nineteenth Century," *Enterprise & Society* 3, no. 1 (2009): 3–37.

76. William Henry Browne, *A Treatise on the Law of Trade-Marks and Analogous Subjects* (1873), v.

77. John Clark Ridpath, ed., "Rules and Forms Adopted by the United Sates Patent Office for the Registration of Trade-Marks under the Act of March 3, 1881," *Standard American Encyclopedia of Arts, Sciences, History, Biography, Geography, Statistics, and General Knowledge*, vol. 8 (1897), 3048.

78. "Phosphorus," *Transactions of the Detroit Medical and Library Association,* July 1879, front advertising material; "Caution to Physicians!," *Ohio Medical Recorder,* Jan. 1879, 1.

79. *Captured and Branded by the Camanche Indians in the Year 1860: A True Narrative* (1876?), 15.

80. In 1874, for example, the New York Court of Appeals found that the combined words "Ferro-Phosphorated Elixir of Calisaya Bark" could not be appropriated as a trademark because the phrase simply indicated the names of the three principal ingredients of the remedy in question. *Caswell v. Davis,* 58 N.Y. 223 (1874).

81. In legal scholarship, the term "thingification" has been used by scholars since the 1920s to refer to the process through which legal relations are hypostatized. For example, Walter Wheeler Cook, "The Logical and Legal Bases of the Conflict of Laws," *Yale Law Journal* 33, no. 5 (Mar. 1924): 476. Following the work of Felix Cohen, scholars interested in the history of

trademark law have used the concept to analyze how during the late nineteenth century the natural rights legal tradition understood trademark law, and the law of unfair competition more broadly, as grounded in property rights; as Cohen put it, " According to the recognized authorities on the law of unfair competition, courts are not *creating* property, but merely *recognizing* a pre-existent Something." Cohen, "Transcendental Nonsense and the Functional Approach," *Columbia Law Review* 35, no. 6 (June 1935): 815. For a recent use of the term in trademark scholarship, see Mark McKenna, "The Normative Foundations of Trademark Law," *Notre Dame Law Review* 82, no. 5 (2006-2007): 1881-1884, 1897n246. For an example of the use of the term in copyright scholarship, see Keith Aoki, "Adrift in the Intertext: Authorship and Audience 'Recording' Rights—Comment on Robert H. Rotstein, 'Beyond Metaphor: Copyright Infringement and the Fiction of the Work,'" *Chicago-Kent Law Review* 68, no. 2 (1992-1993): 807-808. For a discussion of how the late nineteenth-century natural rights legal tradition increasingly viewed trademarks as a form of property, see Robert G. Bone, "Hunting Goodwill: A History of the Concept of Goodwill in Trademark Law," *Boston University Law Review* 86 (2006): 561-567. I use the term "thingification," rather than the term "reification," in order to place myself in this tradition. For my purposes the two terms can be taken as synonyms, although there may be technical differences between the two depending on one's theoretical orientation.

82. *Filkins et al. v. Blackman*, 9 F. 50 (1876).
83. *OGPO*, Jan. 13, 1874, 35.
84. For botanical information on Tonga, see "The Botanical Origin of Tonga," *British Medical Journal*, July 20, 1881, 171.
85. "The 'Tonga' Controversy," *Mississippi Valley Medical Monthly*, Nov. 1881, 530-531.
86. "The Tonga Case," FESP, misc. correspondence, b. 1, p. 46-47.
87. "Tonga," in *Epitome of the Newer Materia Medica,* 23.
88. *Epitome of the Newer Materia Medica*, vi.
89. Report summarized in "Meeting of the Rhode Island Medical Society," *Medical and Surgical Reporter*, Oct. 18, 1879, 339-341.
90. C. A. Lindsley, "Proprietary Medicines," *Proceedings of the Connecticut Medical Society, 1880* (1880), 104-105.
91. "An Evil and Its Remedy," *College and Clinical Record*, Jan. 15, 1881, 12.
92. Francis H. Upton, *A Treatise on the Law of Trade Marks, with a Digest and Review of the English and American Authorities* (1860), 15-16.
93. "Parke, Davis & Co., Manufacturing Chemists," *Cincinnati Medical News*, July 1879, 501.
94. "Poisoned by Mistake," *San Francisco Bulletin*, Jan. 5, 1874, 3; "The Poisoning Case," *San Francisco Bulletin* Jan. 8, 1874, 3; "Culpable Negligence," *San Francisco Bulletin*, Jan. 9, 1874, 1; "Brief Mention," *San Francisco Bulletin*, Jan. 22, 1874, 1; "A Druggist Sued for Damages," *San Francisco Bulletin*, Mar. 30, 1874, 3.
95. Joseph M. Gabriel, "Restricting the Sale of 'Deadly Poisons': Pharmacists,

Drug Regulation and Narratives of Suffering in the Gilded Age," *Journal of the Gilded Age and Progressive Era* 9, no. 3 (July 2010): 313–336; *The General Statutes of the State of Rhode Island and Providence Plantations* (1872), 253–256. Early pharmacy laws were also intended to suppress the consumption of intoxicating drugs for what were sometimes called "degraded" purposes. Following the Civil War, reformers grew increasingly concerned about the frequent and habitual consumption of morphine, opium, hashish, and other intoxicating drugs in ways that seemed both dangerous and transgressive. The early regulation and consumption of these types of intoxicating drugs is beyond the scope of this volume, as is the topic of drug addiction more broadly, but there is a vibrant and growing literature that considers these and related questions. For example, in addition to Gabriel, "Deadly Poisons," see David Courtwright, *Dark Paradise: A History of Opiate Addiction in America* (Cambridge, MA: Harvard University Press, 2009); Diana L. Ahmad, *The Opium Debate and Chinese Exclusion Laws in the Nineteenth-Century American West* (Reno: University of Nevada Press, 2007); Marcus Boon, *The Road of Excess: A History of Writers on Drugs* (Cambridge, MA: Harvard University Press, 2005); Nyan Shah, *Contagious Divides: Epidemics and Race in San Francisco's Chinatown* (Berkeley: University of California Press, 2001), 90–97. For a discussion of the early regulation and use of cocaine, which began in the 1880s, see Spillane, *Cocaine*. For a useful discussion of why it makes sense to treat alcohol and other drugs in similar terms historically, see Sarah W. Tracy and Caroline Jean Acker, "Introduction: Psychoactive Drugs—An American Way of Life," in *Altering American Consciousness: the History of Alcohol and Drug Use in the United States, 1800–2000*, ed. Tracy and Acker (Amherst: University of Massachusetts Press, 2004), 1–32.

96. Amy Dru Stanley, *From Bondage to Contract: Wage Labor, Marriage, and the Market in the Age of Slave Emancipation* (New York: Cambridge University Press, 1998).

97. For example, "Food and Drug Adulteration Law of Massachusetts," reprinted in *Proceedings of the American Pharmaceutical Association* (1882), 492–493. For an example of a law that did not define adulteration, see "An Act to Prevent the Adulteration of Food, Drinks and Medicines," *Session Laws of the State of Wyoming* (1884), 4–5.

98. "Petroleum Jelly Vaseline," *Retrospect of Practical Medicine and Surgery*, July 1879, back advertising material; William Greene, "Supplementary Note on Boric Acid," *Boston Medical and Surgical Journal*, Sept. 9, 1880, 249–250; W. W. Seely, "The New Departure in the Treatment of Purulent Ophthalmia," *Monthly Review of Medicine and Pharmacy*, May 1881, 141.

99. Samuel A. D. Sheppard, "On Vaseline and Petroleum Products," *New Remedies*, April 1881, 105–106.

100. Ibid., 105.

101. *Proceedings of the New York State Pharmaceutical Association, First Annual Meeting, May 21st and 22d, 1879* (1879), 102–104.

102. Sheppard, "Vaseline and Petroleum Products," 106.

103. *Pharmacopoeia of the United States of America,* 6th revision (1882), 248.

104. Sheppard, "Vaseline and Petroleum Products," 106.

CHAPTER 4

1. For an account of one exploration to the American Southwest, see George Bender, "Rough and Ready Research—1887 Style," *Journal of the History of Medicine and Allied Sciences* 23, no. 2 (1968): 159–166.

2. "Historical Financial Data," PDCR, b. 5, f. 32. There is some internal company correspondence from the mid-1880s in Parke, Davis & Company records, Kremers Reference Files, American Institute for the History of Pharmacy, University of Wisconsin, Madison WI, f. "1868–1885" and f. "1886."

3. Quoted in Joseph Spillane, *Cocaine: From Medical Marvel to Modern Menace in the United States, 1884–1920* (Baltimore: Johns Hopkins University Press, 2000), 71–72.

4. George Davis to Francis E. Stewart, Dec. 5, 1889, FESP, b. 7, f. 2. See also Francis E. Stewart, *Working Bulletin for the Scientific Investigation of Cascara Cordial* (1884).

5. USPN 231,236 (Aug. 17, 1880).

6. William E. Upjohn to Henry Upjohn, Mar. 6, 1884, Upjohn Family Papers, 1795–1974, Bentley Historical Library, University of Michigan, b. 1, f. 25.

7. USPN 198,759 (Jan. 1, 1878).

8. Lucius C. West to Henry Upjohn, Dec. 18, 1884, Upjohn Family Papers, 1795–1974, Bentley Historical Library, University of Michigan, b. 1, f. 25.

9. USPN 312,041 (Feb. 10, 1885).

10. "The Upjohn Pill and Granule Co.," *Pharmaceutical Era,* Dec. 31, 1896, 945–947.

11. "Upjohn's Ideal-Coated Pills," *Medical Brief,* Nov. 1887, advertising section, 4.

12. Untitled report, Jan. 13, 1891, Pfizer, Inc., Collection, 1888–1995, 3.2 Conferences and Meetings, Western Michigan University Special Collections, b. 1, f. 2.

13. "Why Certain Pills Should Not Be Prescribed," *Medical Age,* Feb. 11, 1889, 61.

14. Petra Moser has recently suggested that patenting activity is inversely correlated to the degree to which secrecy is protective and that advances in analytic chemistry between 1851 and 1893 made secrecy less protective for chemical products, with a resulting shift toward patenting of chemical goods. Moser, "Innovation without Patents—Evidence from the World Fairs" (Apr. 15, 2011). Available at Social Science Research Network: http://ssrn.com /abstract=930241.

15. 111 U.S. 293 (1884). Although the substance is now typically spelled "alizarin," at the time it was frequently spelled "alizarine."

16. USPN 95,465 (Oct. 5, 1869), reissue no. 4,321 and 4,320 (Apr. 4, 1871).

17. William Callyhan Robinson, *The Law of Patents for Useful Inventions,* vol. 1 (1890), 283n197.

18. Ibid., 282.
19. Ibid., 383.
20. *Glue Co. v. Upton*, 97 US 3 (1877).
21. *Ex Parte Latimer*, 46 O.G. 1638 (1889), in *Decisions of the Commissioner of Patents* (1889), 123–127. For a detailed examination of the decision, see Jon Harkness, "Dicta on Adrenalin(e): Myriad Problems with Learned Hand's Product-of-Nature Pronouncements in *Parke-Davis v. Mulford*," *Journal of the Patent and Trademark Office Society* 93, no. 4 (2011): 363–399. In recent years, legal historians and other scholars have suggested that this was the first decision in which "products of nature" were held to be beyond patentability. Whether or not this is the case, the decision was hardly as remarkable as most commentators have made it out to be. Littlewood's decision was firmly within the long-standing legal tradition that in order to be patentable a previously existing substance must be transformed into something truly new. The decision thus conformed to Willard Phillip's 1837 observation that previously existing things cannot be patented and that this doctrine applies to "known materials, and to such as naturally exist, whether known or not; for the discovery of a new elementary substance or material, by analysis or otherwise, does not give a right of a monopoly of it." Littlewood explicitly explained his decision by pointing out that had he ruled otherwise, "it would be possible for an element or a principle to be secured by patent." On this, see Willard Phillips, *Law of Patents and Invention; Including the Remedies and Legal Proceedings in Relation to Patent Rights* (1837), 113. For an excellent discussion of the products-of-nature doctrine, including *Ex Parte Latimer*, see Christopher Beauchamp, "Patenting Nature: A Problem of History," *Stanford Technology Law Review* 16, no. 2 (Winter 2013): 257–311. Other examples include Michael S. Carolan, "The Mutability of Biotechnology Patents: From Unwieldy Products of Nature to Independent 'Object/s,' " *Theory, Culture & Society* 27, no. 1 (2010): 110–129; Daniel J. Kevles, "Patents, Protections, and Privileges: The Establishment of Intellectual Property in Animals and Plants," *Isis* 98, no. 2 (June 2007): 323–331; John M. Conley and Roberte Makowski, "Back to the Future: Rethinking the Product of Nature Doctrine as a Barrier to Biotechnology (Part I)," *Journal of the Patent Office Society* 85, no. 4 (Apr. 2003): 301–334. See also Harold C. Thorne, "Relation of Patent Law to Natural Products," *Journal of the Patent Office Society* 6, no. 1 (Sept. 1923): 23–28.
22. William Callyhan Robinson, *The Law of Patents for Useful Inventions*, vol. 1 (1890), 411.
23. Only a handful of patents were issued for what were little more than simple assemblages of ingredients. Examples include USPN 303,603 (Aug. 12, 1884), for a medicinal compound made from juniper berries, blossoms of yarrow, balsam of fir, calamus root, and alcohol; USPN 280,281 (June 26, 1883) for a "medicine for scrofula" made from tulip tree bark, sarsaparilla, comfrey root and goldenseal; USPN 234,785 (Nov. 23, 1880), for a cough medicine made from Indian pleurisy root, saltpeter, honey, and brandy.

24. USPN 216,496 (Dec. 27, 1877).
25. USPN 412,838 (Oct. 15, 1889). See also USPN 412,835 (Oct. 15, 1889), for a "liquid digestive compound"; USPN 412,839 (Oct. 15, 1889), for "prepared food for infants and invalids"; USPN 412,836 (Oct. 15, 1889), for a process for preparing digestive compounds.
26. USPN 412,838 (Oct. 15, 1889). Other examples include USPN 409,175 (1889), for a "worm remedy" in which the various ingredients are "evenly and thoroughly mixed with the cream, and the latter is otherwise peculiarly adapted for assimilation with the medicine"; USPN 632,552 (Sept. 5, 1899), for an "internal remedy" made from lupulin powder and hop extract, in which "there is a correlated function between the moist hop extract and the lupulin powder in that the lupulin gives its aromatic values to the hop extract, and the moist hop extract in turn hermetically seals the particles of lupulin . . . [and] by lending its kindred juices to the dried out and depreciated lupulin powder, rejuvenates and restores it and renders it a soluble and useful product"; USPN 260,140 (June 27, 1882), for a cough remedy in which uva ursi is "mingled with the other medicines," including bloodroot and sassafras, and as a result "the compound becomes more certain and effectual in its action upon the mucus secretion of the system." In each of these cases the patent describes a process in which the various ingredients are combined with each other in some way that goes beyond simple assemblage.
27. USPN 364,332 (June 7, 1887).
28. "Sugar Coated Licorice Lozenges," *Pharmaceutical Era*, Nov. 1887, 425.
29. Kara W. Swanson, "The Emergence of the Professional Patent Practitioner," *Technology and Culture* 50, no. 3 (2009): 519–548.
30. *Trade-Mark Cases*, 100 US 82 (1879).
31. On the *Trade-Mark Cases* and the origins of the 1881 Trademark Act, see Zvi S. Rosen, "In Search of the Trademark Cases: The Nascent Treaty Power and the Turbulent Origins of Federal Trademark Law," *St. John's Law Review* 83, no. 3 (Summer 2009): 827–904; Paul Duguid, "French Connections: The International Propagation of Trademarks in the Nineteenth Century," *Enterprise & Society* 10, no. 1 (2009): 3–37.
32. "Patent Medicines," *American Practitioner and News*, Oct. 30, 1886, 319.
33. Harlan P. Hubbard to C. H. Pinkham, Feb. 22, 1882, Lydia E. Pinkham Company Records, 1776–1968, Schlesinger Library, Radcliff Institute for Advanced Study, b. 84, f. 763; label patent no. 536 (Feb. 15, 1876), in *Annual Report of the Commissioner of Patents for the Year 1876* (1877), 442; Pamela Walker Laird, *Advertising Progress: American Business and the Rise of Consumer Marketing* (Baltimore: Johns Hopkins University Press, 1998), 176–178; Sarah Stage, *Female Complaints: Lydia Pinkham and the Business of Women's Medicine* (New York: W. W. Norton, 1979). For a thoughtful discussion of the ingredients in Pinkham's remedy, see Susan Strasser, "Commodifying Lydia Pinkham: A Woman, a Medicine, and a Company in a Developing Consumer Culture,"

Working Paper no. 32, ESRC/AHRC Cultures of Consumption Programme, 11–14, http://www.consume.bbk.ac.uk/publications.html.

34. "Upjohn's Ideal-Coated Pills," *Medical Brief*, Nov. 1887, advertising section, 4.

35. Henry Upjohn to [unclear], May 5, 1885, Upjohn Family Papers, 1795–1974, Bentley Historical Library, University of Michigan, b. 1, f. 25.

36. *Amoskeag Manufacturing Co. v. Trainer*, 101 US 51 (1879). Other examples include *Schneider v. Williams*, 44 N.J. Eq. 391 (1888); *Jennings v. Johnson*, 37 F. 364 (1888); *Marshall v. Pinkham*, 52 Wis. 572 (1881).

37. *Brown Chemical Co. v. Stearns* 37 F. 360 (1889), in *The Federal Reporter*, vol. 37, *Cases Argued and Determined in the Circuit and District Courts of the United States. January–May, 1889* (1889), 360–363.

38. *Avery & Sons v. Meikle & Co.*, 81 Ky. 73 (1883). Other examples include *Leclanche Battery Co. v. Western Electric Co.*, 22 F. 276 (1885); *Armstrong v. Kleinhans*, 82 Ky. 303 (1884).

39. USTN 4,793 (June 26, 1877).

40. "Important Decision of the Supreme Court in the Matter of Imitations of Horsford's Acid Phosphate," *Boston Medical and Surgical Journal*, Dec. 6, 1883, 2.

41. Anthony Polk to Rumford Chemical Works, Oct. 26, 1883, Rumford Chemical Company Records, 1853–1951, Rhode Island Historical Society, b. 5, f. 107.

42. *Rumford Chemical Works v. Muth*, 35 F. 524 (1888).

43. In 1881, the company spent only about $28,000 on advertising and other promotional efforts for acid phosphate. By 1887 the amount had jumped to about $80,000. Data derived from Rumford Chemical Works sales ledgers, vols. 236, 237, 250, and 251, in Rumford Chemical Works Records, 1848–1948, Rutgers University.

44. *Rumford Chemical Works v. Muth*, 35 F. 524 (1888).

45. Data derived from Rumford Chemical Works sales ledgers, vol. 237, in Rumford Chemical Works Records, 1848–1948, Rutgers University.

46. E. M. Hale, "A Colorless Solution of Hydrastis. Its Uses, etc.," *Medical Era*, November 1885, 130–131; Parke, Davis & Company, *Organic Material Medica* (1888), 88; "Lloyd's Hydrastis" promotional flyer, in Lloyd Brothers Pharmacists, Inc., Collection, Lloyd Library and Museum, Cincinnati, b. 3, f. 3; "A Certificate of Character for the President of the American Pharmaceutical Association," *Bulletin of Pharmacy*, June 1888, 148–149. The best treatment of Lloyd, including a useful account of the hydrastis controversy, is Michael Flannery, *John Uri Lloyd: The Great American Eclectic* (Carbondale: Southern Illinois University Press, 1998). In addition to accusations against Lloyd of exaggerating therapeutic claims for the product, Joseph Remington also accused Lloyd of keeping his manufacturing process secret and thereby adopting the methods of quackery. Joseph Remington to John Uri Lloyd, Jan. 10 and 15, 1888, John Uri Lloyd Papers, 1849–1936, Lloyd Library and Museum, Cincinnati, b. 11, f. 223.

47. *A Treatise on Asepsin and Asepsin Soap* (1904), 2, in Lloyd Brothers Pharma-

cists, Inc., Collection, Lloyd Library and Museum, Cincinnati, b. 6, f. 37; USTN 16,519, in *OGPO*, Apr. 23, 1889, 406.

48. *Lloyd Bros. v. Merrill Chemical Co.*, 1891 Ohio Misc. (1891).

49. For example, "Asepsin—the New Antiseptic," *American Medical Journal*, Nov. 1887, 521–523; John Fearn, "Asepsin (Lloyd's)," *California Medial Journal*, Aug. 1892, 361–364.

50. Lloyd Brothers to Cyrus Eison, Jan. 15, 1896, Lloyd Brothers Pharmacists, Inc., Collection, Lloyd Library and Museum, Cincinnati, b. 6, f. 37.

51. Ibid.

52. C. A. Lindsley, "Proprietary Medicines," *Proceedings of the Connecticut Medical Society, 1880* (1880), 104–105.

53. "The Trademark and Patent Medicines," *Medical Bulletin*, Jan. 1882, 22.

54. Francis E. Stewart, "Patent Medicines and Trade-Mark Pharmaceuticals," *Therapeutic Gazette*, March 15, 1882, 109.

55. C. A. Lindsley, "Proprietary Medicines," *Proceedings of the Connecticut Medical Society, 1880* (1880), 104–105.

56. USTN 13,392, 13,393, 13,394, 13,395, in *OGPO*, June 15, 1886, 1232.

57. "Flagrant Abuses Requiring Correction—Proprietary vs. Legitimate Pharmacy," *Kansas City Medical Index*, Apr. 1889, 123.

58. Austin Flint, *Medical Ethics and Etiquette* (New York, 1893), 40–41. On Flint and the controversy, see Ruth R. Faden and Tom L. Beauchamp, *A History and Theory of Informed Consent* (New York: Oxford University Press, 1986), 4, 73.

59. On the origins of medical licensing laws in the decades following the Civil War, see Samuel L Baker, "Physician Licensure Laws in the United States, 1865–1915," *Journal of the History of Medicine and Allied Sciences* 39, no. 2 (Apr. 1984): 173–197; Ronald Hamowy, "The Early Development of Medical Licensing Laws in the United Sates, 1875–1900," *Journal of Libertarian Studies* 3 (1979): 73–119.

60. John Harley Warner, "The 1880s Rebellion against the AMA Code of Ethics: 'Scientific Democracy' and the Dissolution of Orthodoxy," in *The American Medical Ethics Revolution: How the AMA's Code of Ethics Has Transformed Physicians' Relationships to Patients, Professionals, and Society*, ed. Robert B. Baker, Arthur L. Caplan, Linda L. Emanuel, and Stephen R. Latham (Baltimore: Johns Hopkins University Press, 1999), 52–69.

61. David Hunt, "Bigotry in the Medical Profession," *North American Review*, Jan. 1883, 77–87.

62. Warner, "The 1880s Rebellion," 66.

63. For example, George B. H. Swayze, "Comedies of Therapeutics," *Therapeutic Gazette*, Mar. 15, 1882, 83.

64. In 1884, for example, two physicians were denied entry to a county medical society in Mississippi for violating a clause of the society's code of ethics that prohibited physicians from acquiring patents. "Patent Medicine," *Iowa State Medical Reporter*, Sept. 1884, 47.

65. The Ebert Survey of 1885, in E. N. Gathercoal, *The Prescription Ingredient Survey* (American Pharmaceutical Association, 1933), 7.

66. Francis E. Stewart, *Open Pharmacy and Scientific Substitutes for Proprietary Preparations: The Remedy for the Great 'Patent Medicine' Evil* (1882), 3.

67. Francis E. Stewart, "How Can Physicians Aid in Elevating the Profession of Pharmacy," *Maryland Medical Journal*, Aug. 2, 1884, 268.

68. Francis E. Stewart, "Report of the Special Committee on National Legislation," *Proceedings of the American Pharmaceutical Association* (1898), 58. Note that here and in subsequent citations the year given for the *Proceedings of the American Pharmaceutical Association* refers to the year of the meeting, not the year the proceedings were published.

69. Stewart, *Open Pharmacy*, 7.

70. Francis E. Stewart, "On Patent and Trademark Law," *Proceedings of the American Pharmaceutical Association* (1889), 132–154.

71. Francis E. Stewart, "Medicine and the Patent Law," *Medical Herald*, Dec. 1881, 391.

72. Quoted in "Medical Patents and Trade Marks," *Monthly Review of Medicine and Pharmacy*, July 1881, 210.

73. Horatio R. Bigelow, "Trade-Mark Pharmacy," *Medical Brief*, Jan. 1882, 44–45. For other examples, see "Champions of Ethics," *St. Louis Clinical Record*, Jan. 1882, 298–300; "Proprietary Medicines and Quack Medicines," *Sanitarian*, Nov. 1882, 691–693. See also Parke, Davis & Company, "A Word to Those Interested in the Defense of the Rights of the Public, Science, and Medicine," *Denver Medical Times*, Jan. 1883, back material.

74. Examples include Francis E. Stewart, *Legitimate Medicine and Pharmacy vs. Nostrum Vendors* (1881), pamphlet in FESP, b. 17, f. 5; Stewart, "The Materia Medica of the Future," *Transactions of the American Medical Association* (1881), 169–172; Stewart, "The 'Therapeutic Boom' and Unscientific Advertising," *Therapeutic Gazette*, Jan. 15, 1881, 23–25; [Stewart], "The Philadelphia Medical Profession and Copyrighted Medicines," *Therapeutic Gazette*, Jan. 15, 1881, 19; Stewart, "Medicine and the Patent Law," *Medical Herald*, Dec. 1881, 391; [Stewart], "Patent Medicines and Trade-Mark Pharmaceuticals," *Therapeutic Gazette*, Mar. 15, 1882, 106–109; Stewart, "Valoids," *Therapeutic Gazette*, May 15, 1885, 304; Stewart, "Proprietary Medicines Not Protected by Patent-Laws," *Philadelphia Medical Times*, Apr. 17, 1886, 552. Examples of articles published in the pharmaceutical press include Stewart, "Patent Law vs. Patent Medicines," *Bulletin of Pharmacy*, 1889, 49–50; Stewart, "A Code of Pharmaceutical Ethics," *Druggist Circular and Chemical Gazette*, Apr. 1890, 73; Stewart, "On Patent and Trademark Laws," 132–154.

75. Stewart, "Valoids," 304.

76. "Champion of Ethics," *St. Louis Clinical Record*, Jan. 1882, 299.

77. "Akesis Universalis," *JAMA*, (Jan., 1884), 54.

78. Francis E. Stewart to PDC, Jan. 9, 1883, FESP, b. 6, f. 9.

79. PDC to Francis E. Stewart, May 20, 1884, FESP, b. 7, f. 1.

80. PDC to Francis E. Stewart, Apr. 3, 1884, FESP, b. 7, f. 1.
81. Jan R. McTavish, *Pain & Profits: The History of the Headache and Its Remedies in America* (New Brunswick, NJ: Rutgers University Press, 2004), 64–70; B. Zorina Khan, "An Economic History of Patent Institutions" (Feb. 5, 2010), EH.net Encyclopedia, http://eh.net/encyclopedia/article/khan.patents.
82. Wolfgang Wimmer, "Innovation in the German Pharmaceutical Industry," in *The Chemical Industry in Europe, 1850–1914: Industrial Growth, Pollution, and Professionalization,* ed. Ernst Homburg, Anthony S. Travis and Harm G. Schröter (Dordrecht, the Netherlands: Kluwer Academic, 1998), 287–288.
83. "Dr. Stewart's Reply," *Medical Times,* Feb. 1890, 353.
84. *USPN* 307,399 (Oct. 28, 1884).
85. "Antipyrine a Patent Medicine," *St. Louis Medical Journal,* Oct. 1887, 222–223.
86. Quoted in "The Modern Antipyretics," *Medical Record,* June 23, 1888, 697.
87. McTavish, *Pain & Profits,* 77–78; "Constitution of Antipyrine," *American Druggist,* Nov. 1888, 218.
88. H. Reding, "Some Uses of Acetanilid and Antipyrin," *Kansas Medical Journal,* Aug. 1890, 603.
89. "Methozin, Antipyrine, Analgesine," *American Druggist,* Oct. 1889, 200. See also "Antipyrine and Its Substitutes," *Gaillard's Medical Journal,* July 1889, 79–80; "New York Letter," *Western Druggist,* July 1888, 259.
90. "Antifebrin," *Pacific Medical and Surgical Journal and Western Lancet,* Aug. 1888, 502.
91. For example, "Antifebrin," *American Druggist,* Aug. 1887, 156–157.
92. "A Misuse of Scientific Nomenclature," *Druggist's Bulletin,* Jan. 1888, 1.
93. "Phenacetin," *Therapeutic Gazette,* Feb. 15, 1888, 142.
94. USPN 400,086 (Mar. 26, 1889); USTN 18,637.
95. "Phenacetin," *American Practitioner,* June 9, 1888, 381.
96. Quoted in McTavish, *Pain & Profits,* 75. On the impact of the German synthetic drugs, see McTavish, *Pain & Profits,* 64–85, 112–133; McTavish, "What Did Bayer Do before Aspirin? Early Pharmaceutical Marketing Practices in America," *Pharmacy in History* 41, no. 1 (1999): 3–15.
97. "Minutes of the Seventh Decennial Convention for the Revision of the Pharmacopeia of the United State of America" (1890), 355–361, in United States Pharmacopeial Convention Records, 1819–2000, Wisconsin Historical Society, b. 4, f. USP Minutes (1890).
98. PDC to Francis E. Stewart, Oct. 31, 1889, FESP, b. 7, f. 2; George Davis to Francis E. Stewart, Oct. 26, 1889, FESP, b. 7, f. 2; George Davis to Francis E. Stewart, Nov. 29, 1889, FESP, b. 7, f. 2; George Davis to Francis E. Stewart, Nov. 29, 1889, FESP, b. 7, f. 2.
99. "Dr. Stewart's Reply," *New York Medical Times,* Feb. 1890, 353.
100. The revision committee made one mistake and included Salol under that name, which was trademarked, rather than its chemical name of "phenyl salicylate." *Pharmacopeia of the United States of America,* 7th revision (1893), xxxi; "New Remedies in the U.S.P.," *Notes on New Remedies,* April 1894, 170–171.

101. Ibid., 170–171.
102. Stewart, "How Can Physicians Aid in Elevating the Profession of Pharmacy," 267.
103. S. H. Kendall to Lucius Wood, Nov. 14, 1887, Lucius F. Wood Papers, 1882–1889, Rutgers University Special Collections, b. 1, f. 1.
104. Albert B. Prescott, "Nostrums in Their Relation to Public Health," *Physician and Surgeon*, May 1881, 219. The paper was reprinted widely, including in *Annual Report of the Commissioner of the Michigan Department of Health* (1882), 151–160. On Prescott, see Dennis B. Worthen, "Albert Benjamin Prescott (1832–1905): Pharmacy Education's Revolutionary Spark," *Journal of the American Pharmacists Association* 44, no. 3 (May–June 2004): 407–410.
105. Joseph M. Gabriel, "Restricting the Sale of 'Deadly Poisons': Pharmacists, Drug Regulation and Narratives of Suffering in the Gilded Age," *Journal of the Gilded Age and Progressive Era* 9, no. 3 (July 2010): 313–336.
106. Robert H. Cowdrey, "Who Is Responsible for Adulteration?," *Proceedings of the American Pharmaceutical Association* (1883), 363.
107. Ibid., 365.
108. "New York Druggists' Union," *Pharmaceutical Record*, Feb. 15, 1884, 93.
109. Analysis of census data by Matthew Sobek puts the number of practicing pharmacists in 1870 at 18,800 and in 1900 at 50,100. Sobek, "New Statistics on the U.S. Labor Force, 1850–1990," *Historical Methods* 34, no. 2 (Spring 2001), table A1.
110. "The Pharmacists' Trouble," *Pharmaceutical Record*, Mar. 1884, 98.
111. "The Retail Drug Trade at Home and Abroad," *Pharmaceutical Record*, Nov. 1, 1883, 1.
112. Justice, "Trade-Marks in Pharmacy," *Druggist*, Sept. 1883, 206–207. For another example, see "A Pharmaceutical Privilege," *Western Druggist*, Dec. 15, 1886, 451–452.
113. Quoted in *Proceedings of the Fourth Annual Meeting of the New York State Pharmaceutical Association* (1882), 90.
114. *United States v. Braun*, 39 F. 775 (1889). See also "Important Decision Affecting Proprietary Articles," *American Druggist*, Oct. 1889, 197–198.
115. *Battle & Co. v. Finlay et al.*, 45 F. 796 (1891).
116. "The Rebate Plan," *Pharmaceutical Record*, May 1, 1883, 1. For details of other plans, see "The Coming Relief," *Pharmaceutical Record*, Mar. 1, 1884, 99.
117. "Report of Willis G. Tucker, M.D., Ph.D., Analyst of Drugs," in *Annual Report of the State Department of Health of New York* (1890), 454–455, 475.
118. "An Enemy of Science," *Virginia Medical Monthly*, Apr. 1881, 68–69.
119. McTavish, *Pain & Profits*, 37; Albert B. Prescott, "Should Proprietary Medicines Be Required to Give an Account of Contents?," *American Druggist*, July 1885, 123.
120. John Uri Lloyd, "What Is Adulteration in Pharmacy?," *American Druggist*, Dec. 1889, 222–224.
121. Gregory J. Higby, in Lee Anderson and Gregory J. Higby, *The Spirit of Volun-*

tarism, a Legacy of Commitment and Contribution: The United States Pharmacopeia, 1820–1995 (Rockville, MD: United States Pharmacopeial Convention, 1995), 180.

CHAPTER 5

1. Newell Dwight Hillis, *The Quest of Happiness* (1902), 143.
2. Ibid., 475–477.
3. Daniel T. Rogers, "In Search of Progressivism," *Reviews in American History* 10, no. 4 (Dec. 1982): 124.
4. Among other works, my understanding of the corporate transformation of American capitalism during this period has been influenced by Jackson Lears, *Rebirth of a Nation: The Making of Modern America, 1877–1920* (New York: HarperCollins, 2009); Phillip Scranton, *Endless Novelty: Specialty Production and American Industrialization, 1865–1925* (Princeton, NJ: Princeton University Press, 1997); James Livingston, *Pragmatism and the Political Economy of Cultural Revolution, 1850–1940* (Chapel Hill: University of North Carolina Press, 1994); Jackson Lears, *No Place of Grace: Antimodernism and the Transformation of American Culture, 1880–1920* (Chicago: University of Chicago Press, 1994); William R. Leach, *Land of Desire: Merchants, Power, and the Rise of a New American Culture* (New York: Pantheon, 1993); Alfred Chandler, *The Visible Hand: The Managerial Revolution in American Business* (Cambridge, MA: Harvard University Press, 1993); Martin J. Sklar, *The Corporate Reconstruction of American Capitalism, 1890–1916: The Market, the Law, and Politics* (New York: Cambridge University Press, 1988); Alan Trachtenberg, *The Incorporation of America: Culture and Society in the Gilded Age* (New York: Hill and Wang, 1982).
5. J. C. Clayton, "Equity in Patent Cases," *Scientific American*, Feb. 11, 1893, 90.
6. *Annual Report of the Commissioner of Patents for the Year 1902* (1903), v.
7. Mira Wilkins, "The Neglected Intangible Asset: The Influence of the Trademark on the Rise of the Modern Corporation," *Business History* 34, no. 1 (Jan. 1992): 66–95. On advertising more broadly, see Pamela Walker Laird, *Advertising Progress: American Business and the Rise of Consumer Marketing* (Baltimore: Johns Hopkins University Press, 2008); Jackson Lears, *Fables of Abundance: A Cultural History of Advertising in America* (New York: Basic Books, 1994).
8. On the history of the pharmaceutical industry during this period, see Jan R. McTavish, *Pain & Profits: The History of the Headache and its Remedies in America* (New Brunswick, NJ: Rutgers University Press, 2004), 86–133; Joseph F. Spillane, *Cocaine: From Medical Marvel to Modern Menace in the United States, 1884–1920* (Baltimore: Johns Hopkins University Press, 2000); Jonathan Liebenau, *Medical Science and Medical Industry: The Formation of the American Pharmaceutical Industry* (Baltimore: Johns Hopkins University Press, 1987); David E. Lilienfeld, "The First Pharmacoepidemiologic Investigations: National Drug Safety Policy in the United States, 1901–1902," *Perspectives in Biology and Medicine* 51, no. 2 (Spring 2008): 188–198; Jonathan Liebenau,

"Ethical Business: The Formation of the Pharmaceutical Industry in Britain, Germany, and the United States before 1914," *Business History* 30, no. 1 (Jan. 1988): 116–129; John Swann, "The Evolution of the American Pharmaceutical Industry," *Pharmacy in History* 37, no. 2 (1995): 76–86; Liebenau, "Industrial R & D in Pharmaceutical Firms in the Early Twentieth Century," *Business History* 84, no. 26 (Nov. 1984): 329–346; William Haynes, *American Chemical Industry*, vol. 1, *Background and Beginnings* (New York: D. Van Nostrand, 1945), 319–334. See also the company histories in Haynes, *American Chemical Industry*, vol. 6, *Company Histories to 1948* (1954).

9. "The League Plan and Its Work," *Western Druggist*, Nov. 1894, 401. Capital invested in the patent medicine industry increased from about $18.5 million to $37.2 million, while the value of goods increased from $32.6 million to $59.6 million. Bureau of the Census, *Abstract of the Twelfth Census of the United States, 1900* (1904), 315.

10. For an explanation of the impossibility of comparing data between the two censuses for the classifications "druggist's preparations" and "pharmaceutical preparations," see Bureau of the Census, *Manufactures 1905*, part 1, *United States by Industries* (1907), clxxvi.

11. "Historical Financial Data," PDCR, b. 5, f. 32.

12. "McKesson & Robbins' Gelatine-Coated Pills," *Pacific Medical Journal*, Oct. 1880, front advertising material; "The Results of the Quinine Pill Analyses," *Medical News*, Aug. 25, 1883, 209.

13. E. H. Stevenson, "New Remedies," *Annual of Eclectic Medicine & Surgery*, vol. 2 (1891), 42.

14. "McKesson & Robbins," *Pharmaceutical Era*, Dec. 31, 1896, 952.

15. "The Manufacturing Laboratory in the Household of Pharmacy," *Pharmaceutical Review*, Apr. 1897, 64.

16. *Kola* (1894), 1, in Frederick Stearns & Company records, Kremers Reference Files, American Institute for the History of Pharmacy, University of Wisconsin, Madison WI. George Davis's career came to an abrupt end during the silver panic of 1893. Davis was forced out of the company for raiding the firm's treasury to pay off personal debts he had incurred during the crisis. Milton L. Hoefle, "The Early History of Parke-Davis and Company," *Bulletin of the History of Chemistry* 25, no. 1 (2000): 31–32.

17. Liebenau, *Medical Science and Medical Industry*, 42–43.

18. For example, Smith, Kline & French, *Clinical Reports upon the Use of Eskay's Neuro Phosphates* (1902).

19. "The Manufacturing Laboratory in the Household of Pharmacy," *Pharmaceutical Review*, Apr. 1897, 64.

20. *Descriptive Catalogue of the Laboratory Products of Parke, Davis & Company* (1894).

21. E. M. Houghton to J. Z. Hunt, April 5, 1899, PDCR, b. 20, f. 8; unidentified to J. E. Bartlett, Nov. 16, 1899, PDCR, b. 20, f. 17; Houghton to J. N. Martin, May 15, 1899, PDCR, b. 20, f. 10.

22. Biological Department, Memorandum to J. C. Spratt, Oct. 24, 1899, PDCR, b. 20, f. 14.

23. "Chloretone," *New Orleans Medical and Surgical Journal*, (Feb. 1900, 482.

24. USPN 389,485 (Sept. 11, 1888).

25. *Frederick R. Stearns & Co. v. Russell*, 85 F. 218 (1898).

26. For example, "Pill-Dipper Patent Void," *JAMA*, May 7, 1898, 1129.

27. USPN 599,123 (Feb. 15, 1898); George F. Butler, "The Physiological Action and Therapeutics of Guaiamar," *New York Medical Journal*, Sept. 23, 1899, 442.

28. USPN 611,234 (Sept. 27, 1898); USPN 587,278 (July 27, 1897); USPN 565,329 (Aug. 4, 1896).

29. "Death of Mr. John Carnrick," *Medical World*, Feb. 1903, 18.

30. Story B. Ladd, "Patents in Relation to Manufactures," *Census Bulletin*, Aug. 15, 1902, 10–11.

31. USPN 482,108 (Sept. 6, 1892); USPN 492,868 (Mar. 7, 1893); USPN 495,204 (Apr. 11, 1893); USPN 615,970 (Dec. 1, 1898); USPN 607,172 (July 12, 1898).

32. Thomas Martin Reimer, "Bayer & Company in the United States: German Dyes, Drugs, and Cartels in the Progressive Era" (PhD diss., Syracuse University, 1996).

33. USPN 525,823 (Sept. 11, 1894). See also USPN 525,819, USPN 525,820, USPN 525,821, USPN 525,822, USPN 525,824, and USPN 525,825, all issued on Sept. 11, 1894. On Takamine, see T. Yamashima, "Jokichi Takamine (1854–1922), the Samurai Chemist and His Work on Adrenalin," *Journal of Medical Biography*, 11, no. 2 (May 2003): 95–102; H. T. Huang, "Takamine Jokichi and the Transmission of Ancient Chinese Enzyme Technology to the West," in *Historical Perspectives on East Asian Science, Technology and Medicine*, ed. Alan K. L. Chan, Gregory K. Clancey, Hui-Chieh Loy (Singapore: Singapore University Press, 2001), 525–532; Jon Harkness, "Dicta on Adrenalin(e): Myriad Problems with Learned Hand's Product-of-Nature Pronouncements in *Parke-Davis v. Mulford*," *Journal of the Patent and Trademark Office Society* 93, no. 4 (2011): 363–399.

34. USPN 141,072 (July 22, 1873). Other examples include USPN 344,433 (June 29 1886), for an organic ferment; USPN 424,357 (Mar. 25, 1890), for a pepsin made from hog stomach; USPN 441,182 (Nov. 25, 1890), for vegetable pepsin; USPN 493,460 (Mar. 14, 1893), for a "yeast or active ferment"; USPN 565,329 (Aug. 4, 1896), for a "digestive compound."

35. "Taka-Diastase: An Isolated Ferment in Powdered Form for the Treatment of Amylaceous Dyspepsia," *Medical and Surgical Reporter*, June 29, 1895, vii; Reynold Wilcox, "The Treatment of Indigestion of Starchy Foods," *Medical News* (Apr. 11, 1896, 393–397; C. C. Fite, "Diastase in Therapeutics," *Medical News* (Feb. 6, 1897, 167–172.

36. *Indigestion or Superdigestion, with Special Reference to the Employment of Taka-Diastase for Its Relief* (Parke, Davis & Company, 1897). A copy can be found in Medical Trade Ephemera Collection, College of Physicians of Philadelphia, b. "Parke Davis."

37. Evelynn Maxine Hammonds, *Childhood's Deadly Scourge: The Campaign to Control Diphtheria in New York* (Baltimore: Johns Hopkins University Press, 1999). For other descriptions of the early production of the antitoxin by state and municipal boards of health, see Jonathan Liebenau, "Public Health and the Production and Use of Diphtheria Antitoxin in Philadelphia," *Bulletin of the History of Medicine* 61, no. 2 (Summer 1987): 216–236; "An American Antitoxin Laboratory," *American Druggist and Pharmaceutical Record*, Apr. 10, 1895, 227–229.

38. Houghton received a degree in pharmacy in 1893 and a medical degree in 1894, both from the University of Michigan. He is identified as an "Assistant in Pharmacology" at the same institution for the year 1894–1895 in *University of Michigan School of Pharmacy. Announcement for 1900–1901: Register of Alumni for 1869–1900* (1900), 55.

39. On the early history of H. K. Mulford & Co., see Louis Galambos with Jane Eliot Sewell, *Networks of Innovation: Vaccine Development at Merck, Sharp & Dohme, and Mulford, 1895–1995* (New York: Cambridge University Press, 1995), 9–32; Liebenau, *Medical Science and Medical Industry*, 57–78; Clement B. Lowe, "The H. K. Mulford Co.," *Pharmaceutical Era*, Jan. 30, 1908, 133–134.

40. Michael Willrich, *Pox: An American History* (New York: Penguin, 2011), 185–187; Galambos with Sewell, *Networks of Innovation*, 9–17.

41. Willrich, *Pox*, 186.

42. The difficult relationship between Behring and Paul Ehrlich during this period is beyond the scope of this volume, but see Derek S. Linton, *Emil Von Behring: Infectious Disease, Immunology, Serum Therapy* (Philadelphia: American Philosophical Society, 2005).

43. "Diphtheria Antitoxin (Behring)," *American Practitioner*, Mar. 9, 1895, 200.

44. James B. Littlewood to Emil Behring, Jan. 11, 1895, PWFKC no. 60,642.

45. James B. Littlewood to Emil Behring, Dec. 13, 1895, PWFKC no. 60,642.

46. James B. Littlewood to Emil Behring, Jan. 16, 1897, PWFKC no. 60,642.

47. James B. Littlewood to Emil Behring, Dec. 13, 1895, PWFKC no. 60,642.

48. Goepel & Ranger to the Commissioner of Patents, Oct. 14, 1895, PWFKC no. 60,642.

49. Goepel & Ranger to Commissioner of Patents, March 12, 1898, PWFKC no. 60,642.

50. Charles Holland Duell to Emil Behring May 31, 1898, PWFKC no. 60642. Records of the appeal do not appear in the *Official Gazette of the Patent Office*, nor do they appear in the patent wrapper file where one would expect them to. There is, however, a brief indication in the patent wrapper that the appeal was successful. I thank Gene Morris at the National Archives in College Park for his help in trying to locate records related to the appeal.

51. "German Commercialism," *Medical Record*, Aug. 6, 1898, 200. For other statements, see "Serum Antitoxin—A Patent," *American Practitioner and News*, Feb. 9, 1898, 117; "Behring's American Patent on Antitoxin," *Pennsylvania*

Medical Journal, Oct. 1898, 278–279; "Professor Behring to His Critics," *Philadelphia Medical Journal*, Oct. 1, 1898, 636–637.

52. "A Patent in the United States on Diphtheria Antitoxin," *Canada Lancet*, Dec. 1898, 879–880; "Diphtheria Antitoxin Now Patented," *Kansas City Medical Index*, Sept. 1898, 274; "Antitoxin Patented," *ADPR*, Aug. 10, 1898, 82; "The Ethical Standing of the Antitoxin," *JAMA*, Aug. 4, 1900, 283–284.

53. "The Brush Bill and the New York City Board of Health," *Philadelphia Medical Journal*, Feb. 26, 1898, 349.

54. "Code of Medical Ethics and Etiquette of the American Medical Association: Report of Majority of Committee," *JAMA*, Apr. 7, 1894, 508.

55. A. C. Simonton, "Code of Revision," *JAMA*, May 5, 1894, 678–679.

56. Nathan S. Davis, "Proposed Revision of the Code of Ethics of the American Medical Association," *JAMA*, Apr. 14, 1894, 557.

57. Francis E. Stewart, "The Work of the American Pharmaceutical Association in Relation to the Materia Medica of the Future," *JAMA*, Dec. 30, 1899, 1644–1646; Stewart, "Is It Ethical for Medical Men to Patent Medical Inventions?," *JAMA*, Sept. 18, 1897, 583–587; Stewart, "The Eminently Scientific Nature of Our Patent and Copyright Laws," *JAMA*, Aug. 22, 1896, 424–426; Stewart, "The Practice of Pharmacy as a Liberal Profession," *JAMA*, July 11, 1896, 74–80; Stewart, "The Proprietary System and Its Remedy," *JAMA*, Sept. 14, 1895, 450–452. In addition to the numerous other articles cited in this volume, Stewart's reputation was also enhanced by the publication of his *A Compend of Pharmacy* (1886), which went through multiple editions and became a standard text for pharmacy students for several decades. See R. A. Buerki, "American Pharmaceutical Education, 1852–1902," *Journal of the American Pharmaceutical Association* 40, no. 4 (2000): 458–460.

58. Stewart, "Is It Ethical for Medical Men to Patent Medical Inventions?," 586.

59. L. C. Lane, "Address of Welcome to the American Medical Association," *JAMA*, June 23, 1894, 956.

60. "Patents and Trade-Marks," *Medical and Surgical Reporter*, Sept. 12, 1896, 341–343. Other statements include "Patent and Proprietary Medicines," *Sanitarian*, Aug. 1901, 110; "Is It Proper for a Physician to Employ a Patented Drug?," *Medical News* July 8, 1899, 46–47; "The Justification for Medical Patents," *American Medico-Surgical Bulletin* Sept. 25, 1898, 845–847; Arthur H. Cohn, "Proprietary vs. Patent Medicines," *Transactions of the State Medical Society of Wisconsin for the Year 1897* (1897), 603–606; C. Lewis Diehl, "The National Formulary," *JAMA*, Jan. 13, 1894, 48–50; "Doctor's Inventions," *Medical Times and Register*, May 31, 1890, 518. Numerous statements in the pharmaceutical press made similar arguments, such as "Editorial," *Pharmaceutical Review*, Apr. 1897, 61.

61. John Jay Taylor, *The Physician as a Businessman* (1892), 12.

62. "Parke Davis & Co. on the Condition of the Trade," *ADPR*, Nov. 10, 1894, 341.

63. For an example of the grateful response to this type of material, see "The Newer Remedies—A Reference Manual for Physicians, Pharmacists and

Students," *Medical Fortnightly*, Sept. 15, 1896, 520. For a critical perspective, see "To the Illinois Pharmaceutical Association," *Report of Proceedings of the Illinois Pharmaceutical Association* (1898), 136–139.

64. E. L. Priest, "Empiricism in Missouri, and How to Suppress It," *Transactions of the Medical Association of the State of Missouri* (1899), 182–183.

65. B. T. Whitmore, "Corner on Antitoxin: Professor Behring's Attempt to Control the Output of Antitoxin in this Country," *New England Medical Monthly*, Sept. 1898, 435. For other statements, see "German Commercialism," *Medical Record*, Aug. 6, 1898, 200; "Serum Antitoxin—A Patent," *American Practitioner and News*, Feb. 9, 1898, 117; "Diphtheria Antitoxin Now Patented," *Kansas City Medical Index*, Sept. 1898, 274.

66. Jacob R. Johns, "The Antitoxine [*sic*] Patent: Why Refused Five Times Yet Finally Allowed," *Cincinnati Lancet and Clinic*, Sept. 17, 1898, 268; "Diphtheria Antitoxin Patented," *Medical News*, July 30, 1898, 146; "Behring's Patent," *Tri-State Medical Journal*, Aug. 1898, 408; "Diphtheria Antitoxin Patented," *Medical News*, July 30, 1898, 146.

67. "The Patent on Antitoxin," *Cleveland Journal of Medicine*, Aug. 1898, 361.

68. "Diphtheria Antitoxin Patented," *Medical News*, July 30, 1898, 145.

69. "The Monopoly of Antitoxin," *Journal of Medicine and Science*, Aug. 1898, 371.

70. "The Nobel Foundations and Awards," *Medical News*, Feb. 8, 1902, 279.

71. For example, Charles Rice, "Shall Doses and Some of the New Synthetic Remedies Be Introduced into the Next U.S. Pharmacopeia?," *ADPR*, June 10, 1896, 335–226.

72. AMA resolution quoted in McTavish, *Pain & Profits*, 122; "American Medical Association," *Merck's Archive of the Materia Medica and Its Uses*, June 1899, 248–249.

73. Lee Anderson, "The USP from 1900–1990," in Anderson and Gregory J. Higby, *The Spirit of Volunteerism, a Legacy of Commitment and Contribution: The United States Pharmacopeia, 1820–1995* (Rockville, MD: United States Pharmacopoeial Convention, 1995), 213.

74. There were occasional disagreements on this point. For example, "The Introduction of Synthetics," *Proceedings of the American Pharmaceutical Association* (1900), 362. Note that here and in the subsequent citations the year given for the *Proceedings of the American Pharmaceutical Association* refers to the year the meeting took place, not the year the proceedings were published.

75. M. Clayton, "The United States Pharmacopeia," *Medical Bulletin*, Sept. 1905, 327–335.

76. "The Latin of the New Pharmacopeia," *ADPR*, Aug. 14, 1905, 65; Clayton Thrush, "The Eighth Decennial Revision of the U.S.P.," *American Journal of Pharmacy*, Jan. 1906, 36.

77. O. T. Osborne and C. S. N. Hallberg, "Report of Committee on Proprietary Medicines," *JAMA*, Dec. 30, 1905, 2010. Of course, this comment indicates a poor understanding of the distinction between trademarks and copyrights, given that copyright law could not be used to protect the names of phar-

maceuticals. This mistake was not unusual among physicians and pharmacists who debated these issues.

78. For example, see "Signification of Patent Medicine," *Medical News*, Nov. 29, 1902, 1052.

79. "Principle of Medical Ethics," *JAMA*, May 16, 1903, 1380.

80. "The Physician and the Public," *Massachusetts Medical Journal*, Aug. 1903, 381.

81. John S. Billings, "American Inventions and Discoveries in Medicine, Surgery, and Practical Sanitation," in *Annual Report of the Board of Regents of the Smithsonian Institution . . . to July, 1892* (1893), 613–619.

82. William H. Baldwin et al., *The President's Homes Commission: Report of the Committee on Improvement of Existing Houses and Elimination of Insanitary and Alley Houses* in Senate Documents, 60th Cong., 2d Sess., vol. 7 (1909), 263.

83. *OGPO*, Aug. 7, 14, 21, and 28, 1894.

84. Virgil Coblentz, "Can a System of Nomenclature Be Devised for Modern Medical Preparations," *Druggists Circular and Chemical Gazette*, Dec. 1898, 200–201.

85. This claim is based on the results of the Hallberg Prescription Ingredient Survey of 1895, which surveyed twelve thousand prescriptions, made up of more than twenty-eight thousand ingredients, dispensed by ten different pharmacists in Illinois. The results of the survey can be found in E. N. Gathercoal, *The Prescription Ingredient Survey* (American Pharmaceutical Association, 1933). It is not clear how many of these articles were actually trademarked, but given the growing importance of trademarking to drug manufacturers, it is safe to assume that most of them were.

86. E. G. Raeuber, "Are Proprietary Medicines Profitable to Handle in a Retail Drug Store?," *Proceedings of the Wisconsin Pharmaceutical Association* (1900), 73.

87. For example, C. J. Rosenham, "The Fallacy of Substitution, Showing the Demoralizing Effect upon the Employees Who Are Instructed to Substitute," *Proceedings of the Twenty-Fourth Annual Meeting of the Kentucky Pharmaceutical Association* (1901), 79–81; William C. Anderson, "Substitution," *Pharmaceutical Era*, Apr. 24, 1902, 404–406.

88. "Editorial," *Pharmaceutical Review*, May 1897, 83.

89. "The Interstate League and the 'Vaseline' Question," *ADPR*, May 10, 1894, 254.

90. "To the Medical Profession of the United States," *New York Medical Times*, Mar. 1893, xxiv.

91. "Report of Special Committee on National Legislation," *Proceedings of the American Pharmaceutical Association* (1897), 72.

92. *California Fig Syrup Co. v. Frederick Stearns & Co.* 73 F. 812 (1896).

93. Examples include *Elgin National Watch Company v. Illinois Watch Case Company*, 179 US 665 (1901); *American Fibre Chamois Co. v. De Lee et al.*, 67 F. 329 (1895); *Trask Fish Company v. Frank Wooster*, 28 Mo. App. 408 (1888).

94. *Siegert v. Abbott*, 25 NYS 590 (1893). For another example, see *Brown Chemical Company v. Myer*, 31 F. 453 (1887).

95. For example, *Linoleum Manufacturing Co. v. Nairn*, 7 Ch. Div. 834 (1878).

96. *Singer Manufacturing Co. v. June Manufacturing Co.*, 163 US 169 (1896).
97. *Centaur Co. v. Heinsfurter*, 84 F. 955 (1898); *Centaur Co. v. Killenberger*, 87 F. 725 (1898); *Centaur Co. v. Marshall*, 92 F. 605 (1899); *Centaur Co. v. Hughes*, 91 F. 901. Other examples include *Stuart v. F. G. Stewart Co.*, 91 F. 243 (1899); *Dadirrian v. Yacubian et al.*, 98 F. 872 (1900); *Holsapfel's Compositions Company v. Rahtjen's American Composition Company*, 183 US 1 (1901). On the Centaur Company cases, see Abraham S. Greenberg, "The Effect of Patent Expiration on Trade Mark Rights," *Journal of the Patent Office Society* 25, no. 9 (Sept. 1943): 622–623.
98. Stewart, "The Eminently Scientific Nature of Our Patent and Copyright Laws," 425.
99. Francis E. Stewart, "The Use of Word Marks to Protect Science and Legitimate Commerce in Materia Medica Products," *Pacific Medical Journal*, Jan. 1903, 11. Other examples include Stewart, "The Law vs. Proprietary Titles," *Western Druggist*, Mar. 1896, 109–110; Stewart, "An Attempt to Defeat the Object of the Patent-Law," *Philadelphia Medical Journal*, Feb. 19, 1898, 342–344; Stewart, "The Proper Introduction of Materia Medica Products to Science and Brands of the Same to Commerce," *Therapeutic Monthly*, May 1902, 171–178.
100. Stewart, "Use of Word Marks to Protect Science," 11.
101. "Petty Economies Not the Solution," *Paint, Oil, and Drug Review*, May 20, 1896, 10–11.
102. An Act to Protect Trade and Commerce against Unlawful Restrains and Monopolies 26 Stat. 209 (July 2, 1890); Wyatt Wells, *Antitrust and the Formation of the Postwar World* (New York: Columbia University Press, 2002), 81; Sklar, *Corporate Reconstruction*, 133. For an important critique of Sklar's interpretation, see Peter C. Carstensen, "Dubious Dichotomies and Blurred Vistas: The Corporate Reconstruction of American Capitalism," *Reviews in American History* 17, no. 3 (Sept. 1989): 404–411. On the prohibition of "odious" monopolistic behavior by state courts before the Sherman Act, see Peter Karsten, "Supervising the 'Spoiled Children of Legislation': Judicial Judgments Involving Quasi-Public Corporations in the Nineteenth Century U.S.," *American Journal of Legal History* 41, no. 3 (July 1997): 315–367.
103. *Bement v. National Harrow Company*, 186 US 70 (1902).
104. Reimer, "Bayer & Company," 159–168; Ludwig Haber, *The Chemical Industry, 1900–1930: International Growth and Technological Change* (Oxford: Clarendon, 1971), 128–134.
105. In the early 1890s, for example, Bayer and Schering became involved in a complicated dispute in which Bayer claimed that a drug sold by Schering named "Piperazine" was essentially the same drug as its own Lycetol. Rather than battle it out in court, the two companies ended up forming a cartel and pooling profits from the two drugs. The cartel attracted little attention and, even if it had, would probably have been considered legal because it was patent-based. Reimer, "Bayer & Company," 160–161. On Schering, see Christopher Kobrak, *National Cultures and International Competition: The Experience of Schering AG, 1851–1950* (New York: Cambridge University Press, 2002).

106. *Elliman v. Carrington*, 2 Ch. D. 275 (1901); *Garst v. Harris*, 177 Mass. 72 (1900); *Walsh v. Dwight*, 40 NY App. 513 (1899). On this topic, see William J. Shroder, "Price Restriction on the Re-Sale of Chattels," *Harvard Law Review* 25, no. 1 (Nov. 1911): 59–69.

107. Horace R. Lamb, "The Relation of the Patent Law to the Federal Anti-Trust Laws," *Cornell Law Quarterly* 12 (1926–1927): 261–285.

108. John C. Gall, "Resale Price Maintenance," *The Brief* 22, no. 1 (1922–1923): 13.

109. George J. Seabury, *Shall Pharmacists Become Tradesmen?* (1899), 16–17.

110. "Retrospect," *Pharmaceutical Review*, Jan. 1899, 2–3.

111. "The N.A.R.D. Plan of Price Protection," *ADPR*, Oct. 14, 1901, 215–216.

112. *John D. Park & Sons v. National Wholesale Druggists' Association*, 20 A.D. 508 (1898); "Jobbers Can Legally Refuse to Sell to Cutters," *Midland Druggist*, Dec. 1899, 43–44.

113. "President's Address," *Proceedings of the American Pharmaceutical Association* (1897), 10.

114. Examples include Charles Rice, "Ethics of Prescribing: Objectionable and Unobjectionable Proprietaries,. *ADPR*, May 25, 1897, 286–288; "The Ethics and Economics of Proprietary Preparations," *American Journal of Pharmacy*, June 1897, 317–319; "Prescribing Proprietaries," *ADPR*, June 25, 1900, 410; Wilhem Bodeman, "Patent Laws and Medicine," *ADPR*, Sept. 25, 1894, 233; William Bodeman, "Are the Present Patent and Trademark Laws of the United States, as Applied to Medicinal Products, in Conformity with Common Law, Common Sense, Justice, or Equity?," *Bulletin of Pharmacy*, July 1891, 336–337.

115. Wolfgang Wimmer has argued that although the 1891 German patent law prohibited patents on medicines in general, a series of decisions by the Imperial Court between 1888 and 1890 meant that synthetic drugs actually could be patented under the 1891 law. Whether or not this is the case, American druggists had little understanding of the complexities of foreign law and generally believed that medicines could not be patented in Germany. See Wimmer, "Innovation in the German Pharmaceutical Industry," in *The Chemical Industry in Europe, 1850–1914: Industrial Growth, Pollution, and Professionalization*, ed. Ernst Homburg, Anthony S. Travis, and Harm G. Schröter (Dordrecht, the Netherlands: Kluwer Academic, 1998), 287–288.

116. "The New Antipyretic Phenacetine-Bayer," *Buffalo Medical Journal*, July 1888, iii; USPN 400,086 (Mar. 26, 1889).

117. In one case, for example, the company sued a wholesaler named Hugh J. Tinling for importing two thousand ounces of phenacetin that he had bought from another manufacturer. Tinling's defense argued that he had every right to import the drug because he had legally purchased it in a country in which Bayer held no patent rights. In 1897, however, the Circuit Court of Appeals for the District of Colorado disagreed and ruled that Bayer had every right to limit the distribution of the chemical in this country because of its patent. *Dickerson v. Tinling*, 84 F. 192 (1897). On this issue, see Dwight B. Cheever,

"Rights of a Traveler to Use Here Articles Made and Purchased Abroad but Patented Here," *Michigan Law Review*, Jan. 1909, 226–233.

118. Anthony Gref, "The Development of the Legal Department in the United States," internal Bayer Company memo, p. 7, Bayer Company Archives, Leverkusen, Germany, folio "9 A 1 USA Schriftwechsel Besprechungen Historische Entwicklungen 1909–1954 Bd. 3."

119. E. J. Hallock, "Note on Para-Nitro and Para-Amido Phenetol," *American Chemical Journal*, Oct. 1879, 271–272.

120. Anthony Gref, "Brief for Complainants," *Farbenfabriken of Elberfeld Company and Edward N. Dickerson v. Conrad D. Maurer* (1900), 67. Copies of the court documents from the case, including Gref's brief and a transcript of the trial, can be found in Records of the District Courts of the United States, NARA, Philadelphia Branch, RG 21, b. 523, case no. 33. Material from the decision of the Court of Appeals can be found in Records of the United States Courts of Appeals, NARA, Philadelphia Branch, RG 276, b. 21, case no. 49.

121. Quoted in Gref, "Brief for Complainants," 72.

122. *Dickerson et al. v. Maurer*, 108 F. 233 (1901); *Maurer v. Dickerson*, 113 F. 870 (1902); *Maurer v. Dickerson*, 186 US 481 (1902).

123. "The Movement to Revive the Patent Laws," *ADPR*, Oct. 28, 1901, 252; "The Buffalo Convention," *Bulletin of Pharmacy*, Nov. 1901, 457.

124. "The N.A.R.D's Patent Law Bill," *Bulletin of Pharmacy*, Apr. 1904, 137.

125. F. A. Darrin to Thomas C. Platt, Jan. 31, 1905, Records of the US Senate Committee on Patents, NARA, Washington, DC, Branch, RG 46, b. 132, f. 58A-J57.

126. *Hearings before the Committee on Patents of the House of Representatives on H.R. 13679* (Washington, DC: Government Printing Office, 1904).

127. "Hearing on the Mann Bill," *ADPR*, May 28, 1906, 295.

128. The Proprietary Manufacturers Association endorsed the bill, as did some ethical manufacturers. For a statement of support from Sharp & Dohme, see Sharp & Dohme to L. E. McComas, Feb. 14, 1905, Records of the US Senate Committee on Patents, NARA, Washington, DC, Branch, RG 46, b. 132, f. 58A-J57. The extent to which NARD intimidated manufacturers from testifying against the bill is difficult to determine, although circumstantial evidence suggests that this played a role. On this issue, see Reimer, "Bayer & Company," 227.

129. For example, "Resolutions Endorsing the Mann Bill," *Proceedings of the Wisconsin Pharmaceutical Association* (1905), 20–25. Wholesale druggists were unsure about the potential impact of the bill on their trade, and the NWDA declined to endorse it. "National Wholesale Druggists' Association," *Practical Druggist and Pharmaceutical Review of Reviews*, Dec. 1904, 422.

130. Francis E. Stewart, "Report of F. E. Stewart, as Committee of One Appointed for the Purpose of Transmitting the Views of the American Pharmaceutical Association on the Subject of Patents and Trademarks to the Congress," *Proceedings of the American Pharmaceutical Association* (1903), 463–466.

131. For example, George M. Beringer, "Why the Mann Bill Should Be Enacted," *Proceedings of the American Pharmaceutical Association* (1905), 145–152; W. H. Burke, "Why the Mann Bill Should Not Be Enacted," *Proceedings of the American Pharmaceutical Association* (1905), 152–156.

132. "Regarding Trade-Marks," *Bulletin of Pharmacy*, Nov. 1903, 444.

133. For example, one petition was signed by the Chicago Retail Druggists Association, the Retail Druggists of Lafayette, and the Vigo County Druggists Association. "Petition," Records of the US Senate Committee on Patents, NARA, Washington, DC, Branch, RG 46, b. 132, f. SEN58A-J56.

134. "Regarding Trade-Marks," *Bulletin of Pharmacy*, Nov. 1903, 444.

135. Robert Edes, "Shall the Scope of the Pharmacopoeia be Increased?," *JAMA*, Dec. 30, 1899, 1660.

136. Harry Marks, *The Progress of Experiment: Science and Therapeutic Reform in the United States, 1900–1990* (Baltimore: Johns Hopkins University Press, 1997), 17–41.

137. The title of the law was "An Act to Regulate the Sale of Viruses, Serums, Toxins, and Analogous Products in the District of Columbia, to Regulate Interstate Traffic in Said Articles, and for Other Purposes." On the 1902 law, see Willrich, *Pox*; Lilienfeld, "First Pharmacoepidemiological Investigations"; Ramunas A. Kondratas, "Biologics Control Act of 1902," in *The Early Years of Federal Food and Drug Control,* ed. James Harvey Young (Madison, WI: American Institute of the History of Pharmacy, 1982), 8–27; Liebenau, *Medical Science and Medical Industry,* 89–90.

138. *Annual Report of the Surgeon-General of the Public Health and Marine-Hospital Service of the United States for the Fiscal Year 1904* (1904), 372.

139. *Annual Report of the Surgeon-General of the Public Health and Marine-Hospital Service* (1909), 32–33.

140. The first of these was for an improved smallpox vaccine developed by a bacteriologist named LaFayette Parsons, who worked for the company from 1902–1906. USPN 737,656 (Sept. 1, 1903); "LaFayette Russell Parsons," *Seventeenth Annual Catalogue of the University of Idaho 1908–1909* (1910), 20. It is not clear if Parke-Davis commercially introduced vaccines made with Parsons's method, but the company's smallpox "virus" was widely regarded as being of high quality. For example, F. H. Austin, "A Varied Experience in Vaccination," *Journal of Medicine and Science,* July 1903, 292. The second patent was for a vaccine against blackleg, an infectious disease that afflicts livestock, which was developed by Elijah Houghton. USPN 778,667 (Dec. 27, 1904).

141. Examples include patents on antitoxins for hay fever and fatigue. USPN 745,333 (Dec. 1, 1903); USPN 809,347 (Jan. 9, 1906).

142. James F. Curtis, "Importation of Viruses, Serums, Etc.—List of Manufacturing Establishments Licensed," in *Treasury Decisions under Customs and Other Laws,* July–Dec. 1913, 60–61.

143. Parke, Davis & Company, *Complete Catalogue of the Products of the Laborato-*

ries of Parke, Davis & Company, Manufacturing Chemists, Detroit, Mich., U.S.A. Revised to January 1, 1913 (1913), 190–198.

144. For example, see the correspondence between the National Vaccine and Antitoxin Institute and a druggist named O. N. Falk from 1911 in O. N. Falk & Sons Druggists Records, 1904–1946, Wisconsin Historical Society, b. 31, f. 18.

145. For a description of the efforts by Parke-Davis and H. K. Mulford to prevent the passage of a bill in Massachusetts that would have granted the State Board of Health the authority to manufacture and distribute smallpox vaccine, see Frank L. Morse, "The Recent Smallpox Epidemic in Massachusetts," *Journal of the Massachusetts Association of Boards of Health*, July 1903, 57–60.

146. Quoted in Paul Starr, *The Social Transformation of American Medicine: The Rise of a Sovereign Profession and the Making of a Vast Industry* (New York: Basic Books, 1982), 186.

147. "Boards of Health and the Manufacture of Vaccine Virus and Antitoxins," *Boston Medical and Surgical Journal*, Jan. 23, 1902, 99. For other statements of this position, see "The New York City Health Department and the Sale of Prophylactic and Curative Products," *New York Medical Journal and Philadelphia Medical Journal Consolidated*, July 4, 1903, 28–29; "American Medical Association. Minutes of the Section on Materia Medica, Pharmacy and Therapeutics," *JAMA*, July 19, 1902, 153.

148. Joseph Favil Biehn, "Notes on the Use of Chloroform in the Preparation of Vaccine Virus" *Transactions of the Chicago Pathological Society*, April 10, 1905, 254; Robert Johnston, *The Radical Middle Class: Populist Democracy and the Question of Capitalism in Progressive Era Portland, Oregon* (Princeton, NJ: Princeton University Press, 2003), 179–190. As Johnston has argued, skepticism and even hostility was a reasonable response to vaccination during the nineteenth century because of the significant risks associated with the practice. These risks continued following the passage of the 1902 law, albeit in a reduced form. As one physician noted in 1905, for example, "vaccine famines" caused by outbreaks of smallpox and other diseases led manufacturers to put "unripe" or imperfectly purified vaccine on the market, "wherefrom many sore arms and much needless suffering resulted." *Radical Middle Class*, 179–190; Biehn, "Use of Chloroform," 254.

149. Willrich, *Pox*, 264–265.

150. Flora W. Fox, "Vaccination," *Lucifer the Light Bearer*, May 11, 1901, 131.

151. Dudley Tait, "Improper Advertising Methods," *California State Journal of Medicine*, Nov. 1903, 366.

152. William J. Robinson, "The Relation of the Physician to Proprietary Remedies," *Transactions of the Section on Pharmacology of the American Medical Association at the Fifty-Fifth Annual Session* (1904), 52–53. For other examples, see A. Ravogli, "Physicians vs. Medicine Proprietors and Medicine Patentees," *Cincinnati Lancet-Clinic*, Aug. 11, 1900, 141–145; "Patent, Proprietary, and 'Ethpharmal' Preparations," *Medical World*, Apr. 1905, 152–153.

CHAPTER 6

1. George Dock, "Proprietary Medicines and Their Abuses," *Transactions of the Section on Practice of the American Medical Association at the Fifty-Seventh Annual Session* (1906), 49.

2. USPN 637,355 (Nov. 21, 1899), for a compound of nuclein and iron; USPN 637,355 (Nov. 21, 1899), for a compound of nuclein and mercury. See also "Metallic Nucleol Compounds," *ADPR*, Oct. 10, 1899, 202; USTN 84,051 to 84,054, *OGPO*, Jan. 16, 1900, 600.

3. Commissioner of Patents to F. J. Yanes, Apr. 22, 1912, Records of the Patent and Trademark Office, Files of the Office of the Commissioner of Patents, 1850–1949, NARA II, RG 241, file 1–100.

4. Examples include the antiseptics Antiformin and Veroform, introduced in 1902 and 1903 by the American Antiformin Company and the Veroform Hygienic Company, both of New York City. USPN 681,671 (Jan. 21, 1902); USPN 740,424 (Oct. 6, 1903).

5. Perey Frankland, "The Chemical Industries in Germany—II," *Scientific American Supplement*, June 26, 1915, 403.

6. There is some ambiguity about the total number of remedies listed because some products were listed under both generic and proprietary forms and because of other inconsistencies in the text. Eighty-one listed products were currently protected by patent, and another sixteen had previously been protected by patents that had since expired. Based on an analysis of *NNR* (1911).

7. Lewis A. Conner, in "Report of the Committee on the Pharmacopeia," *Transactions of the Section on Practice of Medicine of the American Medical Association at the Sixteenth Annual Session* (1909), 325. Other examples include "Report of the Committee on Patents and Trade-Marks," *JAMA*, June 18, 1910, 2079–2080; "Report of the Committee on Patents and Trade-Marks," *JAMA*, Nov. 25, 1911, 1780–1781.

8. USPN 644,077 (Feb. 27, 1900); USTN 82,805; *OGPO*, May 2, 1899, 889.

9. For example, Farbenfabriken of Elberfeld Company to Milwaukee Drug Company, Mar. 15, 1909, folio 9/K.1, "Pharmazeutischekonferenzen 1909."

10. Diarmuid Jeffreys, *Aspirin: The Remarkable Story of a Wonder Drug* (New York: Bloomsbury, 2005), 90. The story of Aspirin's early history has been told many times, but see also Charles C. Mann and Mark L. Plummer, *The Aspirin Wars: Money, Medicine, and 100 Years of Rampant Competition* (New York: Knopf, 1991).

11. For example, *Frederick R. Stearns & Co. v. Russell*, 85 F. 218 (1898).

12. *Asonia Brass & Copper Co. v. Electrical Supply Co.*, 144 US 11 (1892). Other examples include *Potts v. Creager*, 155 US 597 (1895); *Hobbs v. Beach*, 180 US 383 (1901); *Forsyth v. Garlock*, 142 F. 461 (1905); *Warren Bros. Co. v. City of Owasso*, 166 F. 309 (1909); *General Electric Co. v. Hoskins Mfg. Co.*, 224 F. 464 (1915). On this issue, see Martin A. Ryan, "Patentability of a New Use for an Old Com-

position of Matter," *Journal of the Patent Office Society* 29, no. 11 (Nov. 1947): 787–821.

13. *Badische Anilin & Soda Fabrik v. Kalle & Co.*, 104 F. 802 (1900).

14. *Anheuser-Busch Brewing Association v. United States*, 207 US 556 (1908), citing *Hartranft v. Wiegmann*, 121 US 609 (1887).

15. "In the High Court of Justice—Chancery Division. Before Mr. Justice Joyce. Farbenfabriken Vormals Friedrich Bayer & Co. v. Chemische Fabrik Von Heyden." This is a partial transcript of the hearing, in which Joyce laid out the reasoning behind his decision. It can be found in Records of the Office of Alien Property, Seized Records of the Farbenfabriken Co. Regarding Patent Interference Suits, NARA II, RG 131, b. 1, f. "Interference Cases Involving Aspirin Patents, 1898–1908."

16. Livingston Gifford to Farbenfabriken of Elberfeld Co., July 28, 1905, Records of the Office of Alien Property, Seized Records of the Farbenfabriken Co. Regarding Patent Interference Suits, NARA II, RG 131, b. 1, f. "Interference Cases Involving Aspirin Patents, 1898–1908."

17. Meyer Brothers Drug Company to Farbenfabriken of Elberfeld Co., Aug. 23, 1909, and Aug. 18, 1909; Farbenfabriken of Elberfeld to Meyer Brothers Drug Co., Aug. 25, 1909 and Aug. 20, 1909; PDC to Farbenfabriken of Elberfeld Co., Aug. 11, 1909, and Aug. 6, 1909; Farbenfabriken of Elberfeld Co. to PDC, Aug. 9, 1909, and Aug. 3, 1909, all in Bayer Company Archives, Leverkusen, Germany, folio 9/K.1, "Pharmezeutische Konferenzen 1909–1910." See also Thomas Martin Reimer, "Bayer & Company in the United States: German Dyes, Drugs, and Cartels in the Progressive Era" (PhD diss., Syracuse University, 1996), 170–173, esp. 173n17.

18. *Kuehmsted v. Farbenfabriken of Elberfeld Co.*, 179 F. 701 (1910). For the lower court's decision, which also upheld the patent on similar grounds, see *Farbenfabriken of Elberfeld Co. v. Kuehmsted*, 171 F. 887 (1909).

19. *Kuehmsted v. Farbenfabriken of Elberfeld Co.*, 179 F. 701 (1910).

20. *In re Williams*, 171 F. 2d 319 (1948).

21. "Brooklyn Stable Man and Druggist Fined," *NARD Notes*, Jan. 20, 1913, 1120; "Chemical Importers after the Druggists," *Druggists Circular*, Sept. 1912, 585; Reimer, "Bayer & Company," 205–207.

22. "Aspirin Patent Sustained," *NARD Notes*, Aug. 26, 1909, 971.

23. "Lactophenin and Aspirin," *JAMA*, Sept. 15, 1906, 882.

24. William J. Robinson, "The Relation of the Physician to Proprietary Remedies," *JAMA*, Dec. 3, 1904, 1675–1680.

25. "Acetylsalicylic Acid and Aspirin," *JAMA*, Nov. 2, 1912, 1642.

26. On the development of Salvarsan, see Axel C. Hüntelmann, "Seriality and Standardization in the Production of '606,'" *History of Science* 48 (2010): 435–460; David Greenwood, *Antimicrobial Drugs: Chronicle of a Twentieth Century Medical Triumph* (New York: Oxford University Press, 2008), 52–65. I am grateful to Evan Hepler-Smith for pointing me to Hüntelmann's work.

27. On this issue, see William Kingston, "Streptomycin, *Schatz v. Waksman*, and

the Balance of Credit for Discovery," *Journal of the History of Medicine and Allied Sciences* 59, no. 3 (2004): 441–462; M. Wainwright, "Streptomycin: Discovery and Resultant Controversy," *History and Philosophy of the Life Sciences* 13, no. 1 (1991): 97–124.

28. USPN 986,148 (Mar. 7, 1911); USTN 40,734, *OGPO*, Aug. 23, 1910, 987.

29. "Mississippi Valley Medical Association," *Lancet-Clinic*, Nov. 20, 1915, 462.

30. "Commercial Art Criticism," *Printer's Ink*, Apr. 8, 1903, 22. On brands, trademarks, and advertising at the turn of the century, see Pamela Walker Laird, *Advertising Progress: American Business and the Rise of Consumer Marketing* (Baltimore: Johns Hopkins University Press, 2001) and Susan Strasser, *Satisfaction Guaranteed: The Making of the American Mass Market* (New York: Pantheon Books, 1989).

31. *Warner v. Searle & Hereth Co.*, 191 US 195 (1903).

32. Joseph Campbell Preserve Company to John F. Dryden, Apr. 9, 1904, Records of the US Senate Committee on Patents, NARA, Washington, DC, Branch, RG 46, b. 132, f. SEN58A-J56.

33. An Act to Authorize the Registration of Trade-Marks Used in Commerce with Foreign Nations or Among the Several States or with Indian Tribes, and to protect the Same, Pub. L. no. 58-84, 33 Stat. 727 (1905).

34. Ibid.

35. *Holzapfel's Compositions Co. v. Rahtjen's Co.*, 183 US 1 (1901).

36. Edward S. Rogers, *Good Will, Trade-Marks and Unfair Trading* (1914), 237. For example, *Whann v. Whann*, 116 La. 690 (1906); *Capewell Horse Nail Co. v. Mooney*, 167 F. 575 (1909); *United States v. Seventy-Five Boxes of Alleged Pepper*, 198 F. 934 (1912); *N. Wolf & Sons v. Lord & Taylor*, 41 App. D.C. 514 (1914).

37. For example, *Thaddeus Davids Company v. Davids Manufacturing Company*, 233 US 461 (1914); *Hutchinson, Pierce & Co. v. Loewy*, 217 US 457 (1910).

38. *Standard Paint Company v. Trinidad Asphalt Manufacturing Company*, 220 US 446 (1911).

39. For example, in 1907 the manufacturers of Glycozone successfully prevented the manufacturers of a competing product from registering the name "Liquozone" because of the similarity between the names. *Medical Times*, Mar. 1907, xviii.

40. *Jacobs v. Beecham*, 221 US 263 (1911).

41. *Pharmaceutical Review*, Dec., 1906, back advertising material.

42. *USP*, 8th revision (1905), 233.

43. "Elixir No. 154: Uritone Compound," *Therapeutic Notes*, Nov. 1906, 118.

44. "Uritone, Urotropin, Formin, etc. in Commerce," *Practical Druggist and Review of Reviews*, Dec. 1906, 566.

45. "Phenacetin Patent Will Expire in March," *Pharmaceutical Era*, Feb. 15, 1906, 158.

46. On Lehn & Fink's advertising, see "Why Pay More?," *Pharmaceutical Review*, June 1907, back page.

47. "Farbenfabriken Trade-Mark Suits Discontinued," *Druggists Circular*, Apr. 1913, 221.

48. Farbenfabriken of Elberfeld Company to Monsanto Chemical Works, Apr. 3 1909, Bayer Company Archives, Leverkusen, Germany, folio 9/K.1, "Pharmazeutischekonferenzen 1909." See also Farbenfabriken of Elberfeld Company to Monsanto Chemical Works, Mar. 25, 1909, and Monsanto's response, Mar. 27 1909, both in ibid.

49. "Farbenfabriken Trade-Mark Suits Discontinued," *Druggists Circular*, Apr. 1913, 221.

50. "Brief of John F. Queeny, President of Monsanto Chemical Works, St. Louis, MO," *Tariff Schedules Hearings before the Committee on Ways and Means*, H.R. Doc. No. 1447, vol. 1, 62d Cong., 3d Sess. (1913), 79.

51. On this question, see "Phenacetin and Acetphenetidin," *Pharmaceutical Era*, Nov. 1912, 696.

52. "Why Not 'Made in the United States?,'" *Pharmaceutical Era*, Oct. 1912, 619.

53. See Margaret C. Levenstein, "Price Wars and the Stability of Collusion: A Study of the Pre–World War I Bromine Industry," *Journal of Industrial Economics* 45, no. 2 (June 1997): 118–121.

54. Reimer, "Bayer & Company," 160–163; William Haynes, *American Chemical Industry*, vol. 4, *The Merger Era* (New York: Van Nostrand, 1945), 246–248; USPN 775,810 (Nov. 22, 1904); USPN 782739 (Feb. 14, 1905). On the early development of the barbiturates, see Francisco López-Munoz, Ronaldo Ucha-Udabe, and Cecilio Alamo, "The History of Barbiturates a Century after Their Clinical Introduction," *Neuropsychiatric Disease and Treatment* 1, no. 4 (Dec. 2005): 329–343.

55. Efforts to modify the patent law continued through the outbreak of World War I, but these bills were more modest in scope and pursued with significantly less enthusiasm. For example, "A New Patent Bill," *Bulletin of Pharmacy*, Jan. 1907, 10; "Currier Patent Bill Impracticable," *Practical Druggist and Pharmaceutical Review*, July 1908, 446.

56. *Loder v. Jayne*, 142 F. 1010 (1906).

57. "Tripartite Plan Attacked by Government," *Practical Druggist and Pharmaceutical Review of Reviews*, June 1906, 442.

58. "Judge McFarlane's Views of Patent Medicines," *Indiana Medical Journal*, Apr. 1906, 403.

59. *Dr. Miles Medical Co. v. Goldthwaite*, 133 F. 794 (1904); *Dr. Miles Medical Co. v. Jaynes Drug Co.*, 149 F. 838 (1906); *Dr. Miles Medical Co. v. Platt*, 142 F. 606 (1906); *Dr. Miles Medical Co. v. Snellenburg*, 152 F. 661 (1907).

60. *Dr. Miles v. Platt*, 142 F. 606 (1906).

61. *Dr. Miles Medical Co. v. John D. Park & Sons Co.*, 220 US 373 (1911). On the decision, see Rudolph J. R. Peritz, "'Nervine' and Knavery: The Life and Times of *Dr. Miles Medical Company*," in *Antitrust Stories*, ed. Eleanor M. Fox and Daniel A. Crane (St. Paul: West, 2007).

62. *Standard Oil Co. of New Jersey v. United States*, 221 US 1 (1911); *United States v. American Tobacco Company*, 221 US 106 (1911). My interpretation of the 1911 decisions follows Martin J. Sklar, *The Corporate Reconstruction of American Capitalism, 1890–1916: The Market, the Law, and Politics* (New York: Cambridge University Press, 1988),127–154.

63. *Standard Sanitary Manufacturing Co. v. United States*, 226 US 20 (1912).

64. USPN 601,995 (Apr. 5, 1898).

65. *Bauer & Cie v. O'Donnell*, 229 US 1 (1913).

66. Sklar, *Corporate Reconstruction*, 127–154; Steven Wilf, "The Making of the Post-War Paradigm in American Intellectual Property Law," *Columbia Journal of Law & the Arts* 31, no. 2 (2008): 192–194, esp. 194n279.

67. Harry B. Mason, "The Supreme Court and Price Restriction," *Bulletin of Pharmacy*, July 1913, 271.

68. "Statement of Mr. Louis D. Brandeis," in *Hearings before the Committee on Interstate and Foreign Commerce House of Representatives Sixty-Third Congress Second and Third Sessions on H.R. 13305* (Jan. 9, 1915), 198.

69. Federal Trade Commission, *Conference Report*, H.R. 1142, 63d Cong., 2d Sess., (Sept. 4, 1914), 18–19.

70. Lewis S. McMurtry, "The American Medical Association: Its Origin, Progress and Purpose," *Medical News*, July 15, 1905, 101–102. On the early history of the Council on Pharmacy and Chemistry, see Eric W. Boyle, *Quack Medicine: A History of Combating Health Fraud in Twentieth-Century America* (Santa Barbara, CA: Praeger, 2013), 17–33; Harry Marks, *The Progress of Experiment: Science and Therapeutic Reform in the United States, 1900–1990* (New York: Cambridge University Press, 1997), 17–33; James G. Burrow, "The Prescription-Drug Policies of the American Medical Association in the Progressive Era," in *Safeguarding the Public: Historical Aspects of Medicinal Drug Control*, ed. John Blake (Baltimore: Johns Hopkins Press, 1970), 112–122.

71. Francis E. Stewart, "The Working Bulletin System, National Pharmacological Association and Proposed Investigation of the World under the Auspices of the Government of the United States," *Addresses, Papers and Discussions in the Section of Materia Medica and Pharmacy, at the Forty-Second Annual Meeting of the American Medical Association* (1891), 35–52.

72. Francis E. Stewart, "Proposed National Bureau of Materia Medica," *JAMA*, Apr. 27, 1901, 1175–1176. For a similar proposal, see H. Bert. Ellis, "Necessity for a National Bureau of Medicines and Foods," and comments by T. D. Davis in "Discussion," both in *Bulletin of the American Academy of Medicine*, Dec. 3, 1903, 486–494 and 495–496.

73. George H. Simmons to Francis E. Stewart, Mar. 19, 1901, FESP, b. 6, f. 17.

74. Francis E. Stewart to H. K. Mulford & Co., Oct. 2, 1903, Robert P. Fischelis Papers, 1821–1981, Wisconsin Historical Society, b. 41, f. 15; "National Medicine and Food Bureau" *JAMA*, Apr. 11, 1903, 1002–1003; "Report of Committee on National Bureau of Materia Medica," *Transactions of the American Therapeutic Society 1900–1902* (1903), 62–66. In 1910, the *Practical Druggist*

and Pharmaceutical Review noted the role Stewart played in the origins of
the Council on Pharmacy and Chemistry, pointing out that his 1901 article
"became the basis upon which was organized the Council on Pharmacy
and Chemistry of the American Medical Association." See "Francis Edward
Stewart, Ph.G., M.D.," *Practical Druggist and Pharmaceutical Review*, May 1910,
897–898.

75. The initial set of drugs approved by the council was published by the *Journal
of the American Medical Association* in a series of articles in January 1907. The
list was then updated in the first issue of the journal each month.

76. *NNR* (1907).

77. "Report of the Council on Pharmacy and Chemistry," *JAMA*, July 22, 1905, 263.

78. *NNR* (1909), 8.

79. "Catchy Names and Their Dangers," *JAMA*, Sept. 22, 1906, 944; *Reports of
the President's Homes Commission*, S. Doc. No. 644, 60th Cong., 2d Sess.,
(1909), 263.

80. "German Greed and American Gold," *American Journal of Clinical Medicine*,
July 1912, 690–691.

81. Francis E. Stewart, "The Proper Introduction of Materia Medical Products
to Science and Brands of the Same to Commerce," *Therapeutic Monthly*, May
1902, 175.

82. Solomon Solis-Cohen, "The Limit of Proprietorship in Materia Medica,"
JAMA, Jan. 19, 1907, 197.

83. USPN 812,554 (Feb. 13, 1906); USTN 14,207, *OGPO*, Mar. 27, 1906, 1325.

84. *NNR* (1909), 11–19.

85. "The Council on Pharmacy and Chemistry," *JAMA*, Sept. 1905, 110.

86. For examples of the influence of the policy, see "The Advisability of the
Faculty Publishing Its Own Journal," *Maryland Medical Journal*, Sept. 1906,
352–355; "Concerning Our Advertisements," *Pennsylvania Medical Journal*,
Feb. 1906, 356–359; "Stick to the Council on Pharmacy and Chemistry,"
Journal of the Missouri State Medical Association, Nov. 1913, 173.

87. "Support the Council on Pharmacy and Chemistry," *Journal of the Indiana
State Medical Association*, Feb. 15, 1908, 66.

88. For a contrasting interpretation, which suggests that the council had little
impact on the market, see Jan R. McTavish, *Pain & Profits: The History of Headache and its Remedies in America* (New Brunswick, NJ: Rutgers University Press,
2004), 103–105.

89. For example, Wallace C. Abbott, *A Brief Therapeutics of Some of the Principle Alkaloidal Medicaments* (1901); Abbott, *Abbott's Alkaloidal Digest* (1904). Abbott
also published a medical journal, *The American Journal of Clinical Medicine*, to
popularize his ideas.

90. Wallace C. Abbott, "Hyoscine, Morphine and Cactin Anesthesia," *Chicago
Medical Recorder*, Sept. 1907, 559.

91. "H.M.C. Tablets," *California State Journal of Medicine*, Feb. 1908, 71.

92. "The 'Hyoscin-Morphin-Cactin' Anesthesia," *JAMA*, Dec. 21, 1907, 203.

93. "The Abbott Alkaloidal Company Modern High Finance and Methods of Working the Medical Profession," *JAMA*, Mar. 14, 1908, 895–900. See also "The Abbot Alkaloidal Company's Reply," *Texas State Journal of Medicine*, Feb. 1908, 268; "Dr. Abbott's Controversy," *Texas Sate Journal of Medicine*, Oct. 1908, 141–142.

94. *NNR* (1914).

95. Untitled memo, Records of the Office of Alien Property, Records of the Bureau of Investigations, Records of Investigations, 1917–1921, NARA II, RG 131, b. 161, f. 6.

96. Francis E. Stewart, "Dioxydiamidoarsenobenzol as a Patented Product," *Monthly Cyclopedia and Medical Bulletin*, Sept. 1911, 523–524. Other examples of this concern include Burnside Foster, "Some Thoughts on the Ethics of Medical Journalism," *Saint Paul Medical Journal*, July 1901, 478–481; "We Write Our Own Editorials," *Medico-Pharmaceutical Critic and Guide*, Oct. 1908, 335; "The Blight on Medical Journalism," *JAMA*, Aug. 8, 1908, 504; Francis M. Pottenger, "Some Observations on the Present State of Medical Journalism," *California State Journal of Medicine*, Sept. 1915, 339–341. For an example of a physician discussing the difficulties of publishing his scientific work because of opposition by vaccine manufacturers, see H. M. Alexander to Joseph McFarland, Dec. 4, 1901, Joseph McFarland Papers, 1900–1943, College of Physicians of Philadelphia, b. 1, f. 8. For an analysis of the marketing of Salvarsan using a publication strategy, see Axel Hüntelmann, "A Different Mode of Marketing? The Importance of Scientific Articles in the Marketing of Salvarsan," *History and Technology* 29, no. 2 (2013): 116–134. See also Jean-Paul Gaudillière and Ulrike Thomas, "Pharmaceutical Firms and the Construction of Drug Markets: From Branding to Scientific Marketing," *History and Technology*, 29, no. 2 (2013): 105–115.

97. There is a large literature on the 1906 Pure Food and Drug Act, including Philip J. Hilts, *Protecting America's Health: The FDA, Business, and One Hundred Years of Regulation* (Chapel Hill: University of North Carolina Press, 2003), 3–71; Lorine Swainston Goodwin, *The Pure Food, Drink, and Drug Crusaders, 1879–1914* (Jefferson, NC: McFarland, 1999); James Harvey Young, *Pure Food: Securing the Federal Food and Drugs Act of 1906* (Princeton, NJ: Princeton University Press, 1989). See also Kara Swanson, "Food and Drug Law as Intellectual Property Law: Historical Reflections," *Wisconsin Law Review* 2011, no. 2 (2011): 331–398.

98. USPN 673,070 (Apr. 30, 1901). See also USPN 673,347 (Apr. 30, 1901), for a method for manufacturing explosives; USPN 673,069 (Apr. 30, 1901), for a method of treating vegetable matter; USPN 615,376 (Dec. 6, 1898), for a method of manufacturing alcohol. In 1901 Wiley was involved in an interference suit with another inventor who claimed that his patent on smokeless gunpowder was invalid. See "Harvey W. Wiley vs. Hudson Maxim Interference No. 22,822," in Harvey Washington Wiley Papers, Library of Congress, b. 207, f. "Patents."

99. Harvey W. Wiley, "Federal Control of Drugs," *Transactions of the Section on Pharmacology of the American Medical Association at the Fifty-Fifth Annual Session* (1904), 38. The extent to which Wiley was directly influenced by Stewart is unclear, but the two moved in many of the same circles, and Wiley was clearly aware of his arguments. For example, see "Eighth Annual Meeting of the American Therapeutic Society," *Boston Medical and Surgical Journal*, Oct. 31, 1907, 606–608; "Eighth Annual Meeting of the American Therapeutic Society," *Boston Medical and Surgical Journal*, Nov. 14, 1907, 671–675.

100. For example, Wiley, "Federal Control of Drugs," 37.

101. Harvey Wiley to Editor, *Wine Trade Review*, Mar. 6, 1905, clipping in Records of the Bureau of Chemistry, Miscellaneous Records, 1877–1910, NARA II, RG 88, b. 2, unlabeled volume of loose-leaf letters.

102. Harvey W. Wiley, "Drugs and Their Adulterations and the Laws Relating Thereto," 208–210, pamphlet in Records of the Bureau of Chemistry, Miscellaneous Records, 1877–1910, NARA II, RG 88, b. 3, loose pamphlets.

103. In addition to the *USP*, the 1906 law and later regulations also required drugs listed in the *National Formulary* to conform to the standards promulgated by that text. I have excluded discussion of the formulary in this volume for simplicity's sake, but see Gregory J. Higby, ed., *One Hundred Years of the National Formulary: A Symposium* (Madison, WI: American Institute of the History of Pharmacy, 1989).

104. An Act of June 30, 1906, Public Law 59-384, for Preventing the Manufacture, Sale, or Transportation of Adulterated or Misbranded or Poisonous or Deleterious Foods, Drugs, Medicines, and Liquors, and for Regulating Traffic Therein, And For Other Purposes, Pub. L. 59-384, 34 Stat. 768 (1906).

105. United States Department of Agriculture, "Rules and Regulations for the Enforcement of the Food and Drug Act" (Oct. 16, 1906), in S. Doc. No. 252, 59th Cong.. 2d Sess. (1907).

106. William Wheeler Thorton, *The Law of Pure Food and Drugs* (1912), 852.

107. Ibid., 325.

108. "A New Danger to Trade Marks and Trade Names," *Practical Druggist and Pharmaceutical Review*, Dec. 1906, 566.

109. *United States v. Johnson*, 221 US 488 (1911).

110. *The Pure Food and Drugs Act. Hearings before the Committee on Interstate and Foreign Commerce House of Representatives Sixty Second Congress, Second Session, Part 1*, 62d Cong., 2nd Sess. (1912), 40, 181, 358.

111. In relation to this and similar issues, see the testimony of F. E. Holiday, ibid., 445.

112. The text of the Sherley Amendment can be found in "The Sherley Amendment," *ADPR*, Nov. 1912, 33.

113. "'Joker' in Amendment to Pure Food Law," *Journal of the Medical Society of New Jersey*, Oct. 1912, 272.

114. George H. Simmons to Joseph P. Remington, July 20, 1912, United States

Pharmacopeial Convention Records, 1819–2000, Wisconsin Historical Society, b. 17, f. 6. See also McTavish, *Pain & Profits*, 128–129.

115. "Memorandum from Mallinckrodt Chemical Works, St. Louis. Inorganic Chemicals," Edward J. Mallinckrodt Jr. Papers, 1798–1981, University of Missouri–St. Louis, Western History Manuscript Collection, b. 65, f. 1308. A note indicating that the memorandum was sent to the revision committee is on the front page. On this issue, see also *Report Submitted by the Delegates of the American Chemical Society to the Convention for Revising the United States Pharmacopoeia* (May 10, 1910), 2, in Edward J. Mallinckrodt Jr. Papers, 1798–1981, University of Missouri–St. Louis, Western History Manuscript Collection, b. 65, f. 1307.

116. *USP*, 9th revision (1916), 5.

117. The company's first working bulletin was for a product sold under the trade name "Somnos." *Working Bulletin for the Co-Operative Investigation of the Somnos Brand of Trichlorethidene Propenyl Ether* (1906). A copy can be found in Medical Trade Ephemera Collection, College of Physicians of Philadelphia, b. "H. K. Mulford."

118. Francis E. Stewart, "The Relation of the Patent and Trade-Mark Laws to Materia Medica Nomenclature," *Transactions of the Section on Pharmacology and Therapeutics of the American Medical Association at the Sixty-Third Annual Session* (1912), 30.

119. Francis E. Stewart, "Some Objections to Materia Medica Standardization, with Reference to the U.S. Pharmacopeia," *Medical Record*, Nov. 22, 1913, 941.

120. Brian B. Hoffman, *Adrenaline* (Cambridge, MA: Harvard University Press, 2013), 22–47; Elliot Valenstein, *The War of the Soups and the Sparks: The Discovery of Neurotransmitters and the Dispute over How Nerves Communicate* (New York: Columbia University Press, 2005), 13–17; T. B. Aldrich, "Adrenalin, the Active Principle of the Suprarenal Glands," *Journal of the American Chemical Society*, Sept. 1905, 1075; Armour and Company, *Suprarenalin (Crystalline Powder and Solution) and the Suprarenal Preparations* (1900). On Abel, see John Parascandola, *The Development of American Pharmacology: John J. Abel and the Shaping of a Discipline* (Baltimore: Johns Hopkins University Press, 1992).

121. Takamine worked with a young chemist named Keizo Wooyenaka. I have left aside the question of whether or not Wooyenaka deserves credit as the inventor or coinventor of Adrenalin, but see Jon M. Harkness, "Dicta on Adrenalin(e): Myriad Problems with Learned Hand's Product-of-Nature Pronouncements in *Parke-Davis v. Mulford*," *Journal of the Patent and Trademark Office Society* (2011), 374n79.

122. Jokichi Takamine to John J. Abel, Nov. 28, 1900, John Jacob Abel Collection, Alfred Mason Chesney Medical Archives, Johns Hopkins University, b. 54, f. 5. See also Jokichi Takamine, "Adrenalin: The Active Principle of the Suprarenal Glands and Its Mode of Preparation," *American Journal of Pharmacy*, Nov. 1901), 523–531.

123. Correspondence between employees of Parke-Davis and a number of physi-

cians about the clinical testing of the drug can be found in PDCR, b. 23, f. 9. For examples of published writings about such experiments, see Emil Mayer, "Clinical Experience with Adrenalin," *Philadelphia Medical Journal*, Apr. 1901, 819–821; "A Discussion on the Practical Value of Adrenalin," *Therapeutic Notes*, Oct. 1901, 63–64.

124. *Therapeutic Notes*, April, 1901, back page.

125. Little correspondence related to the debate survives, but in a 1900 memo one employee of the company argued against securing a patent on Takamine's discovery. "You know very well the attitude of the Medical Profession in regard to patents and particularly the product patents," the employee wrote. "I feel sure that even were such a patent granted, it would do harm to the house." See "Memo to Mr. Fink," May 23, 1900, PDCR, b. 9, f. 8.

126. Takamine obtained the following process patents for manufacturing Adrenalin: USPN 730,175 (June 2, 1903); USPN 730,196 (June 2, 1903); USPN 730,197 (June 2, 1903); USPN 730,198 (June 2, 1903). There is a large amount of correspondence in the wrapper files for these patents about these issues. These wrappers are available in PWFKC.

127. James B. Littlewood to Jokichi Takamine, Dec. 7, 1900, PWFKC no. 730,175.

128. Ibid.

129. Knight Bros. [for Jokichi Takamine], "Amendment," Oct. 22, 1901, PWFKC no. 730,175.

130. James B. Littlewood to Jokichi Takamine, Nov. 7, 1901, PWFKC no. 730,175.

131. Christopher Beauchamp, "Patenting Nature: A Problem of History," *Stanford Technology Law Review* 16, no. 2 (Winter 2013): 287–289, quote on 288; Knight Bros. [for Jokichi Takamine], "Amendment" (filed Sept. 1902); USPN 730,175 (June 2, 1903). I follow Beauchamp's general argument here. For a different interpretation and a much closer reading of the exchange between Takamine (and his lawyers) and Littlewood, see Harkness, "Dicta on Adrenalin(e)," 369–383. Takamine obtained an additional patent on the product in 1904. USPN 753,177 (Feb. 23, 1904).

132. Beauchamp, "Patenting Nature," 287–289.

133. "The Suprarenal Wars," *Practical Druggist and Pharmaceutical Review*, Aug. 1904, 328.

134. PDC to Charles F. Chandler, Mar. 16, 1904, Charles Frederick Chandler Papers, Rare Book and Manuscript Library, Columbia University, b. 54, f. 1.

135. F. E. Stewart to C. S. Bacon, May 27, 1910, FESP, b. 1, f. 9.

136. "The Suprarenal Situation," *Medical Times*, July 1911, viii.

137. E. M. Tansey, "What's in a Name? Henry Dale and Adrenaline, 1906," *Medical History* 39, no. 4 (1995): 459–476.

138. For a statement of the company's position on this issue, see "The Adrenalin Controversy," Aug. 12, 1904, PDCR, b. 23, f. 22.

139. USTN 11,909, *OGPO*, Jan. 16, 1906, 885.

140. Stewart, "Dioxydiamidoarsenobenzol as a Patented Product," 526–528.

141. For example, Philip Mills Jones, "The Active Principle of the Adrenal Gland:

What Name Shall Be Given To It?," *California Sate Journal of Medicine*, June 1904, 178.

142. "New and Non-Official Remedies," *JAMA*, (Sept. 15, 1906, 856–857.

143. *NNR* (1909).

144. "The Name 'Epinephrin' versus the Name 'Adrenalin,' " *JAMA*, Mar. 25, 1911, 901.

145. "Proprietary versus Unprotected Names," *JAMA*, Mar. 25, 1911, 910–915. See also William Henry Schultz, "Quantitative Pharmacologic Studies: Adrenalin and Adrenalin-like Bodies," *Bulletin of the Hygienic Laboratory* 55 (1909).

146. Livingston Gifford to Charles F. Chandler, June 19, 1906, Charles Frederick Chandler Papers, Rare Book and Manuscript Library, Columbia University, b. 54, f. 1.

147. *Parke-Davis & Co. v. H. K. Mulford Co.*, 189 F. 95 (1911).

148. Beauchamp, "Patenting Nature," 63.

149. *Parke-Davis & Co. v. H. K. Mulford Co.* 189 F. 95 (1911). Extensive documentation relating to the case can be found in Records of the District Courts of the United States, NARA, New York City Branch, RG 21, b. 1567; Records of the United States Courts of Appeals, NARA, New York City Branch, RG 276, b. 1684. The latter includes an extensive transcript of the trial.

150. Statement reprinted in "The Adrenalin Patents Valid," *Practical Druggist and Review of Reviews*, June 1911, 54.

151. For example, "Adrenalin Patent Sustained," *ADPR*, May 1911, 60; "Patents on Suprarenal Active Principle Upheld," *Druggists Circular*, June 1911, 328.

152. Hoffman, *Adrenaline*, 7.

153. USPN 1,271,111 (July 3, 1918). The company also obtained at least two patents on new anesthetic compounds in 1916 and a patent on an arsenical compound in 1914. USPN 1,193,651 (Aug. 8, 1916); USPN 1,193,634 (Aug. 8, 1916); USPN 759,038 (Dec. 1, 1914).

154. In 1912, for example, John Uri Lloyd patented an alkaloidal preparation and licensed it to Eli Lilly & Company, which marketed it under the name "Alcresta." See John Uri Lloyd Papers, Lloyd Library and Museum, Cincinnati, b. 18, f. 354, f. 355. Eli Lilly also obtained patent rights to a number of other chemicals around this time. For example, USPN 1,150,252 (Aug. 17, 1915); USPN 1,150,253 (Aug. 17, 1915); USPN 1,138,936 (May 11, 1915); USPN 1,138,937 (May 11, 1915); USPN 1,114,734 (Oct. 27, 1914); USPN 1,098,022 (May 26, 1914); USPN 1,079,693 (Nov. 25, 1913).

CONCLUSION

1. "Address of Dr. Carl L. Alsberg, Chief of the Bureau of Chemistry, Department Agriculture," *NARD Notes*, Feb. 27, 1913, 1329–1330.

2. Charles A. Catlin, "Alsberg Interview," Sept. 23, 1913, Rumford Chemical Company Records, 1853–1951, Rhode Island Historical Society, b. 6, f. 111.

See also "In the Matter of the Rumford Chemical Works, I.S. No. 2230," ibid, b. 6, f. 111.

3.   *Alien Property Custodian Report*, Mar. 1, 1919, 7–20.

4.   Julius Stieglitz, "Synthetic Drugs," *JAMA*, Feb. 23, 1918, 536. There is significant correspondence and other material related to this issue in Records of the Federal Trade Commission, Trading with the Enemy Files, 1916–1924, NARA II, RG 122, b. 4, f. 30A.

5.   In 1919 A. Mitchell Palmer and Francis Garvan, who had been appointed by Woodrow Wilson to oversee German property seized during the war, arranged for the sale of the patents to a private holding company that they established for this purpose. The transfer was controversial, and in the early 1920s critics filed a lawsuit to determine the legality of the move. On this issue, see Kathryn Steen, "Patents, Patriotism, and 'Skilled in the Art': *USA v. The Chemical Foundation, Inc.*, 1923–1926," *Isis: Journal of the History of Science in Society* 92, no. 1 (Mar. 2001): 96–102.

6.   Milton Campbell to Robert Fischelis, Apr. 6, 1921, Robert P. Fischelis Papers, 1821–1981, Wisconsin Historical Society, b. 111, f. 3; "Announcement of the H. K. Mulford Company in Regard to the Patenting of Materia Medica," n.d., ibid.; Robert Fischelis to Milton Campbell, Mar. 31, 1921, ibid.

7.   A. M. Lewers, "Composition of Matter," *Journal of the Patent Office Society* 4, no. 11 (July 1922): 551; F. O. Taylor to O. W. Smith, Nov. 18, 1926, PDCR, b. 37, f. 13.

8.   "Principles of Medical Ethics," *Journal of the Indiana State Medical Association*, Sept. 15, 1912, 407.

9.   "Report of the Judicial Council," *JAMA*, July 4, 1914, 106.

10. On the discovery and early history of insulin, see Maurice Cassier and Christiane Sinding, "'Patenting in the Public Interest': Administration of Insulin Patents by the University of Toronto," *History and Technology* 24, no. 2 (June 2008): 153–171; Chris Feudtner, *Bittersweet: Diabetes, Insulin, and the Transformation of Illness* (Chapel Hill: University of North Carolina Press, 2003); Christiane Sinding, "Making the Unit of Insulin: Standards, Clinical Work, and Industry, 1920–1925," *Bulletin of the History of Medicine* 76, no. 2 (Summer 2002): 231–270; Michael Bliss, *The Discovery of Insulin* (Chicago: University of Chicago Press, 1982). On Harry Steenbock and the patenting of his method for producing Vitamin D, see Robert Bud, "Upheaval in the Moral Economy of Science? Patenting, Teamwork, and the World War II Experience of Penicillin," *History and Technology* 24, no. 2 (June 2008): 173–190; Rima D. Apple, *Vitamania: Vitamins in American Culture* (New Brunswick, NJ: Rutgers University Press, 1996), 33–53; Rima D. Apple, "Patenting University Research: Harry Steenbock and the Wisconsin Alumni Research Foundation" *Isis: Journal of the History of Science in Society* 80, no. 3 (Sept. 1989): 375–394. For an examination of some other examples, see Charles Weiner, "Patenting and Academic Research: Historical Case Studies," *Science, Technology &*

Human Values 12, no. 1 (Winter 1987): 50–62. On collaboration between the pharmaceutical industry, academic scientists, and clinicians in the interwar period more generally, see Joseph M. Gabriel, "The Testing of Sanocrysin: Science, Profit, and Innovation in Clinical Trial Design, 1926–1931," *Journal of the History of Medicine and Allied Sciences* (forthcoming); Nicolas Rasmussen, *On Speed: The Many Lives of Amphetamine* (New York: New York University Press, 2008), 1–52; Jeffrey L. Furman and Megan J. MacGarvie, "Academic Science and the Birth of Industrial Research Laboratories in the U.S. Pharmaceutical Industry," *Journal of Economic and Behavior Organization* 63, no. 4 (Aug. 2007): 756–776; Nicolas Rasmussen, "The Drug Industry and Clinical Research in Interwar America: Three Types of Physician Collaborator," *Bulletin of the History of Medicine* 79, no. 1 (Spring 2005): 50–80; John Parascandola, *The Development of American Pharmacology: John J. Abel and the Shaping of a Discipline* (Baltimore: Johns Hopkins University Press, 1992), 91–125; John P. Swann, *Academic Scientists and the Pharmaceutical Industry: Cooperative Research in Twentieth-Century America* (Baltimore: Johns Hopkins University Press, 1988).

11. For example, see *Wisconsin Alumni Research Foundation v. George A. Breon & Co.* 85 F.2d 166 (1936); *New Discoveries, Inc. v. Wisconsin Alumni Research Foundation* 15 F. Supp. 596 (1936). Also see *General Baking Co. v. Grocers' Baking Co.* 3 F. Supp. 146 (1933).

12. For example, "Report of the Committee on Clinical Investigation and Scientific Research of the American Academy of Pediatrics, 1935" *Journal of Pediatrics*, Jan. 1936, 124–130.

13. Morris Fishbein to Claude E. Forkner, Dec. 28, 1937, Morris Fishbein Papers, 1912–1976, University of Chicago, Special Collections and Research Center, b. 92, f. 12.

14. "Problem of Medical Patents," *JAMA*, July 22, 1933, 284–285.

15. Morris Fishbein, "Copper-Iron Patent," Apr. 20, 1933, Morris Fishbein Papers, 1912–1976, University of Chicago, Special Collections and Research Center, b. 92, f. 13.

16. Morris Fishbein to Ludvig Kast, May 23, 1934, Morris Fishbein Papers, 1912–1976, University of Chicago, Special Collections and Research Center, b. 92, f. 12. See also Morris Fishbein to James Randolph, Sept. 25, 1937, ibid., b. 92, f. 13.

17. "The Insulin Monopoly," *JAMA*, July 12, 1941, 112.

18. Comments printed in "An Institute for Cooperative Research as an Aid to the American Drug Industry," *Journal of Industrial and Engineering Chemistry*, Jan. 1919, 59–69.

19. On the history of the Food and Drug Administration, see Dominique A. Tobbell, *Pills, Power, and Policy: The Struggle for Drug Reform in Cold War America and Its Consequences* (Berkeley: University of California Press, 2012); Daniel Carpenter, *Reputation and Power: Organizational Image and Pharmaceutical Regulation at the FDA* (Princeton, NJ: Princeton University Press, 2010);

Arthur A. Daemmrich, *Pharmacopolitics: Drug Regulation in the United States and Germany* (Chapel Hill: University of North Carolina Press, 2004); Philip J. Hilts, *Protecting America's Health: The FDA, Business, and One Hundred Years of Regulation* (New York: Alfred A. Knopf, 2003); Harry M. Marks, *The Progress of Experiment: Science and Therapeutic Reform in the United States, 1900–1990* (New York: Cambridge University Press, 1997), 71–97.

20. Robert Fischelis to Milton Campbell, Nov. 24, 1919, Robert P. Fischelis Papers, 1821–1981, Wisconsin Historical Society, b. 111, f. 3; Henry C. Thomson, "Trade-Mark Origination and Protection," *Eighth Annual Meeting American Drug Manufacturers' Association* (1919), 124–145. On the turn toward psychology, advertising, and trademarks, see Michael Pettit, *The Science of Deception: Psychology and Commerce in America* (Chicago: University of Chicago Press, 2013), 121–155.

21. On the history of generic drugs in the twentieth century, see Jeremy A. Greene, *Generic: The Unbranding of Modern Medicine* (Baltimore: Johns Hopkins University Press, 2014); Greene, "The Materiality of the Brand: Form, Function, and the Pharmaceutical Trademark," *History and Technology* 29, no. 2 (June 2013): 210–226; Greene, "What's in a Name: Generics and the Persistence of the Pharmaceutical Brand in American Medicine," *Journal of the History of Medicine and Allied Sciences* 66, no. 4 (Oct. 2011): 468–506; Tobbell, *Pills, Power, and Policy*; Daniel Carpenter and Dominique A. Tobbell, "Bioequivalence: The Regulatory Career of a Pharmaceutical Concept," *Bulletin of the History of Medicine* 85, no. 11 (Spring 2011): 93–131.

22. For example, see *J. W. Kobi Co. v. Federal Trade Commission*, 23 F. 2d 41 (1927).

23. Myron W. Watkins, "An Appraisal of the Work of the Federal Trade Commission," *Columbia Law Review* 32, no. 2 (Feb. 1932): 277.

24. For example, *Sears, Roebuck & Co. v. Federal Trade Commission*, 258 F. 307 (1919); *Federal Trade Commission v. Raladam Company*, 283 US 643 (1931).

25. Inger L. Stole, *Advertising on Trial: Consumer Activism and Corporate Public Relations in the 1930s* (Urbana: University of Illinois Press, 2006), 144–158. On the history of consumer activism, see Lawrence B. Glickman, *Buying Power: A History of Consumer Activism in America* (Chicago: University of Chicago Press, 2009); Charles F. McGovern, *Sold American: Consumption and Citizenship, 1890–1945* (Chapel Hill: University of North Carolina Press, 2006).

26. Alan Brinkley, *The End of Reform: New Deal Liberalism in Recession and War* (New York: Alfred A. Knopf, 1995). See also Wyatt Wells, *Antitrust and the Formation of the Postwar World* (New York: Columbia University Press, 2002), 37–42.

27. Mira Wilkins, *The History of Foreign Investment in the United States, 1914–1945* (Cambridge, MA: Harvard University Press, 2004), 457–470.

28. Tobbell, *Pills, Power, and Policy*, 89–120; Jeremy A. Greene and Scott H. Podolsky, "Reform, Regulation, and Pharmaceuticals–the Kefauver-Harris Amendments at 50," *New England Journal of Medicine* 367 (Oct. 18, 2012): 1481–1483; Carpenter, *Reputation and Power*, 231–238; Hilts, *Protecting America's Health*,

129–143; Richard Edward McFadyen, "Estes Kefauver and the Drug Industry" (PhD diss., Emory University, 1973); Richard Harris, *The Real Voice* (New York: Macmillan, 1964).

29. *Federal Trade Commission v. Actavis* 133 S. Ct. 1630 (2013). For a brief overview, see "Supreme Court Lets Regulators Sue over Generic Drug Deals," *New York Times,* June 17, 2013. See also Federal Trade Commission, "Pay-for-Delay: When Drug Companies Agree Not to Compete," http://www.ftc.gov/opa /reporter/competition/payfordelay.shtml.

30. In 2011 the Leahy-Smith America Invents Act was signed into law, transform-ing the United States from a "first to invent" to a "first to file" system and thereby eliminating interference proceedings. Debates about the reform and its potential effects are beyond the scope of this volume.

31. "Top Pharma Companies by 2012 Revenue," *FiercePharma,* Mar. 26, 2013, http://www.fiercepharma.com/special-reports/top-pharma-companies-2012 -revenues?utm_medium=nl&utm_source=internal

32. There is a vast literature on these and related issues. Examples include Thomas Pogge, Matthew Rimmer, and Kim Rubenstein, eds., *Incentives for Global Public Health: Patent Law and Access to Essential Medicines* (New York: Cambridge University Press, 2010); Frederick M. Abbott and Graham Dukes, *Global Pharmaceutical Policy: Ensuring Medicines for Tomorrow's World* (Northampton, MA: Edward Elgar, 2009); Frank A. Sloan and Chee-Ruey Hsieh, eds., *Pharmaceutical Innovation: Incentives, Competition, and Cost-Benefit Analysis in International Perspective* (New York: Cambridge University Press, 2007); Michael A. Santoro and Thomas M. Gorrie, eds., *Ethics and the Pharmaceutical Industry* (New York: Cambridge University Press, 2005); Alfonso Gambardella, *Science and Innovation: The US Pharmaceutical Indus-try during the 1980s* (New York: Cambridge University Press, 1995). I have refrained from discussing the complex issues surrounding the patenting of genetic material, including the 2013 Supreme Court decision *Association for Molecular Pathology v. Myriad Genetics* 569 US 12-398 (2013) because of the complexity of the issues involved and the fact that a significant amount of attention has been focused on the topic elsewhere. However, my analysis of *Parke-Davis & Co. v. H. K. Mulford* (1911) and the broader history of legal doctrines related to purity, utility, and related issues in patent case law sug-gests that the issues involved are far from new. On the patenting of genetic material, see Christopher Beauchamp, "Patenting Nature: A Problem of His-tory" *Stanford Technology Law Review* 16, no. 2 (Winter 2013): 257–311; Luigi Palombi, *Gene Cartels: Biotech Patents in the Age of Free Trade* (Northampton, MA: Edward Elgar, 2009); David B. Resnik, *Owning the Genome: A Moral Anal-ysis of DNA Patenting* (Albany: State University of New York Press, 2004).

33. Jane Hamsher, "House Health Care Bill: A Death Sentence for My Fellow Breast Cancer Survivors," Oct. 29, 2009, http://fdlaction.firedoglake.com /2009/10/29/house-health-care-bill-a-death-sentence-for-my-fellow-breast -cancer-survivors/.

34. Federal Trade Commission, "Emerging Health Care Issues: Follow-On Biologic Drug Competition," June 2009, http://www.ftc.gov/os/2009/06 /P083901biologicsreport.pdf

35. As the Office of the US Trade Representative put it, "Biologic drugs need data protection because those drugs require enormous amounts of time and money to develop. Before entrepreneurs are willing to make the investment in new therapies, they want to know that they will have the rights to their own research for a certain period of time in order to see a return on their investments." "Stakeholder Input Sharpens, Focuses U.S. Work on Pharmaceutical Intellectual Property Rights in the Trans-Pacific Partnership," Nov. 29, 2013, http://www.ustr.gov/about-us/press-office/blog/2013/November /stakeholder-input-sharpen-US-work-on-pharmaceutical-IP-in-TPP.

36. Since 2004 there have been more than fifteen major settlements between the US Department of Justice (DOJ) and drug manufacturers for illegal marketing practices. This suggests that there are structural forces in place that drive the industry to violate federal law as a routine part of its efforts to create markets. In July 2012, for example, the DOJ announced that the pharmaceutical giant GlaxoSmithKline had agreed to settle criminal and civil complaints related to its illegal marketing of the popular antidepressants paroxetine (sold under the brand name "Paxil") and bupropion (sold under the brand mane "Wellbutrin") for $3 billion. In addition to a number of other offenses, the settlement covered allegations that the company had promoted these drugs for unapproved uses and that it had failed to report safety data to the FDA for another of its products, the popular diabetes drug rosiglitazone (sold under the brand name "Avandia"). Three billion dollars may seem like a lot of money, but given the fact that paroxetine, buproprion, and rosiglitazone earned GlaxoSmithKline at least $28 billion during the period covered by the settlement, it can also, as one analyst commented, "be rationalized as the cost of doing business." US Department of Justice, "GlaxoSmithKline to Plead Guilty and Pay $3 Billion to Resolve Fraud Allegations and Failure to Report Safety Data," July 2, 2012, http://www.justice.gov/opa/pr/2012/July /12-civ-842.html; Patrick Burns quoted in K. Thomas and M. S. Schmidt, "Glaxo Agrees to Pay $3 Billion in Fraud Settlement," *New York Times*, July 3, 2012.

37. Sergio Sismondo, "Ghost Management: How Much of the Medical Literature Is Shaped behind the Scenes by the Pharmaceutical Industry?," *PLoS Medicine* 4, no. 9 (Sept. 2007): 1429–1433.

38. Examples include G. Caleb Alexander, "Seeding Trials and the Subordination of Science," *Archives of Internal Medicine* 171, no. 12 (June 27, 2011): 1107–1108; Howard Brody and Donald W. Light, "The Inverse Benefit Law: How Drug Marketing Undermines Patient Safety and Public Health," *American Journal of Public Health* 101, no. 3 (Mar. 2011): 399–404; Carl Elliot, *White Coat, Black Hat: Adventures on the Dark Side of Medicine* (Boston: Beacon, 2010); Howard Brody, *Hooked: Ethics, the Medical Profession, and the Phar-*

maceutical Industry (Lanham, MD: Rowman and Littlefield, 2007); Barton Moffatt and Carl Elliott, "Ghost Marketing: Pharmaceutical Companies and Ghostwritten Journal Articles," *Perspectives in Biology and Medicine* 50, no. 1 (Winter 2007): 18–31; John Abramson, *Overdosed America: The Broken Promise of American Medicine* (New York: HarperCollins, 2005); Marcia Angell, *The Truth about the Drug Companies: How They Deceive Us and What to Do about It* (New York: Random House, 2004).

39.   There is a large and growing literature on the impact of the pharmaceutical industry on biomedical knowledge, health politics, subjective experience, and related issues. For example, Joseph Dumit, *Drugs for Life: How Pharmaceutical Companies Define Our Health* (Durham, NC: Duke University Press, 2012); Susan E. Bell and Anne E. Figert, "Medicalization and Pharmaceuticalization at the Intersections: Looking Backward, Sideways and Forward" *Social Science and Medicine* 75, no. 5 (Sept. 2012): 775–783; Keith Wailoo, Julie Livingston, Steven Epstein, and Robert Aronowitz, eds., *Three Shots at Prevention: The HPV Vaccine and the Politics of Medicine's Simple Solutions* (Baltimore: Johns Hopkins University Press, 2010); Simon J. Williams, Jonathan Gabe, and Peter Davis, eds., *Pharmaceuticals and Society: Critical Discourses and Debates* (New York: Wiley-Blackwell, 2009); Jeremy A. Greene, *Prescribing by Numbers: Drugs and the Definition of Disease* (Baltimore: Johns Hopkins University Press, 2007); Nikolas Rose, *The Politics of Life Itself: Biomedicine, Power, and Subjectivity in the Twenty-First Century* (Princeton, NJ: Princeton University Press, 2007); Andrew Lakoff, *Pharmaceutical Reason: Knowledge and Value in Global Psychiatry* (New York: Cambridge University Press, 2006); Steven Epstein, *Impure Science: AIDS, Activism, and the Politics of Knowledge* (Berkeley: University of California Press, 1996).

Index